HOMOLOGY

RICHARD OWEN
A lithograph (1850) by T. H. Maguire (Wellcome Institute
Library, London). Reproduced from Nicholas A. Rupke (1993).
Richard Owen's Vertebrate Archetype. *Isis* **84**, p. 232.

HOMOLOGY

THE HIERARCHICAL BASIS
OF COMPARATIVE BIOLOGY

Edited by

Brian K. Hall

Department of Biology
Dalhousie University
Halifax, Nova Scotia
Canada

ACADEMIC PRESS

A Division of Harcourt Brace & Company

San Diego   New York   Boston   London   Sydney   Tokyo   Toronto

This book is printed on acid-free paper. ⊚

Academic Press, Inc.
525 B Street, Suite 1900, San Diego, California 92101-4495

United Kingdom Edition published by
Academic Press Limited
24–28 Oval Road, London NW1 7DX

Library of Congress Cataloging-in-Publication Data

Homology : the hierarchical basis of comparative biology / edited by
    Brian K. Hall.
        p.   cm.
    Includes bibliographical references and index.
    ISBN 0-12-318920-9
    1. Homology (Biology)   I. Hall, Brian Keith,   Date,
QH367.5.H66   1994
574--dc20   93-41646
                                                        CIP

PRINTED IN THE UNITED STATES OF AMERICA
94 95 96 97 98 99   BB   9 8 7 6 5 4 3 2 1

This volume commemorates the 150th anniversary of Richard Owen's landmark paper on homology, "Lectures on Comparative Anatomy and Physiology of the Invertebrate Animals," delivered at the Royal College of Surgeons in 1843.

CONTENTS

2. HOMOLOGY, TOPOLOGY, AND TYPOLOGY: THE HISTORY OF MODERN DEBATES

Olivier Rieppel

3. HOMOLOGY AND SYSTEMATICS

G. Nelson

4. HOMOLOGY, FORM, AND FUNCTION

George V. Lauder

9. WITHIN AND BETWEEN ORGANISMS: REPLICATORS, LINEAGES, AND HOMOLOGUES
V. Louise Roth

10. HOMOLOGY IN MOLECULAR BIOLOGY
David M. Hillis

CONTRIBUTORS

Numbers in parentheses indicate the pages on which the authors' contributions begin.

Fred L. Bookstein (197), Center for Human Growth and Development, University of Michigan, Ann Arbor, Michigan 48109

Michael J. Donoghue (393), Department of Organismic and Evolutionary Biology, Harvard University, Cambridge, Massachusetts 02138

Brian Goodwin (229), Department of Biology, The Open University, Milton Keynes, MK7 6AA, England

Harry W. Greene (369), Museum of Vertebrate Zoology, and Department of Integrative Biology, University of California, Berkeley, Berkeley, California 94720

Brian K. Hall (1), Department of Biology, Dalhousie University, Halifax, Nova Scotia B3H 4J1, Canada, and Department of Anatomy and Human Biology, The University of Western Australia, Nedlands, Perth, Western Australia 6009, Australia

David M. Hillis (339), Department of Zoology, The University of Texas, Austin, Texas 78712

George V. Lauder (151), School of Biological Sciences, University of California, Irvine, Irvine, California 92717

G. Nelson (101), Department of Herpetology and Ichthyology, American Museum of Natural History, New York, New York 10024, and School of Botany, The University of Melbourne, Parkville, Victoria 3052, Australia

Alec L. Panchen (21), Department of Marine Sciences, University of Newcastle upon Tyne, Newcastle upon Tyne NE1 7RU, England

Olivier Rieppel (63), Department of Geology, Field Museum of Natural History, Chicago, Illinois 60605

V. Louise Roth (301), Department of Zoology, Duke University, Durham, North Carolina 27708

Michael J. Sanderson (393), Department of Biology, University of Nevada, Reno, Nevada 89557

Rolf Sattler (423), Department of Biology, McGill University, Montreal, Quebec, Canada H3A 1B1

Neil H. Shubin (249), Department of Biology, University of Pennsylvania, Philadelphia, Pennsylvania 19104

Günter P. Wagner (273), Department of Biology, Yale University, New Haven, Connecticut 06511

INTRODUCTION

Brian K. Hall

Department of Biology
Dalhousie University
Halifax, N.S. B3H 4J1
Canada and
Department of Anatomy and Human Biology
The University of Western Australia
Nedlands, Perth, WA 6009
Australia

The subtitle of this volume summarizes the motivation for producing it; homology is the hierarchical basis of comparative biology. Homology can be and is studied at all levels of biological organization from molecules through genes, cells, organs, embryos, organisms, populations, communi-

Homology: The Hierarchical Basis of Comparative Biology
Copyright © 1994 by Academic Press, Inc.

ties, behavior, even biogeographical regions. Homology therefore informs gene regulation, ontogeny and phylogeny, morphology and physiology, molecular and cell biology, botany and zoology, systematics and classification, ecology and biogeography. As Gareth Nelson tells us in Chapter 3, "If homology pervades biological data then all of biology relates to homology." Homology is indeed morphology's central conception as Julian Huxley described it.

In many respects, however, homology remains in 1994 where it was 45 years ago when Szarski concluded:

> After examining the present status of the concept of homology, one arrives at disquieting results. A basic term of one of the most important zoological sciences, that of comparative anatomy, cannot be exactly defined. (Szarski, 1949, p. 127)

As Günter Wagner put it: "among evolutionary biologists, homology has a firm reputation as an elusive concept" (1989, p. 51). The present volume has been produced, in part, because of such ongoing difficulties. That homology informs the entire hierarchy of biological patterns and processes is both its ultimate strength as a unifying concept and a major stumbling block to its application. The following questions and the ambivalence of some of the answers, illustrate this conundrum.

I. SOME UNANSWERED QUESTIONS

1. Do we possess a single concept of homology transcending (uniting?) all of biology? No.
2. Can a single definition of homology apply to all elements and all levels in the hierarchy of biological complexity? No.
3. If homology is defined differently at different levels in the biological hierarchy, then how do (or should?) such definitions relate to one another? Opinions will vary on the answer to this one.
4. Should the definition(s) incorporate criteria to recognize homology, and/or provide explanations of mechanisms through which homology can, or should be, established? No.

5. Should homology encompass explanations of proximate causation (i.e., ontogeny or physiology) *and* explanations of ultimate causation (i.e., phylogenetic change through evolutionary time)? Again, answers will vary.

6. Must homology always be tied to knowledge of evolutionary origins (descent with modification) of the cells, organs, organisms, species, populations, communities, under consideration? Yes, but some would say no.

7. Must homology always be tied to structure, even when behavioral or physiological characters are under investigation? No, but some would say yes.

8. Must homologous features always share common embryonic development, or can homologous features arise from nonhomologous (nonequivalent) developmental processes? No and yes, but some would say yes and no.

Answers to some of these questions are equivocal and have been for as long as biologists have posed them. This volume provides a compilation of current opinions on, and approaches to, answers to these and other unresolved questions concerning homology.

II. RICHARD OWEN AND HOMOLOGY

Timing of the production of these 13 essays is not accidental. Nineteen ninety-three marked the 150th anniversary of the delineation of *homology* ("*homologue*"[1]) from *analogy* ("*analogue*") by Richard Owen (1804-1892). He became Sir Richard and (the first) Superintendent (Director) of the British Museum (Natural History), a position he held for 27 years beginning in 1856. Boyden (1943) produced a seminal review of Owen's role to mark the centennial of his definitions of homology and analogy. A volume commemorating the centennial of his death was published recently (Gruber and Thackray, 1992). Other thoughtful

---

[1]The British rather than the U.S. spelling of homologue(s) [U. S. homolog(s)] is used throughout this book. In part, this is to conform to the original spellings. It is also because homologue has become the accepted scientific usage; none of the U. S.-based authors used the U. S. spelling in their chapters.

analyses of Owen and his work are those by Huxley (1894),
Clark (1900),[2] Ghiselin (1976), and Richards (1987).

Owen delineated homology from analogy in the glossary
to the published version of his *Lectures on Comparative
Anatomy and Physiology of the Invertebrate Animals,
Delivered at the Royal College of Surgeons in 1843*:

> Homologue...The same organ in different animals under every vari-
> ety of form and function....Analogue...A part or organ in one animal
> which has the same function as another part or organ in a different
> animal. (Owen, 1843, pp. 379, 374)

Owen's definitions are deceptively simple, especially as
they are limited to structures, and within structures to
organs. Organs that are the same in different animals (or
plants) are homologous, even if they serve different functions in those organisms. Organs that are not the same in
different organisms, but that serve the same function, are
analogous.

III. HISTORY OF THE TERM AND CONCEPT

Owen's delineation of homology from analogy was
neither the first recognition of these concepts, nor the first
use of the terms. These two issues are discussed by Alec
Panchen in Chapter 1 in an analysis of Richard Owen's contribution to homology.

Aristotle was familiar with homologous structures. He
used them in his descriptions and classifications of animal
forms and relationships. Homology, therefore, has been tied
to morphology and classification for millennia. Belon presented a new view of animal forms (what we would now call
homologues) in his representation of equivalent structures
(skeletons) across different animal groups (vertebrates). His
famous 1555 depiction of comparisons between the skeletons of birds and humans is reproduced in Chapter 1 and
discussed by Rieppel in Chapter 2.

---

2 Clark's book, *Old Friends at Cambridge and Elsewhere*, consists of
"biographical notices of distinguished Cambridge men" (p. v). Owen is the
only "non-Cambridge" man included in the book, an inclusion justified by
Clark, who argued "that the Senate coopted Owen by selecting him, in 1859,
as the first recipient of an honorary degree under the new statutes" (p. vii).

Etienne Geoffroy Saint-Hilaire developed his *principle des connections* in a series of publications from 1796 onwards, most fully in his two-volume work, *Philosophie Anatomique* (1818-1822). For Geoffroy, shape or composition of structures might differ between organisms but constancy of relationships and of connections to adjacent structures enabled homologous (what he termed analogous) structures to be identified in two organisms under investigation.

Although we celebrated 1993 as the sesquicentennial of Owen having delineated homology from analogy, both the terms and the concepts were in use well before 1843. Homology was in use as an Anglicized term for agreement about things from at least the mid-1700s onwards. It has been used as an anatomical term from early in the 19th century (Huxley, 1894).

Panchen discusses the legacy acknowledged by Owen to earlier workers in proffering "his" definitions. Owen acknowledged Strickland as having precedence but, as Panchen convincingly argues, it was William Sharp MacLeay who drew the distinction between homology and analogy in 1821. Panchen's discussion reminds us of the complexities and difficulties involved in tracing the history of ideas, terms, and definitions. We celebrate in this volume therefore, the 172nd anniversary of Macleay's elucidation and the sesquicentennial of Owen's.

Owen, as had Oken and Goethe before him, made much use of the vertebral origin of the skull (skulls as a linear series of modified vertebrae) to illustrate how homology could be used to trace structures across a natural group of organisms, in this case, across the vertebrates. Owen, the last of the grand idealistic morphologists,[3] made much use of the *archetype* in his concept of homology. His 1848 book, produced under commission from the British Association for the Advancement of Science, was titled *On the Archetype and Homologies of the Vertebrate Skeleton*.

---

3 Although Russell (1916) regarded Owen as the last of the grand idealistic morphologists, Rolf Sattler (personal communication) author of Chapter 13 in the present volume, reminded me of my less than adequate background in plant morphology by claiming that position for Wilhelm Troll, "perhaps the most influential plant morphologist in the 20th century."

Russell's wonderful history of animal morphology contains a thorough analysis of Owen's contribution (Russell, 1916).

Currently, definitions of homology that embody typology are often called *idealistic homology*. In his chapter Panchen has included II from Owen's treatise, in which the archetype is beautifully illustrated. Most writers — Russell, Patterson, Ruse, Rudwick, Appel, and I — have used this depiction, but only reproduced Fig. 1 (*Archetypus*) from Owen's Plate II. Panchen has done the considerable service of reproducing and discussing the other 18 figures found in Owen's original plate. These illustrate the vertebrate archetype in fish, reptiles, birds, mammals, and man, and depict representations of archetypal vertebral columns and limbs.

In his treatment of definitions and concepts, Panchen discusses Lankester's introduction of the terms *homoplastic* (homoplasy) and *homogeny*, the latter used by Lankester to incorporate descent from a common ancestor into a concept of homology. Lankester thought the term homology too loaded with residues of Platonic idealism and notions of the archetype, and so substituted homogeny. [Recall that homology developed, not as an evolutionary concept, but as an aid to classification and systematics. The definition of homology proposed by Owen and Macleay preceded the concept of descent with modification from a common ancestor published by Charles Darwin in *The Origin of Species*].

Panchen introduces a further topic that recurs throughout this volume. This is the relationship(s) between embryonic development and homology. In issues going to the heart of the questions posed in Section I, Owen argued that embryology, which can be a *criterion of homology*, is also, along with phylogenetic history and the archetype, a possible *explanation* for homology. For Darwin, community of embryonic development provided powerful, but not over-riding, evidence for commonality (homology) of structures through descent from a common ancestor.

> Thus, community in embryonic structure reveals community of descent; but dissimilarity in embryonic development does not prove discommunity of descent, for in one of two groups the developmental stages may have been suppressed, or may have been so greatly modified through adaptation to new habits of life, as to be no longer recognizable. (Darwin, 1910, pp. 371-372)

The precise relationships between embryonic development and the *definition(s), criteria,* and *explanation(s)* of homology are an active area of enquiry amply represented in this book; see also Hall (1992, 1994).

Panchen argues for a clear threefold separation of "definitions of homology, criteria for testing proposals of homology, and explanations of … homology." Such a separation goes a long way toward clarifying current difficulties. In the end Panchen concludes that, however unsatisfying, "the same organ in different animals under every variety of form and function" (Owen, 1843) is the appropriate definition of homology.

IV. A PHYLOGENETIC DEFINITION OF HOMOLOGY

Chapters 2 to 4 evaluate the phylogenetic definition of homology, using specific examples spanning several levels across the biological hierarchy. They provide a telling testament to homology as the hierarchical basis of comparative biology.

In Chapter 2, Olivier Rieppel takes us into a history of modern debates on relationships between homology, topology, and typology. Those who reject typology as Platonic idealism, might consider Tomlinson's first homily, "do not eschew typology, it is the first step in any morphological analysis" (1984, p. 380).

Rieppel's concern is with homology as similarity[4] due to common descent (the *phylogenetic, evolutionary,* or *his-*

---

[4] Homologous features need not be structurally similar (Moment, 1945; Cracraft, 1967; Smith, 1967; Gans, 1985). Bones of the middle ears of mammals are homologus with the bones of lower jaws of reptiles but certainly not similar to them. Analogy and homoplasy, not homology, are identified on the basis of similarity of structure or function, similarity being some commonality, not necessarily of the entire features. [See Young (1993) for the contrary view of homology as fundamentally a similarity-based concept]. Potential (oftentimes actual) independence of similarity of structure, embryonic development, and function/behavior has been discussed by Gans (1985). Sattler (1984) argued that homology should be adandoned for "resemblances" based on similarity, structural correspondence, or structural relationships. The issue of similarity versus complexity and homology assessment in plants is dealt with by Donoghue and Sanderson in Chapter 12. Whether sequence data provide evidence of homology or similarity is discussed in Chapters 9 to 12.

torical definition of homology). He is also concerned with relationships between ontogeny, homology, homoplasy, and types. Types in the context of cladistic analysis are "groups of organisms tied together by shared mechanisms preserving relative topological relations of constituent elements in a biological structure." Homology is "the potential expressed by shared ontogenetic primordia" (see also Chapters 8 and 9).

Rieppel comes to grips with the relationship between *proximate* (ontogenetic) and *ultimate* (evolutionary) causes of homology. He discusses *developmental constraints* as a proximate explanation for homology. Shared developmental mechanisms or constraints (proximate causes) are seen as the essence of monophyletic taxa; homology is always only an hypothesis of similarity due to common ancestry (ultimate causation). As Rieppel concluded a recent review on logical underpinnings of the homology concept: "Homology is a relation bearing on recency of common ancestry, not on commonality of developmental pathways or constraints" (1992, p. 713).

I share Rieppel's view that homology need not require commonality of embryonic development. My assessment was that "homology is a statement about pattern, and should not be conflated with a concept about processes and mechanisms" (Hall, 1992, p. 194). I would much prefer to have come to a different conclusion. There are, however, sufficient instances of clearly homologous characters that do not share common developmental pathways that commonality of development cannot be used as a mandatory criterion of homology. These examples, which have been discussed in detail elsewhere (Hall, 1992, 1994), include mechanisms of gastrulation, origin of the alimentary canal and germ cells, induction of the eye and Meckel's cartilage, development of internal or external cheek pouches in squirrels, mice, and geomyoid rodents, selection for increased tail length in mice, and regeneration of tails and lenses.

Is the lens that regenerates from the iris in a urodele amphibian not homologous with the lens in the other eye because one arose from dedifferentiated iris cells and the other arose *de novo* through the mechanisms of embryonic development? If increased tail length obtained through selection in mice occurs by *addition* of a vertebra in one line but by increased growth of *preexisting* vertebrae in another, are the tails (or vertebrae) not homologous because differing processes brought them into being? Are regenerated tails not homologous with the tail they replaced? Regenerates arise from a blastema of dedifferentiated cells and depend on nervous innervation while the original tails arose from embryonic mesenchyme independently of innervation. If a lizard regenerates the distal third of its tail, should we regard that third as not homologous with any portion of the original tail? In all cases the answer to these and similar questions is that homology is retained even when developmental pathways vary.

Relationships between homology, cladistics, and systematics are discussed fully and forthrightly by Gareth Nelson in Chapter 3. As noted by Gegenbaur 115 years ago, "Blood relationship becomes dubious exactly in proportion as the proof of homologies is uncertain." Homology is considered by Nelson to be a "set of shared derived characters indicative of phylogenetic relationships" — a *synapomorphy*. This is the relation which characterizes monophyletic groups (Patterson, 1982). With this definition, taxa relate whole organisms, while homology relates their parts. This position is echoed by Lauder in Chapter 4: "It is the phylogenetic relationships among taxa that allow us to assess the homology or analogy of individual characters".

As noted at the outset, homology can apply to population and community levels of biological organization. Nelson pursues this notion in a section devoted to biogeography — Croizat's panbiogeography — built on a basis established by the British botanist John Christopher Willis. In his discussion, Nelson includes the important concept of time as an independent criterion of homology.

Application of the concept of homology to populations, communities, and biogeographic regions (briefly) as well as to function and behavior (in detail) is the concern of George Lauder in Chapter 4. (Homology and behavior is dealt with in detail by Harry Greene in Chapter 11). Lauder decries requirements for prior determination of structural homologies in studies of behavior. Nonstructural characters, such as behavior or function, stand as homologies in their own right, requiring no *a priori* assessment of a common structural basis of structural homology. Lauder notes that Owen's definition of homology was restricted to structure ("the same *organ* in different animals...".

Lauder's chapter is explicitly hierarchical, as is his view of homology; see, for example, his Table II. Like Rieppel and Nelson, he advocates a phylogenetic definition of homology, recognizing homologous traits "*a posteriori* as a consequence of a global phylogenetic analysis of many characters of all kinds." For Lauder, recognition of two characters as homologous requires analysis of the phylogenetic distribution "not only of those characters, but also of many other characters...". No single set of criteria (genes, development, connections, structure) is adequate for determining homology. And no criterion is adequate unless framed in a phylogenetic context. "The claim of homology...stands or falls on the basis of the phylogeny as a whole....The result of the phylogenetic analysis alone determines if two characters are homologous." Lauder also provides the first consideration in this volume of serial or *iterative homology*, the class of homology that provided so much trouble for de Beer in his classic 1971 reader on homology, for more of which see below.

Adoption of a phylogenetic definition of homology forced Fred Bookstein, author of Chapter 5, to an uncomfortable conclusion. Morphometric characters (biometrical shape) cannot be homologous characters. Bookstein's penetrating analysis is rooted in studies of Karl Pearson, originator of biometrics and cofounder of the journal *Biometrika*. Bookstein cogently argues that "similarity" in biometrics (regression coefficients among correlations) is not, and cannot, be the same as "similarity" in homology ("the same organ in every animal..."). Consequently, morphometric data do not contribute evidence that can be used in assessing homology. There is no biometrical homology.

Bookstein discusses morphometrics and homology by examining the method of Cartesian transformations introduced by D'Arcy Wentworth Thompson. He also utilizes his own "partial warps," which are a means for analyzing biometric criteria, independent of homology. Incongruence between biometrics and homology occurs because "morphometric shape variables cannot possibly form a hierarchy...Owen's homology, at root, is thus not a metric concept, biometric or otherwise". As Bookstein sees it: "The languages of homology and of morphometrics are mutually incomprehensible."

V. EMBRYONIC DEVELOPMENT AND HOMOLOGY

With Chapter 6, Brian Goodwin leads us into three chapters devoted to relationships between embryonic development and homology, what has been called *biological homology*.

As did Panchen, Rieppel, Lauder, Roth and Shubin, Goodwin introduces de Beer's insightful 1971 reader *Homology: an Unsolved Problem*. Gavin de Beer, like Owen before him, was knighted for services as Director of the British Museum (Natural History). de Beer's seminal treatment dealt deftly and succinctly with problems associated with each of the criteria by which homology might be identified. I have recently analyzed these in relation to embryonic development and homology (Hall, 1992, 1994), and my position on the relationship between homology and embryonic development was summarized in Section IV.

Goodwin takes up a discussion of whether homology has a genetic basis, utilizing recent studies on homeobox (especially *Hox*) genes in vertebrate limbs. Homology for Goodwin is equivalence, specifically "an equivalence relation over the members of a set, defined by a transformation that takes any member into any other member within the set." Hierarchical relationships between developmental processes and generative rules of embryonic development determine the members of a set. Morphogenetic processes therefore can be used to produce a hierarchical taxonomy of biological forms. Goodwin develops this concept by utilizing a model for the generation of patterns of vertebrate limbs, the organ system used by Owen in his classic studies on the archetype and homology. Neil Shubin also uses the vertebrate limb as the major example in his chapter. Other recent analyses of the archetype in vertebrate limb development, evolution, and homology may be found in Hinchliffe (1989, 1991) and Müller (1991).

Methodologies of developmental biologists such as Goodwin vary considerably from the cladistic approach of many systematists outlined by Gary Nelson in Chapter 3 and in Section I above. Nevertheless, in both of these subdisciplines, embryonic development is linked to systematics through homology.

Like all "new" ideas, the notion of a link between systematics and homology is not new, but rooted in earlier morphological analysis. As Thomas Henry Huxley noted in an essay on Richard Owen's position in the history of anatomical sciences, published in 1894:

> Thus, in course of time, there arose in the minds of thoughtful systematists a distinction between 'analogies' and 'affinities;' and, in those of the philosophical anatomists, a corresponding discrimination between 'analogous' and 'homologous' structures. (Huxley, 1894, Vol. II, p. 301).

Neil Shubin explores the tetrapod limb in some detail in Chapter 7. He especially uses the digits of avian wings and urodele hands in an analysis of relationships between homology, ontogeny and the archetype. Shubin makes explicit three steps involved in the analysis of homology; formulate an hypothesis; test the hypothesis; determine the mechanistic explanation of that hypothesis. Shubin argues persuasively for the utility, indeed the necessity, of both historical and ahistorical approaches to homology.

In Chapter 8, Günter Wagner develops the concept of biological homology by distinguishing the developmental processes that generate structures (what he calls *morphogenetic mechanisms*) from those processes that maintain structures that are already formed (what he calls *morphostatic mechanisms*). Wagner argues that greater emphasis on morphostatic mechanisms will have important implications for any study of relationships between homology and intraspecific variation.

Wagner's approach to homology is rooted in structure: "structural identity is conceptually required for the historical homology concept to make sense and thus has to be the starting point for a theory of homology." He provides a modification of Owen's definition that illuminates the utility of the biological homology concept. Lack of heritable variation seen in conserved structures is discussed as due to effects of *developmental constraints*. Coexistence of developmental constraints with variable developmental pathways[5] is discussed in relationship to stabilizing selection, and the separation of generative and morphostatic mechanisms by generative and morphostatic constraints. Wagner ends with

---

5 These are homologous structures that can arise by different developmental processes. I termed them "nonequivalent developmental processes" (Hall, 1992).

the potential importance of genes, such as the *Hox* series, that establish the structural identity of morphological characters. [This aspect is developed more fully by Brian Goodwin in Chapter 6]. Homeobox genes may provide the genetic basis for homologous features of phylotypic stages (zootypes) uniting natural groups of organisms (Hall, 1992; Slack *et al.*, 1993), an exciting prospect, with the potential to open-up evolutionary developmental biology to genetic analysis.

VI. CONTINUITY OF INFORMATION AND HOMOLOGY

Louise Roth takes a broadly synthetic approach in her analysis of replicators and lineages in Chapter 9. She examines homology both between and within organisms — homologue and serial homology as defined by Owen — and comes to grips with one of the more difficult problems confronting those who study homology; does homology describe pattern or process? Because Roth follows Van Valen in equating homology with correspondence and continuity of information,[6] she takes a hierarchical approach to homology and so is able to incorporate both pattern and process in her analysis. Her chapter includes an explicit discussion of the sources of continuity in the context of the replicators responsible for maintaining continuity of information. An extensive argument for homology as hierarchical (presented in the context of the differing levels at which natural selection acts) is presented as the means "to discover what the most fundamental properties of biological lineages and replicators are that cause differences in their behavior at different levels." Roth presents what she sees as a means to reconcile, on the one hand, analysis of pattern and process in population biology, systematics, genetics, and developmental biology, and on the other, taxic and transformational approaches in comparative biology.

---

6 "Homology is resemblance caused by a continuity of information. In biology it is a unified developmental phenomenon" (van Valen, 1982, p. 305).

VII. HOMOLOGY IN MOLECULAR BIOLOGY

David Hillis begins Chapter 10 by asking whether it is sensible to speak of genes sharing 50% homology when what is really meant is 50% similarity.[7] He sets out a detailed argument for homology in molecular biology as having to fit a definition of evolutionary homology. In so doing he distinguishes between and among homology, similarity, homoplasy, analogy, isology, ortology, paralogy and xenology. Hillis addresses mechanisms producing molecular variation among species, emphasizing relationships between homology at molecular and morphological levels. This is an important analysis. Molecular biology may not "solve the problem of homology," but any rational approach to homology as the hierarchical basis of comparative biology must come to grips with homology in molecular biology.

VIII. HOMOLOGY AND BEHAVIOR

Including behavior in analyses of homology has been alluded to in several places in this introduction. The hierarchical view outlined in the opening paragraph included behavior as one of the levels of biological organization at which homology may be studied. In posing the list of unanswered questions about homology in Section I, I asked whether homology must always be tied to structure, even when behavioral or physiological characters are being investigated. In Chapter 4, George Lauder emphatically argued that nonstructural characters (behavior, function) stand as homologues in their own right. This issue of homology and behavioral repertoires is discussed more fully by Harry Greene in Chapter 11, who provides the most thoughtful analysis of the topic since the chapters by Atz published almost 25 years ago (Atz, 1970) and by Lauder almost a decade ago (Lauder, 1986).

Greene dismisses notions that behavior is more variable or more difficult to measure and compare than morphology. He discusses the problem of lack of a "behavioral fossil record," but then provides some intriguing examples from the fossil record. His analysis is primarily illustrated with

7 See footnote 4 and Chapter 12.

behavior associated with the capture of prey by snakes, in an approach to the homology of behavior that is overtly phylogenetic and optimistic, despite the paucity of data, Greene places behavior squarely in the realm of hierarchical analysis of comparative biology.

IX. HOMOLOGY AND PLANT BIOLOGY

The issue of similarity as a test (the most powerful test?) of homology is discussed in depth by Michael Donoghue and Michael Sanderson in Chapter 12. This and the subsequent chapter by Rolf Sattler is devoted explicitly to detection of homology in plants. Nevertheless, themes developed in the "animal" chapters — complexity, the role of development as a criterion for homology, and relationships between biological and evolutionary homology — emerge as common themes in the "plant" chapters.

Donoghue and Sanderson deal particularly with real or perceived differences between morphology and molecules in assessing homology. They also deal with complexity. What is it? How can it be measured? How does complexity relate to the determination of homology? They emphasize a shift away from the "morphology versus molecules" debate toward determination of "the relative simplicity or complexity of the features under consideration." Complexity is essentially the number of parts and the irregularity of their arrangement. They trace their approach back to Remane's classic 1952 treatment of homology *Die Grundlagen des natürlichen Systems, der vergleichenden Anatomie und der Phylogenetik*. Donoghue and Sanderson propose a test of complexity and apply it to morphological cladistic analyses of angiosperms. This test is capable of integrating morphological and molecular data as a function of structural complexity. The potential integration of biological and historical approaches to homology, using such an approach, is emphasized. Again, homology emerges as the hierarchical basis of comparative biology.

Detection of homology in plants is also the theme of the final chapter by Rolf Sattler. He begins with a comprehensive introduction to the kinds of homology, squaring the circle begun by Panchen in Chapter 1. Sattler pays special attention to correspondence ("the common basis of all ver-

sions of homology") and to partial correspondence; his
dynamic view of homology requires recognition of partial
correspondences. Partial correspondence is where "a structure does not share all relevant properties with another
one." These two aspects of homology were considered in
some depth by Owen when he delineated criteria for detect-
ing and classifying homology. Although previously skeptical
about partial homology as a viable concept before reading
Sattler's chapter, I found myself convinced by the logic and
cogency of his arguments.

Sattler also provides an extensive discussion of the
relationships between *homeosis* and homology. He uses
replacements of stamens by petals or by branchlets to illustrate the roles of position and one-to-one correspondence
in the determination of homology. Partial correspondence
nicely gets around the twofold problems of intermediate
forms and homeosis, where similarity and corresponding
connections are broken. Sattler uses principal component
analysis, continuum and process morphology, and morphological distance to provide a dynamic approach to structure
as process and to homology as a statement about "properties
of structures," in an approach essentially similar to that
taken by Rieppel, Goodwin, and Wagner. The challenge for
the future is to determine how morphological distances
relate to phylogenetic distances, and how degree of similarity relates to phylogenetic reconstruction and/or phylogenetic relationships. What is the hierarchical relationship
between biological and phylogenetic homology?

Similarity and complexity, ontogeny and phylogeny,
structure and behavior, form and function, pattern and
process — relationships between and among these pairs are
the foundation of homology as the hierarchical basis of comparative biology.

ACKNOWLEDGMENTS

It is a pleasure to thank all the contributors for their
thorough, scholarly and up-to-date contributions and for the
timely submission of their chapters. David Carlson provided
input during the development of the themes discussed in
the volume. Charles Crumly of Academic Press steered the
volume through production. June Hall provided invaluable

help and support with the production of the camera-ready copy. Much of the editorial work was carried out during a sabbatical leave in the Department of Anatomy and Human Biology of the University of Western Australia. I thank the members of that department for their generous welcome, the Raine Foundation of The University of Western Australia for the awarding of a Visiting Professorship, and the Killam Trust of Dalhousie University for the provision of a research professorship. My own research is supported by NSERC (Canada) and NIH (USA).

REFERENCES

Atz, J. W. (1970). The application of the idea of homology to behavior. *In* "Development and Evolution of Behavior" (L. R. Aronson, E. Tobach, D. S. Lehrman, and J. S. Rosenblatt, eds.), pp. 53-74. W. H. Freeman, San Francisco.

Belon, P. (1555). "L'Histoire de la Nature des Oyseaux." Guillaume Cavellat, Paris.

Boyden, A. (1943). Homology and analogy: A century after the definitions of "homologue" and "analogue" of Richard Owen. *Q.. Rev. Biol.* **18**, 228-241.

Clark, J. W. (1900). "Old Friends at Cambridge and Elsewhere." Macmillan, London.

Cracraft, J. (1967). Comments on homology and analogy. *Syst. Zool.* **16**, 355-359.

Darwin, C. (1910). "The Origin of Species by Means of Natural Selection." John Murray, London.

de Beer, G. R. (1971). "Homology: An Unsolved Problem." Oxford University Press, Oxford.

Gans, C. (1985). Differences and similarities: comparative methods in mastication. *Am. Zool.* **25**, 291-301.

Gegenbaur, C. (1878). "Elements of Comparative Anatomy" (F. J. Bell, translator), Macmillan, London.

Geoffroy Saint-Hilaire, E. (1818-1822). *Philosophie Anatomique.* Vol. 1. *Des Organes respiratoires sous le rapport de la détermination et de l'identité de leurs pièces osseuses.* Vol. 2. *Des monstruosités humaines.* J.-B. Baillière, Paris.

Ghiselin, M. T. (1976). The nomenclature of correspondence: A new look at "homology" and "analogy." *In*

"Evolution, Brain and Behavior: Persistent Problems" (R. B. Masterton, W. Hodos, and H. Jerison, eds.), pp. 129-142. Lawrence Erlbaum, Hillsdale, New Jersey.

Gruber, J. W., and Thackray, J. C. (eds.) (1992). *Richard Owen Commemoration.* Cambridge University. Press, Cambridge.

Hall, B. K. (1992). "Evolutionary Developmental Biology." Chapman and Hall, London.

Hall, B. K. (1994). Homology and embryonic development. *Evol. Biol.* **28** (in press).

Hinchliffe, J. R. (1989). Reconstructing the archetype: Innovation and conservatism in the evolution an development of the pentadactyl limb. *In* "Complex Organismal Functions: Integration and Evolution in Vertebrates" (D. B. Wake and G. Roth, eds.), pp. 171-189. John Wiley & Sons, New York.

Hinchliffe, J. R. (1991). Developmental approaches to the problem of transformation of limb structure in evolution. *In* "Developmental Patterning of the Vertebrate Limb" (J. R. Hinchliffe, J. M. Hurle and D. Summerbell, eds.), NATO ASI Series, Series A: Life Sciences, pp. 313-324. Plenum, London.

Huxley, T. H. (1894). Owen's position in the history of anatomical science. *In* The Life of Richard Owen by His Grandson the Rev. Richard Owen, M. A." Vol. 11, pp. 273-332. John Murray, London (Reprinted in 1970 by Gregg International Publishers, Westmead, England.)

Lauder, G. V. (1986). Homology, analogy, and the evolution of behavior. *In* "Evolution of Animal Behavior" (M. H. Nitecki and J. A. Kitchell, eds.), pp. 9-40. Oxford University Press, New York.

Moment, G. B. (1945). The relationship between serial and special homology and organic similarities. *Am Nat.* **79,** 445-455.

Müller, G. B. (1991). Evolutionary transformations of limb pattern: Heterochrony and secondary fusion. *In* "Developmental Patterning of the Vertebrate Limb" (J. R. Hinchliffe, J. M. Hurle, and D. Summerbell, eds.), NATO ASI Series, Series A: Life Sciences, pp. 395-405. Plenum, New York.

Owen, R. (1843). "Lectures on Comparative Anatomy and Physiology of the Invertebrate Animals, Delivered at the

Royal College of Surgeons in 1843." Longman, Brown, Green, and Longman, London.

Owen, R. (1848). "On the Archetype and Homologies of the Vertebrate Skeleton." Richard and John E. Taylor, London.

Patterson, C. (1982). Morphological characters and homology. *In* "Problems of Phylogenetic Reconstruction" (K. A. Joysey and A. E. Friday, eds.), Systematics Association Special Volume 25, pp. 21-74. Academic Press, London.

Patterson, C. (1988). Homology in classical and molecular biology. *Mol. Biol. Evol.* **5**, 603-625.

Richards, E. (1987). A question of property rights: Richard Owen's evolutionism reassessed. *Br J. Hist. Sci.* **20**, 129-171.

Rieppel, O. (1992). Homology and logical fallacy. *J. Evol. Biol.* **5**, 701-715.

Russell, E. S. (1916). "Form and Function. A Contribution to the History of Animal Morphology." John Murray, London. (Reprinted in 1972 by Gregg International Publishers, Westmead, England, and in 1982 by the University of Chicago Press, Chicago, with a new introduction by George V. Lauder.)

Sattler, R. (1984). Homology — a continuing challenge. *Syst. Bot.* **9**, 382-394.

Slack, J. M. W., Holland, P. W. H., and Graham, C. F. (1993). The zootype and the phylotypic stage. *Nature* (*London*) **361**, 490-492.

Smith, H. M. (1967). Biological similarities and homologies. *Syst. Zool.* **16**, 101-102.

Szarski, H. (1949). The concept of homology in the light of the comparative anatomy of vertebrates. *Q Rev. Biol.* **24**, 124-131.

Tomlinson, P. B. (1984). Homology: An empirical view. *Syst. Bot.* **9**, 374-381.

Van Valen, L. (1982). Homology and causes. *J. Morphol.* **173**, 305-312.

Wagner, G. P. (1989). The biological homology concept. *Annu. Rev. Ecol. Syst.* **20**, 51-69.

Young, B. A. (1993). On the necessity of an Archetypal Concept in morphology: with special reference to the concepts of "Structure" and "Homology." *Biol. Philos.* **8**, 225-248.

1

RICHARD OWEN AND THE CONCEPT OF HOMOLOGY

Alec L. Panchen

Department of Marine Sciences
University of Newcastle upon Tyne
Newcastle upon Tyne
NE1 7RU
England

I. INTRODUCTION

In 1971, subsequent to his retirement from the direc-
torship of the British Museum (Natural History) in London,
Sir Gavin de Beer published a sixteen-page booklet in a
series entitled "Oxford Biology Readers". These little
reviews, aimed at high school or undergraduate audiences,
were written by eminent biologists on problems or lines of
research of contemporary interest. de Beer was much involved in this series: he wrote "Some General Biological
Principles Illustrated by the Evolution of Man" (1971a),
"Adaptation" (1972), and "Evolution of Flying and Flightless
Birds" (1975). But on this occasion his subject was
"Homology, an Unsolved Problem" (de Beer, 1971b).

As an eminent comparative anatomist and embryologist
and also as a biographer and editor of Darwin and his work,
it is not surprising that de Beer attributed the phenomenon
of homology to community of descent:

> In other words it is homologous organs that provide evidence of
> affinity between organisms that have undergone descent with modi-
> fication from a common ancestor, i.e. evolution. Furthermore since
> evolution is the explanation of the 'agreement' between homologous
> organs, their study, if they are hard parts susceptible of fossilization,
> is not restricted to the morphology of living organisms, but the entire
> range of palaeontology is available for it. (de Beer, 1971b, p. 4)

So the problem was not the explanation of the
phenomenon of homology — that was evolution — but some
of the criteria that had been proposed for adjudicating on
hypotheses of homology. de Beer was led to end the introductory section of his review thus:

> So, provided with a cast-iron explanation in terms of affinity,
> of inheritance in evolution from a common ancestor, it looked as if
> the concept of homology was at last soundly based and presented no
> more problems of principle; however, as will be seen below, it unfor-
> tunately does. (de Beer, 1971b, pp. 4-5)

The body of the review leads off with the idea that the
parts of the flowers of angiosperms, that is, carpels,
stamens, petals, and sepals, are modified leaves, a theory
first proposed by Goethe (1807). This is followed by the
cliché example of transformational homology, the homology

of two mammalian ear ossicles, incus and malleus, with the two cartilage bones (quadrate and articular, respectively) that form the jaw joint in reptiles — and, either as cartilage or bone, that of all other jawed vertebrates. Then de Beer goes on the talk about "conservative effects of homology," taking the recurrent laryngeal nerve, "an example of how the topology of homologous structures determines some curious anomalies in adult anatomy."

But after these examples trouble sets in. The first problem for the concept of homology is characterized by de Beer as "the displacement of homologous structures."

> There is no doubt whatever that the forelimb in the newt and the lizard and the arm of man are strictly homologous, inherited with modification from the pectoral fin of fishes 500 million years ago. They have identical elbow and wrist joints and their hands end in five fingers [not true for the newt which has only four, the maximum for extant Amphibia]. The bones and muscles that they contain also correspond. But a minute examination of their comparative anatomy reveals the astonishing fact that they do not occupy the same positions in the body. (de Beer, 1971b, p. 8)

He goes on to explain that vertebrate limbs — bones, muscles, nerves — are formed from several adjacent trunk segments. But the forelimb of the newt is formed from segments 2-5, that of the lizard from segments 6-9, and that in humans from segments 13-18. "The limb is a pattern which has been transposed over the long axis of the vertebrate body, like a tune that can be transposed over the keys...." A similar example, cited by de Beer, is the position of the occipital arch, which marks the hind end of the skull. But a different number of segments is incorporated into the head, thus lying in front of the arch, in sharks, newts, frogs, reptiles, and mammals.

de Beer then goes on to talk about serial homology, after which his next anomaly is what he refers to as *latent homology*. This is where a feature occurs commonly in some high-ranking animal taxon, a large taxonomic group of animals, but is not of general occurrence in some or all of the subgroups into which each of the major groups is divided. Thus ants, bees, and wasps all include species that have complex innate social organization, but each group also contains solitary species and there are some intermediate forms. It is therefore assumed that the inclusive group, the

Hymenoptera, shows a tendency to evolve social instincts, that "there is a genetically based homology which provides some evidence of affinity between the groups that show it." de Beer argues similarly for spiral cleavage of the fertilized egg in the development of some platyhelminth worms, nemertines, annelids, and molluscs, and which is taken to unite the animal phyla that they represent.

His next anomaly concerns differences in individual development of apparently homologous structures. Thus, the alimentary canal in vertebrates can be formed from the roof of the embryonic gut cavity (sharks), the floor (lampreys, newts), both (frogs), or from the lower layer of the blastoderm (reptiles, birds). The lens of the eye has to be induced by the underlying optic cup in the frog species *Rana fusca*, but in the closely related *Rana esculenta* [as we now know, not a true species but a hybrid (Uzzell *et al.*, 1977)] the lens develops from the epidermis with or without induction. These two cases move de Beer to the following aphorisms (his italics):

> ... *correspondence between homologous structures cannot be pressed back to similarity of position of the cells of the embryo or the parts of the egg out of which these structures are ultimately differentiated.*
> ... *homologous structures can owe their origin and stimulus to differentiate to different organiser-induction processes without forfeiting their homology.* (de Beer, 1971b, p. 13)

de Beer's final anomaly, concerning homology and genetics, inspires another aphorism:

> ... *homologous structures need not be controlled by identical genes, and homology of phenotypes does not imply similarity of genotypes.* (de Beer, 1971b, p. 15, emphasis de Beer's)

The next example is not in de Beer's review.

The butterfly *Papilio dardanus* occurs in all those parts of Africa south of the Sahara that are not desert, and also as isolated populations in Ethiopia, Madagascar and the Comoro islands (Ford, 1975; Turner, 1962). In the last two localities the females are male-like, with a pattern of bright lemon yellow and black on the wings; this is also the case with most females in the Ethiopian population. But the butterfly is famous because in the main African stock, divided into a series of races (Vane-Wright and Smith, 1991), all females are strikingly unlike the males and are usually polymorphic, with each common morph a mimic of a species of poisonous and/or distasteful butterfly from either the subfamily Danainae or the Acraeinae. In the 1960s Clarke and Sheppard investigated the genetics of the polymorphism, using a technique of hand-pairing to make all possible crosses. Although all males from the main African stock look similar, each belongs genetically to one of the female morphs and thus, if its history is known, can be used in meaningful crosses — the morph appearance is sex-limited to the females. Figure 1 shows a male and three females, the latter with their respective models.

Clarke and Sheppard investigated, *inter alia*, the relationship of genetic dominance between different morphs from the same race within the main African stock. The three female morphs shown, all mimics of danaine species, (Fig. 1 e-g) proved to represent a dominance series; *hippocoonides* (Fig. 1e) is recessive to the other two. Thus, all the first generation female offspring between *hippocoonides* and either *cenea*, or *trophonius*, would look like pure-bred *cenea* or *trophonius*, respectively. Then while *cenea* is dominant to *hippocoonides* it is recessive to *trophonius*, so that *trophonius* is dominant to the other two. Thus, these complex wing patterns behave genetically as though each were controlled by a single Mendelian gene (in fact a switch gene within a supergene complex), all the genes being alleles at the same locus. This simple dominance relationship is obviously adaptive — if "hybrids" (heterozygotes) between two morphs were of intermediate appearance they would not look like any model butterfly: their mimetic protection would be lost.

But there are two more mimetic female morphs in some populations, each of which mimics a different acraeine butterfly (Fig. 2). These morphs are *planemoides* and *niobe*, and they might be expected to continue (or at least fit into) the dominance series. But if *planemoides* is crossed with *trophonius* the result (if female) looks identical to *niobe* (Clarke and Sheppard, 1960). It is not, however, "true" *niobe* because there is a *niobe* gene quite distinct from those for *planemoides* and *trophonius*. So are all the features of wing and body pattern and color homologous between true and false *niobe*, or not? The answer is yes, if one accepts de

Fig. 1. The mimetic swallowtail *Papilio dardanus* Brown and its models. (a) nonmimetic male; (b-d) danaine model species, (e-g) mimetic *P. dardanus* females: (e) *hippocoonides*, (f) *cenea*, (g) *trophonius*. [After Panchen (1993)]

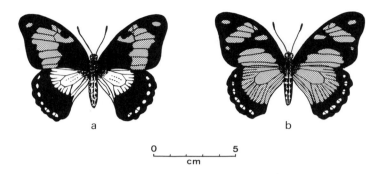

Fig. 2. Female morphs of *Papilio dardanus* Brown. (a) *planamoides*, (b) *niobe.*

Beer's reasoning — "homologous structures need not be controlled by identical genes...."

So, de Beer demonstrated that homologous structures in two or more related species may differ in their relationship to the segments or other markers of position in the body, in their origin within the developing embryo, in their mode of induction in embryology, and in their genetic origin. Furthermore, features apparently homologous between animal groups may not be of universal occurrence within those groups (as is the case with eusociality in hymenopterous insects).

So, while admitting that topographical, embryological, genetic, and taxonomic criteria of homology may all fail, de Beer seems sure of two things, the first explicit, the second not overtly expressed. The first is that the explanation of homology is evolutionary, or rather phylogenetic, community of descent. The second is that homology is a term properly applied to anatomical features: anatomical criteria override those from embryology or genetics.

In setting out the orthodox position, that evolution is the explanation of homology, de Beer quotes a favorite passage from *The Origin of Species* (Darwin, 1859, pp. 434-5):

> What can be more curious than that the hand of a man, formed for grasping, that of a mole for digging, the leg of the horse, the paddle of the porpoise, and the wing of the bat, should all be constructed on the same pattern, and should include the same bones, in the same relative positions?

Darwin then goes on:

> Geoffroy St. Hilaire has insisted strongly on the high importance of relative connexion in homologous organs: the parts may change to almost any extent in form and size and yet they always remain connected together in the same order.

Then the next paragraph:

> Nothing can be more hopeless than to attempt to explain this similarity of pattern in members of the same class, by utility or by the doctrine of final causes. The hopelessness of the attempt has been expressly admitted by Owen in his most interesting work on the 'Nature of Limbs'. On the ordinary view of the independent creation of each being, we can only say that so it is; — that it has so pleased the Creator to construct each animal and plant.

Darwin explains the variation seen in homologous structures in different organisms as due to the action of natural selection on a common ancestral pattern. But as quoted above he has already referred to other earlier traditions in comparative anatomy, to the "principle of connexions" of Etienne Geoffroy Saint-Hilaire (1818), to the teleology of Cuvier (1817) ("by utility or by the doctrine of final causes"), and to his rival Richard Owen. In the following paragraph, Darwin misuses a concept that Owen had made his own:

> If we suppose that the ancient progenitor, the *archetype* as it may be called, of all mammals, had its limbs constructed on the existing general pattern, for whatever purpose they served, we can at once perceive the plain signification of the *homologous* construction of the limbs throughout the whole class. (italics added) (Darwin, 1859, p. 435)

II. THE ARCHETYPE

Richard Owen's most famous contributions to theoretical comparative anatomy were to distinguish between homologous and analogous features in organisms and to present the concept of the archetype. To understand Owen's concept of homology it is necessary to understand that of the archetype. Both concepts were presented by Owen in their definitive form in *On the Archetype and Homologies of the Vertebrate Skeleton* (1848), but each had a long and convoluted history both in Owen's thought and in that of other comparative anatomists before him. The work *On the Archetype* was first presented, not in its final form as a book, but as a report in the published proceedings of the sixteenth meeting of the British Association for the Advancement of Science, held in Southampton in September 1846 (Owen, 1847). Owen had given two lectures at the meeting — "On the Homologies of the Bones Collectively Called 'Temporal' in Human Anatomy" on September 17th, and "On the Vertebrate Structure of the Skull" on September 21st (Rupke, 1993). These were then elaborated together as Owen's published report.

The final book differed from the published report in one important particular. While the latter contained many woodcuts, mostly of skull bones of a variety of vertebrates, the book contained for the first time a large foldout diagram illustrating "the archetype of the vertebrate skeleton" (reproduced herein as Figure 3) together with illustrations of the whole skeletons of a bony fish (unnamed; "typical skeleton of a fish"), a reptile (crocodile), a bird (unnamed), a mammal (dog), and a human, with corresponding bones labelled throughout. The diagram of the archetype alone has been reproduced many times, e.g., in Russell's (1916) classic *Form and Function: A Contribution to the History of Animal Morphology* (his Fig. 6) and in de Beer (1971b, see above); the rest of the Owen foldout diagram (his Plate II) has not. This seems to undermine Owen's intention. The whole plate and its lengthy caption (Owen, 1848, pp. 175-203) embody his archetypal theory much more successfully than the preceding text. Rupke (1993) quite rightly distin-

---

Fig. 3, (overleaf). Richard Owen's archetype of the vertebrate skeleton and its modifications characteristic of the four great divisions of the vertebrate subkingdom; after Owen (1848). See text for further details. (Photographed from the original by permission of the Robinson Library, University of Newcastle upon Tyne.)

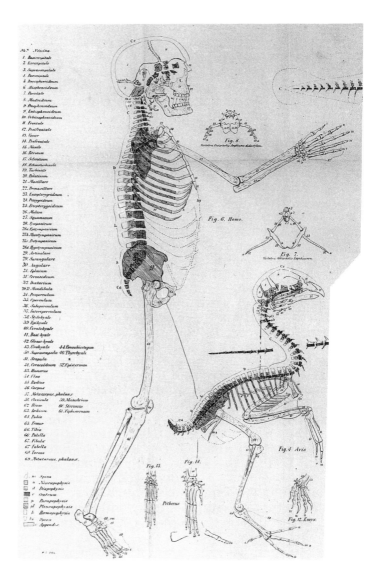

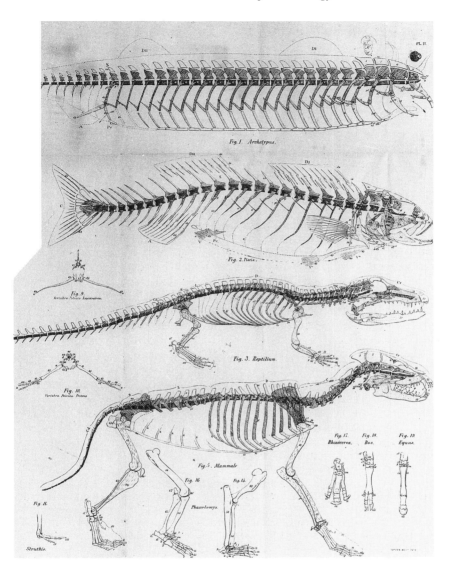

Fig. 1. Archetypus.

Fig. 2. Pisces.

Fig. 9.
Vertebra Pelvica Lepidosteus.

Fig. 3. Reptilium.

Fig. 10.
Vertebra Pelvica Proteus.

Fig. 5. Mammale.

Fig. 17. Fig. 18. Fig. 19.
Rhinoceros. Bos. Equus.

Fig. 16. Fig. 15.
Phascolomys.

Fig. 11.

Struthio.

guishes between the *visual representation* and the *philosophical interpretation* of Owen's archetype. In the whole plate, the fish, reptile, bird, and mammal skeletons represented the four taxonomic classes of vertebrates then recognized (Amphibia were conflated with reptiles, and Agnatha — jawless vertebrates — with sharks and bony fish), and the human skeleton the highest manifestation of the archetypal plan (as well as the source of most of the listed technical names of the bones). In addition to the six complete skeletons (Owen's Figs. 1-6; see Fig. 3), the plate displays a series of smaller drawings of various modifications of the "hands" and feet of tetrapod vertebrates (Owen's Figs. 11-19; see Fig. 3), and four drawings of "vertebrae" (Owen's Figs. 7-10; see Fig. 3). It is these vertebrae that give a clue to the whole enterprise. They are not vertebrae alone, as the term would be understood by a modern anatomist, but include (where present) ribs, and, as "diverging appendages," the backward (uncinate) rib processes and parts of the appendicular skeleton (limbs and limb girdles) which attach to the vertebra in the same body segment.

The vertebrate archetype, or at least that of the vertebrate skeleton, consists of a linear series of Owen's "vertebrae" and "appendages," little modified from a single basic plan, and little different from one another. Each vertebra of the archetype is then a serial homologue or *homotype* of every other vertebra of the archetype, just as every vertebra of a real organism is a homotype of the other vertebrae of that organism. But two corresponding vertebrae, each from a different animal, are *special homologues* of one another, and *general homologues* of the corresponding vertebra of the archetype:

> Relations of homology are of three kinds: the first is that above defined, viz. the correspondency of a part or organ, determined by its relative position and connections, with a part or organ of a different animal; the determination of which homology indicates that such animals are constructed on a common type: when for example the correspondence of the basilar process of the human occipital bone with the distinct bone called 'basi-occipital' in a fish or crocodile is shown, the *special homology* of that process is determined.
>
> A higher relation of homology is that in which a part or series of parts stands to the fundamental or general type, and its enuncia-

tion involves and implies a knowledge of the type on which a natural group of animals, the vertebrate for example, is constructed. Thus when the basilar process of the human occipital bone is determined to be the 'centrum' or 'body of the last cranial vertebra', its *general homology* is enunciated.

If it be admitted that the general type of the vertebrate endo-skeleton is rightly represented by the idea of a series of essentially similar segments succeeding each other longitudinally from one end of the body to the other .... then any given part of one segment may be repeated in the rest of the series, just as one bone may be reproduced in the skeletons of different species, and this kind of repetition or representative relation in the segments of the same skeleton I call 'serial homology'. As, however, the parts can be namesakes only in a general sense, as centrums, neurapophyses, ribs, &c.; and since they must be distinguished by different special names according to their particular modifications in the same skeleton, as e.g. mandible, coracoid, pubis &c., I call such serially related or repeated parts 'homotypes' .... when the basi-occipital is said to repeat in its vertebra or natural segment of the skeleton the basi-sphenoid or body of the parietal vertebra, or the bodies of the atlas and succeeding vertebrae, its *serial homology* is indicated. (Owen, 1848, pp. 7-8; his italics)

Thus, these kinds of homologous relationship are all assumed by Owen to be dependent on the archetype. Whatever the *criteria* used to suggest an hypothesis of homology, the *reason* for (or explanation of) a true case of homology is that the two structures, each in a different animal, are homologous because they are derivations of the same structure in the achetype.

While it was implicit in Owen's thought that archetypes existed for organisms other than vertebrates and systems other than the vertebrate skeleton, the theory was only developed fully for vertebrates. It is not surprising, therefore, that components of the archetypal vertebra appear relatively early in his writings, in a short paper describing a fossil plesiosaur (Owen, 1838, 1840). In this case, however, he was simply setting out and defining the parts of a generalized vertebra, which had not yet attained the transcendental significance it later acquired.

In *On the Archetype* [and before that in his published vertebrate lectures (Owen, 1846)] this generalised vertebra was figured with its labelled parts, followed as an example by its actual realization in the thoracic "vertebra" of a bird (Figs. 4 and 5). But the most controversial part of Owen's theory — the vertebral theory of the vertebrate skull — did not originate with him. Owen credits Oken as originator of this idea:

> The gifted and deep-thinking naturalist, OKEN, obtained the first clew [sic] to this discovery by the idea of the arrangement of the cranial bones of the skull into segments, like the vertebrae of the trunk. He informs us that walking one day in the Hartz forest, he stumbled upon the blanched skull of a deer, picked up the partially dislocated bones, and contemplating them for a while, the truth flashed across his mind, and he exclaimed "It is a vertebral column!" (Owen, 1848, pp. 73-74, with a footnote giving the original German which dates the incident in August 1806 — but publication in 1818)

Russell (1916, p. 96) gives Goethe priority for the skull theory with a very similar anecdote: "He tells us that the idea flashed into his mind when contemplating in the Jewish cemetery at Venice a dried sheep's skull. The discovery was made in 1790, but not published till 1820". (Goethe, 1817-1824, Vol. 1, Part 2, p. 250). (Subsequently

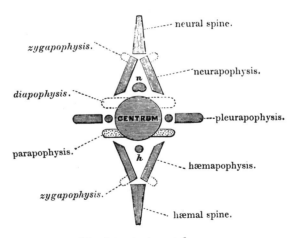

Ideal typical vertebra.

Fig. 4. The archetype vertebra, after Owen (1849). (From the original by permission of the Robinson Library, University of Newcastle upon Tyne.)

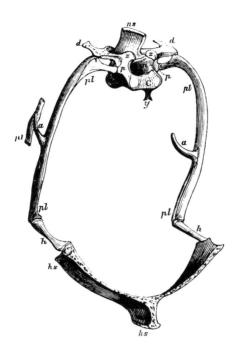

Fig. 5. "Natural skeleton-segment, 'osteocomma' or 'vertebra'. Thorax of Bird", after Owen (1849). (From the original by permission of the Robinson Library, University of Newcastle upon Tyne.)

Oken claimed priority; for a nineteenth century account, see Lewes, 1864, pp. 357-364). By the time of publication of Owen's *On the Archetype* the vertebral theory of the skull was widely accepted, particularly by the German *Naturphilosophen* but with wide disagreement as to the number of vertebrae represented (Owen, 1848, pp. 73-79; Russell, 1916, pp. 96-97). Oken and Owen, however, settled for what seems to have been the majority opinion — four vertebrae (Fig. 6).

In *special homology*, equivalent parts of the skeleton in two different vertebrates are homologues because each is equivalent to the same structure in the archetype.

Cranial Vertebræ.[1]   (After Owen, 1848, p. 165.)

Vertebræ.	Occipital.	Parietal.	Frontal.	Nasal.
Centra.	Basioccipital.	Basisphenoid.	Presphenoid.	Vomer.
Neurapophyses.	Exoccipital.	Alisphenoid.	Orbitosphenoid.	Prefrontal.
Neural Spines.	Supraoccipital.	Parietal.	Frontal.	Nasal.
Parapophyses.	Paroccipital.	Mastoid.	Postfrontal.	None.
Pleurapophyses.	Scapular.	Stylohyal.	Tympanic.	Palatal.
Hæmapophyses.	Coracoid.	Ceratohyal.	Articular.	Maxillary.
Hæmal Spines.	Episternum.	Basihyal.	Dentary.	Premaxillary.
Diverging Appendage.	Fore-limb or Fin.	Branchiostegals.	Operculum.	Pterygoid and Zygoma.

Fig. 6. Table of cranial vertebrae and their components [after Owen (1848), from Russell (1916).]

In *general homology* the comparison is made directly between organism and archetype.

In *serial homology* different structures are equivalent ('homotypes') because the archetypal equivalents have the same basic structure.

But how did Owen view 'the archetype'? The development of Owen's concept, and its pre-Owenite history, are reviewed by Rupke (1993).

Neither the term "archetype", nor the concept, or even the diagram, originated with Owen, but, as Rupke says at the end of his essay, "Exclusively Owen's own was all the hard work."(p. 251). In the long description of his "archetype plate" Owen (1848, p. 177) explains his use of the term in a footnote:

> I have used this word ... in the sense which it bears in such classical works of our own language as Glanville's Scepsis and Watt's Logic, and agreeably with its definition in Johnson's and other

dictionaries, as the original or pattern of which any resemblance is made: and as equivalent to the terms 'general type' and 'fundamental type' as they occur in my "Lectures on the Vertebrate Animals" 8vo. 1846, p. 41, and *passim.*

The footnote continues by giving credit to "Joseph Maclise, Esq." for the use of the term "archetype" in his *Comparative Osteology* together with two remarkably opaque quotations from Maclise, but without any critical comment from Owen.

Rupke uses Maclise's (1846, 1847) concept of the archetype to illuminate the changing use that Owen made of the term. To Maclise, the archetype (again of the vertebrate skeleton) represented a perfect embodiment from which all mundane vertebrate skeletons could be derived by subtraction and/or degradation of parts. The human skeleton was, of course, closest to the archetype. Owen's archetype on the other hand represented a generalized and primitive condition from which the skeletons of real vertebrates were derived — the human skeleton, representing the nearest approach to perfection, was furthest removed from the archetype. As Rupke points out, this somewhat arcane distinction is of considerable theoretical importance in relation not only to Owen's views, but to nonevolutionary explanations of homology.

The standard story (including that in Panchen, 1992) is that Owen's concept is a Platonic one: Plato's idealism refers to his theory of *forms* or *ideas* (ἰδέαι) (Oldroyd, 1986). At first these were generalizations from classes of objects. A triangle, as drawn, may be imperfect, with sides that are not straight lines, so that its three angles do not add up to precisely 180°. It is, however, fairly easy to have a concept of a perfect triangle congruent with the imperfectly drawn one(s). With a little more difficulty one could picture an ideal table or chair. Later Plato was to suggest that abstract concepts such as beauty or justice had realities of their own, to be apprehended, if dimly, by discovering what all (e.g.) beautiful objects had in common. Eventually, in the *Timaeus,* Plato was to claim not only that his *ideas* were real, but that they were the only real entities. The same stance is represented by his famous metaphor of the Cave in the *Republic.* Prisoners in the cave are constrained so that they can see only the wall opposite the entrance; their view of the world outside (the world of *ideas*) is simply that of shadows cast by objects in the real world. The shadows are the objects of sensory perception: the real world consists of *ideas*.

It seems that to some degree there is a parallel between Plato's hardening of attitude toward his eventual position of radical idealism (Oldroyd,1986) and Owen's attitude toward the archetype. As Rupke says, Owen was no philosopher, and at first probably attributed no metaphysical significance to the "fundamental plan" of the vertebrate skeleton. The concept seems to have come to him from a number of sources (Rupke,1993; Sloan, in Owen,1992) and even the famous diagram was almost certainly adapted from one in *Von den Ur-Theilen des Knochen- und Schalengerüstes* by Carus (1828), as Rupke has shown.

Then, in *On the Archetype* itself, Owen is thoroughly confused. He introduces the Platonic concept of an *idea*, but in a way directly opposed to that of Plato. Owen postulates a "polarising force," responsible for repetition of similar segments in a vertebrate skeleton and is even more prevalent in invertebrates "as for example in the rings of the centipede and the worm," so that "the principle of vegetative repetition prevail[s] more and more as we descend in the scale of animal life...." This, in the development of an animal, is opposed to a greater or lesser degree, by an "organising principle, vital property, or force" that results in the various adaptations that characterize individual animal species. The greater the effect of this "organising principle," the higher the resultant animal is in the scale of being. But Owen identifies the Platonic ιδεαι [sic] not with the "polarising force" responsible for serial homology and thus at its purest in the archetype, but with the "organising principle" that produces adaptation, advancement over the archetype, and the individuality of species (Owen, 1848, pp. 171-172).

Almost immediately after the publication of *On the Archetype* in 1848, Owen produced another important theoretical work, edited from "a discourse delivered on Friday, February 9th, at an evening meeting of the Royal Institution of Great Britain," and entitled *On the Nature of Limbs* (Owen,1849). Owen starts his discourse by demonstrating the special homology of the forelimbs of the dugong, mole, bat and horse and resemblances and differences from fish fins, and then turns to serial homology between fore- and hindlimbs in vertebrates, comparing carpals and tarsals in considerable detail. Owen then addresses himself to the general homology of the limbs and limb girdles, homologizing parts of the pectoral girdle with components of his last cranial (occipital) vertebra, the forelimb or fin with the "diverging appendage" (Fig. 3), and the pelvic girdle and limb similarly with sacral vertebrae.

But then, at the end of his discourse, Owen appears to indulge in a *volte face* from his previous opinions as expressed in *On the Archetype*, as Rupke (1993) points out. Instead of the Platonic *idea* being inherent in the "organising principle," the archetype itself has become the *idea*:

> Now, however, the recognition of an ideal Exemplar for the Vertebrated animals proves that the knowledge of such a being as Man must have existed before man appeared. For the Divine mind which planned the Archetype also foreknew all its modifications.
>
> The Archetypal idea was manifested in the flesh, under divers such modifications, upon this planet, long prior to the existence of those animal species that actually exemplify it. (Owen, 1849, pp. 85-86)

Rupke discusses the reason for Owen's change of opinion and rejects as improbable any direct influence from philosophy. He does, however, suggest that the influence of Sedgwick, Whewell, and Conybeare persuaded Owen to establish the archetype as in the mind of the Creator. This contrasts with the view of Desmond (1989), who, assuming that Owen's archetype had always corresponded to a Platonic *idea*, saw it as an establishment ploy against radical medical practitioners in London, who were advocates of Lamarckian evolution and of the suspect transcendentalism of Geoffroy Saint-Hilaire (see below).

III. THE DEFINITION OF HOMOLOGY

Received opinion is that Richard Owen was the first to distinguish clearly between homology and analogy, at least in English, and that he made the distinction by defining both in the published version of his 1843 lectures on invertebrates delivered at the Royal College of Surgeons (Owen, 1843). It comes as something of a surprise, therefore, that neither analogy nor homology (or corresponding adjectives) appear in the extensive index to the original published volume. The famous definitions appear only in the Glossary:

> ANALOGUE A part or organ in one animal which has the same function as another part or organ in a different animal. *See* HOMOLOGUE. (p. 374).
>
> HOMOLOGUE (Gr. *homos; logos*, speech.) The same organ in different animals under every variety of form and function. (p. 379).

This gives the impression strongly that in 1843 Owen did not see himself as recording new concepts or even coining new definitions, but simply as recording current usage. For "homologue" and "homology," this was certainly the case. If one turns to the recently published transcriptions of the surviving lectures from Owen's first series of Hunterian lectures, delivered in 1837, one finds him using "homologies," as though it were a familiar term, in a discussion of the rival schools of comparative anatomy of Cuvier and the "Transcendentalists":

> Now the inquiry into the homologies or signification of the different parts of the animal frame is replete with beautiful results when pursued in accordance with those rules of investigation, by which alone the Human Understanding can attain to any certain truth. (Owen, 1992, p. 190)

On the next page we find Owen equating "signification" with the German word *Bedeutung*. By the time of publication of *On the Nature of Limbs, Bedeutung* has, according to Owen, become the most apt term for general homology, i.e., validation by comparison with his now fully Platonic archetype.

But even before the publication of *On the Nature of Limbs*, Owen appears to accord greater importance to his definitions of *analogue* and *homologue* than in 1843. He repeats the 1843 glossary definitions near the beginning of *On the Archetype* (Owen, 1848, p. 7) and then adds a footnote:

> ... My ingenious and learned friend Mr Hugh Strickland has made a strong and able appeal to the good sense of comparative anatomists in favour of the restriction of these terms to the senses in which they are here defined. *Phil. Mag.* 1846, pp. 358-362. (Owen 1848, p. 7, footnote)

Reference to Strickland's (1846) paper is replete with irony in the light of Owen's footnote. First, there is no reference whatsoever to Owen, nor any quotation of Owen's definitions. But, more importantly Strickland gives credit for contrasting the *concepts* of homology and analogy to William Sharp MacLeay:

> Zoologists had long been aware that certain sets of characters produced an arbitrary or artificial method if employed for classification, while others seemed to lead to a natural system, but the question was involved in obscurity till the time of MacLeay, who was the first to give us clear definitions on the distinction between AFFINITY and ANALOGY. (Strickland, 1846, p. 356)

On p. 358, Strickland does indeed suggest that homology should be used in the same sense as does Owen, "as the relation between equivalent organs is one of real *affinity*," and on p. 362 recommends the restriction of analogy to the Owenite sense. But he is correct in giving MacLeay the credit.

Macleay is known principally as the inventor of the Quinarian system of classification (for details, see Panchen, 1992, pp. 23-25) in which animal taxa were arranged in groups of five, represented in a circle, with individual taxa subdivided into five and so on. But MacLeay's system started with his attempts to arrange the 10 then recognized major groups of insects. He ordered them into two groups of five, the Mandibulata and Haustellata, using the nature of their mouth parts (after Cuvier and Lamarck). Each group of five could then be arranged in a linear series, depending chiefly on the completeness of metamorphosis, but each member of the series was paralleled by a member of the other series (Fig. 7) so that they were coupled by *relations of analogy* in contrast to the *affinity* that united members of the same taxon.

Thus, at least in English, the distinction between affinity (recognized by homology) and parallelism (recognized by analogy) was first made by MacLeay (1821, pp. 365-367), but reinforced by Owen (1843, 1847) and Strickland (1846). The *concept* of homology itself is much more ancient. Russell (1916) traces it back to Aristotle. In the sixteenth century there is a famous figure by Belon (1555) showing correspondence between the skeleton of a man and that of a bird, with the bones that he regarded as equivalent

MANDIBULATA.		HAUSTELLATA.
	Relations of Analogy.	
1. HYMENOPTERA *Linn.*	Metamorphosis incompleta vel coarctata. Larva apoda.	1. DIPTERA *Arist.*
Strepsiptera ? *Kirby.*		Homaloptera *Leach.*
2. COLEOPTERA *Arist.*	Metamorphosis incompleta.	2. APTERA *Lam.*
Dermaptera *Degeer,* 3. ORTHOPTERA *Oliv.* Dictyoptera *Leach.*	Metamorphosis semicompleta.	3. HEMIPTERA *Linn.*
4. NEUROPTERA *Linn.*	Metamorphosis subsemicompleta.	4 HOMOPTERA *Degeer.*
5. TRICHOPTERA *Kirby.* Tenthredina.	Metamorphosis obtecta. Larva pedibus membranaceis. Imaginis os mandibulis abbreviatis incompletis, labio et maxillis ad basin saltem coalitis.	5. LEPIDOPTERA *Linn.*

Fig. 7. Affinity (vertical columns) and analogy (horizontal rows) among 10 orders of insects. (after MacLeay, 1821).

in the two labelled (Fig. 8). The concept of homology is also implicit in the work of Goethe, Carus, and Oken, and Owen gives credit for the use of the term *homology* to "the philosophical cultivators of that science [anatomy] in Germany."

Confusion between the terms *homology* and *analogy* seems largely to have been engendered by the great French transcendental anatomist Etienne Geoffroy Saint-Hilaire, who used both "homologue" and "analogue" as synonyms. Thus, Geoffroy's theory of homologues is described by him in the *Discourse preliminaire* of his *Philosophie Anatomique* (1818) as the *Théorie des analogues*, whereas, as quoted by

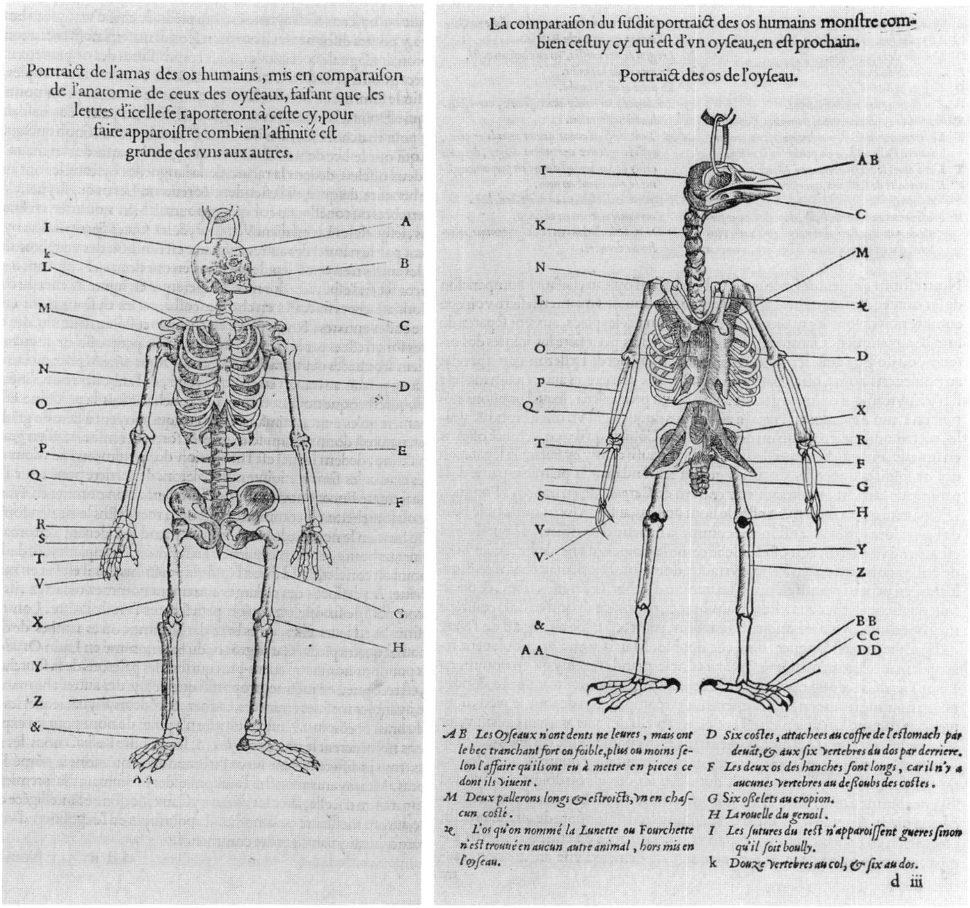
Portraict de l'amas des os humains, mis en comparaiſon de l'anatomie de ceux des oyſeaux, faiſant que les lettres d'icelle ſe raporteront à ceſte cy, pour faire apparoiſtre combien l'affinité eſt grande des vns aux autres.
La comparaiſon du ſuſdit portraict des os humains monſtre combien ceſtuy cy qui eſt d'vn oyſeau, en eſt prochain.
Portraict des os de l'oyſeau.
AB Les Oyſeaux n'ont dents ne leures, mais ont le bec tranchant fort ou foible, plus ou moins ſelon l'affaire qu'ils ont eu à mettre en pieces ce dont ils viuent.
M Deux pallerons longs & eſtroicts, vn en chaſcun coſté.
L'os qu'on nommé la Lunette ou Fourchette n'eſt trouué en aucun autre animal, hors mis en l'oyſeau.
D Six coſtes, attachees au coffre de l'eſtomach par deuāt, & aux ſix vertebres du dos par derriere.
F Les deux os des hanches ſont longs, car il n'y a aucunes vertebres au deſſoubs des coſtes.
G Six oſſelets au cropion.
H La rouelle du genoil.
I Les ſutures du teſt n'apparoiſſent gueres ſinon qu'il ſoit boully.
k Douze vertebres au col, & ſix au dos.
d iii

Fig. 8, (overleaf). Skeletons of a man and a bird to show homologous bones [after Belon (1555). (From the original, by permission of the Natural History Society of Northumbria).]

Owen (1848, p. 5), Geoffroy uses "homologue" in the sense that "serial homology" is used by the German school, Owen and de Beer:

> Les organs des sens sont *homologues*, comme s'exprimerait la philosophie Allemande; c'est-à-dire qu'ils sont analogues dans leur mode de développement, s'il existe véritablement en eux une même principe de formation, une tendence uniforme à se répéter, à se reproduire de la même façon. (Geoffroy Saint-Hilaire, 1825, p. 341)

There is an important and common misapprehension about the terms homologue and analogue. Ill-taught schoolchildren and students (and others who should know better) treat them as antonyms and thus mutually exclusive. Owen was quite clear in saying that this was not the case:

> But homologous parts may be, and often are, also analogous parts in a fuller sense, viz. as performing the same functions: thus the fin or pectoral limb of a Porpoise is homologous with that of a Fish, inasmuch as it is composed of the same or answerable parts: and they are analogues of each other, inasmuch as they have the same relation of subserviency to swimming. (Owen, 1848, p. 7)

Owen's usage is current. There is therefore need for a third term to describe a structure that is analogous between two organisms, but not homologous, a phenomenon usually attributed to convergent evolution. The term is *homoplastic*, the phenomenon *homoplasy*. Both terms were coined by Lankester (1870) although not in quite the same sense in which they are used today. Ray Lankester was an ardent evolutionist and suggested as a criterion for homology that the structure apparently shared by two organisms should have been present in their nearest common ancestor. But he did not like the term homology, — it was tainted with pre-evolutionary Platonism — so he suggested it be replaced by *homogeny*. All other resemblances were then homoplastic. Thus, examples of Owen's serial homology, such as the resemblances between successive vertebrae, were homoplastic and while the heart of a bird and that of a mammal were homogenous as vertebrate hearts, the four chambers into which each is divided (assuming independent origin) were homoplastic between the two animals. Finally, two structures analogous but not homologous by any definition were also homoplastic. It is in the last two senses that homoplastic is used today.

IV. DEFINITIONS, CRITERIA, AND EXPLANATIONS

A. *Richard Owen*

In looking at the work of Richard Owen we have seen his *definition* of homology and what, up to approximately 1850, was his principal *explanation* of the phenomenon — two structures, each in a separate organism, were special homologues of one another because each was a general homologue of, and thus deriveable from, the same structure in the archetype. It is further necessary to distinguish criteria of homology. It was the failure of traditional *criteria* that de Beer apparently found so disturbing and discussed in his little pamphlet on the subject.

Owen, in a work later than those considered so far, his *On the Anatomy of Vertebrates* (1866-1868), expresses his criteria thus:

> 'Homological Anatomy' seeks in the characters of an organ and part those, chiefly of relative position and connections, that guide to a conclusion manifested by applying the *same name* to such part or organ, so far as the determination of the namesakeism [*sic!*] or homology has been carried out in the animal kingdom. This aim of anatomy concerns itself little, if at all, with function, and has led to generalisations of high import, beyond the reach of one who rests on final causes. (Owen, 1866-8, Vol. I, p. vii)

So the criteria are "relative position and connections." The latter represents the most important contribution of Etienne Geoffroy Saint-Hilaire to comparative anatomy, the *principe des connections* . Geoffroy's theoretical ideas were developed in a series of *mémoires* from 1796 onward and given definitive expression in his *Philosophie Anatomique* (1818). Homologous structures might differ in shape, or even composition, but they were to be recognized by the constancy of relationships to surrounding organs and structures. Whatever the subsequent definitions and explanations of homology, anatomical similarity, including that of relative position and connections, is still the principal criterion used today (but see discussions in Patterson, 1982; Rieppel, 1988; Panchen, 1992; and see below).

B. Similar Embryonic Development

Another criterion of homology, similarity of embryological development, was one of those found wanting by de Beer, just as too literal an interpretation was also rejected by Owen:

> M. Agassiz seems, in like manner, to give undue importance to similarity of development in the determination of homologies, where he repudiates the general homology of the basi-sphenoid with the vertebral centrum, and consequently its serial homology with the basi-occipital, because the pointed end of the chorda dorsalis has not been traced further forwards along the basis of the cranium in the embryo osseous fish than the basi-occipital....
>
> There exists doubtless a close general resemblance in the mode of development of homologous parts; but this is subject to modification, like the forms, proportions, functions and very substance of such parts, without their essential homological relationships being thereby obliterated. (Owen, 1848, p. 6)
>
> The femur of the cow is not the less homologous with the femur of the crocodile, because in the one it is developed from four separate ossific centres, and the other from only one such centre. (Owen, 1848, p. 5)

Owen was also well aware that the development of the embryo was subject to modification for functional (teleological) reasons, but that

> embryology affords no criterion between the ossific centres that have a 'homological' and those that have a 'teleological' signification. A knowledge of the archetype skeleton is requisite to teach how many and which of the separate centres that appear and coalesce in the human, mammalian, or avian skeleton, represent and are to be reckoned as distinct bones, or elements of the archetype vertebra. For want of this guide great and estimable anatomists have gone astray. Thus CUVIER, commenting on the arbitrary enumeration of the single bones in the human skeleton, affirmed that to learn their true number in any given species we must go to the first osseous centres as these are manifested in the foetus; and GEOFFROY ST. HILAIRE concurred in this view. (Owen, 1866-1868, Vol. I, pp. xxv-xxvi)

Embryology is a (fallible) criterion of homology, but it is also possible to regard it as the *explanation* of the phenomenon. While Owen's "official" explanation of homology, at least up to about 1850, was one of derivation of homologous structures from corresponding features in the archetype, recent historical research makes it clear that Owen was toying with the idea of transmutation from the late 1830s onward (Richards, 1987, and references therein) so that a phylogenetic explanation of homology would have been acceptable. Owen objected, not to transmutation itself, but to the origin of man from apelike ancestors, to the picture of phylogeny as a linear rather than a divergent pattern, and, most strongly (after 1859) to the theory of natural selection (Owen, 1860a).

C. Owen's Polarizing Force and Organizing Principle

A divergent pattern of organisms, whether regarded as modifications of the archetype, or interpreted as transmutation, is implicit in Owen's views on the interaction of his "polarising force" and "organising principle" (see Section II above and Panchen, 1992, pp. 27-28), but is also explained by divergence from similar initial stages of different species during their embryological development as set out in four laws by von Baer (1828) in his *Über Entwickelungsgeschichte der Thiere: Beobachtung und Reflexion.* As quoted in English translation by Gould (1977, p. 56) these are as follows:

(1) The general features of a large group of animals appear earlier in the embryo than the special features.
(2) Less general characters are developed from the most general, and so forth, until finally the most specialised appear.
(3) Each embryo of a given species [*Thierform*], instead of passing through the stages of other animals, departs more and more from them.
(4) Fundamentally therefore, the embryo of a higher animal is never like [the adult of] a lower animal, but only like its embryo.

In the peroration of the last of his 1843 lectures on invertebrates (whence the glossary definitions of homologue and analogue) Owen gives what could be regarded as a succinct summary of von Baer's third and fourth laws:

> The extent to which the resemblance, expressed by the term 'Unity of Organisation', may be traced between the higher and lower organised animals, bears an inverse ratio to their approximation to maturity. (Owen, 1843, p. 368)

In late 1844 Owen wrote, via the publisher, to the anonymous author of the (1844) *Vestiges of the Natural History of Creation* (Robert Chambers), drawing attention to his, Owen's, 1843 "Law" (Richards, 1987). Richards shows how Owen fought a fierce battle to claim priority for the principle of divergence. In the first edition of *Vestiges*, Chambers set out two versions of embryological development, first the law of parallel development and second that of divergence. Parallel development is based on the ancient idea that all organisms can be ranged on a *scala naturae*, the "Great Chain of Being" (Lovejoy, 1936) and that higher forms pass through an approximation of the adult stages of lower forms in their embryological development (Gould, 1977; Panchen, 1992, pp. 11-15). Chambers produced a diagram (Fig. 9b) to illustrate divergence, at least for the major groups of vertebrates, that he had copied and modified, unacknowledged, from Carpenter (1841) (Fig. 9a). Richards traces this back to von Baer via a more complicated diagram published by Martin Barry (1837).

According to Richards, Owen failed to acknowledge Barry's priority, fought Carpenter to a standstill and also took on Milne-Edwards, who expressed von Baer-like views (Milne-Edwards, 1844), claiming priority over Owen. In an anonymous review Owen (1851, p. 430) even claimed priority, or at least greater precision, over von Baer himself. But Richards (1987, p. 143) suggests that Owen had not really read von Baer (1928) with understanding until an English translation by T. H. Huxley of the appropriate "Fifth Scholion" of *Entwickelungsgeschichte* became available in 1853.

But why was Owen apparently so desperate to establish his priority in enunciating the principle of divergence? There seem to be two reasons, one political, one scientific.

Desmond (1989) describes the political reason. In the late 1830s Owen, as Professor at the Royal College of Surgeons, saw himself, and was seen by others, as representing the medical and scientific establishment in its battle against the upstart private medical schools in London, whose guru was Robert Grant of the newly founded University College. Their creed was a combination of Lamarckian evolution and the transcendental morphology of Geoffroy. Owen rejected the first, largely on religious grounds, and criticized the scientific excesses of the second (e.g. Owen, 1992, p. 191; and see Sloan's note 95-16 on page 202).

Owen's scientific stance was a reconciliation of the teleology of Cuvier, the morphology of Geoffroy, and the German *Naturphilosophen*. But Geoffroy insisted on the unity of the animal kingdom. Homologies could be traced among all animals. The vertebra of a vertebrate was the homologue of one unit in the segmental exoskeleton of a lobster, and of all other annelids and arthopods ("dermo-vertébrés") (Russell, 1916, pp. 60-64). A cephalopod could be compared to a vertebrate bent back at the level of the umbilicus (Russell, 1916, pp. 64-65; Appel, 1987, p. 145ff.).

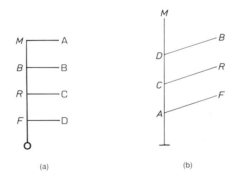

Fig. 9.   Diagrams representing von Baer's principle of the divergence of vertebrate embryos in ontogeny. (a) After Carpenter, (1841): F, R, B, M stages at which a fish, a reptile, a bird, and a mammal, respectively, diverge from the common course of development, to reach their definitive states D, C, B, A. (b) After Chambers (1844): the definitive states are now labelled F, R, B, M respectively. (Redrawn).

Cuvier, von Baer, and Owen rejected the concept of the universal homology of adult structures, with its implication of a single animal series. From 1812 onward Cuvier divided the animal kingdom into four *embranchements*, vertebrates, molluscs, articulates, and zoophytes plus radiates (Cuvier, 1812, 1817; Russell, 1916, pp. 39-41; Outram, 1986, p. 359); von Baer, apparently independently, settled on the same arrangement. Owen (and Agassiz, 1857) followed Cuvier. Thus, according to Owen it is pointless to look for homologies between the adults of members of different *embranchements*. They resemble one another only at the very earliest stage of development, but thereafter homologies should be sought only within each major group:

> Thus every animal in the course of its development typifies or represents some of the permanent forms of animals inferior to itself; but it does not represent all the inferior forms, nor acquire the organisation of any of the forms which it transitorily represents. Had the animal kingdom formed, as was once supposed, a single and continuous chain of being progressively ascending from the Monad to Man, unity of organisation might then have been demonstrated to the extent in which the theory has been maintained by the Geoffroyan school.(Owen, 1843, p. 370)

But even within an *embranchement* the natural pattern is one of radiation, branching, rather than a *scala naturae*:

> The *Vertebrated* ovum ... [eventually] assumes ... the form and condition of the finless cartilaginous fish, from which fundamental form development radiates in as many and diversified directions and extents, and attains more extraordinary heights of complication and perfection than any of the lower secondary types appear to be susceptible of. (Owen, 1843, p. 371)

These *dicta* are repeated in the second edition of *Invertebrate Lectures* (Owen, 1855) with only minor rewording. Thus, Owen continues to insist on the divergence of embryos and thus, by implication a hierarchical pattern of homologies within each *embranchement*, after the archetype had reached its metaphysical peak in *On the Nature of Limbs* (Owen, 1849).

So, whatever the failings of comparative embryology as a *criterion* for homology, it could still be an *explanation* of homology, even a sufficient explanation. With or without the explanation of derivation from the archetype and/or that of common ancestry, homologous structures could be homologous because of their basic resemblance but divergence from one another in ontogeny. Both resemblance and divergence are empirical phenomena.

We now have three possible explanations of special homology, in Owen's sense of correspondence between similar structures in two or more different organisms. They are phylogeny, Platonic idealism (the archetype), and ontogeny (using comparative embryology). While Owen is famous for his advocacy of the archetype he would, at different stages of his career, have assented to the others.

D. Teleology and Essentialism

But there are two further possible explanations of homology — teleology and Aristotelian essentialism. Owen emphatically rejected the first; it is difficult to judge what his reaction might have been to the second. One approach to the teleological explanation is to ask the question, "Under what circumstances would there be no distinction between homology and analogy (as defined and characterized by Owen)?" The answer would be if similarity of structure implied nothing but similarity of function; and, conversely, if differences in structure of features having the same broad function (say the wing of a bat compared to that of a bird) were entirely to be explained as functional differences. Such extreme teleology was rejected by Owen. Having considered, in *the Nature of Limbs*, the variations on a theme represented by the limbs of various mammals — swimming and diving dugongs, burrowing moles, flying bats, running horses, and climbing monkeys, Owen goes on:

> And consider the various devices that human ingenuity has conceived and human skill and perseverance have put into practice in order to obtain corresponding results!
> To break his ocean bounds the islander fabricates his craft, and glides over the water by means of the oar, the sail or the paddle wheel. To quit the dull earth man inflates the balloon.... With the arched shield and the spade or pick he bores the tunnel: and his modes of accelerating his speed in moving over the surface of the ground are many and various. (Owen, 1849, p. 9)

Human machines for locomotion, or any other purpose, are designed solely for the purpose in hand. They do not have to be produced as variants of some common archetypal plan:

> There is no community of plan or structure between the boat and the balloon, between Stephenson's locomotive engine and Brunel's tunnelling machinery. (Owen, 1849, p. 10)

Homologous structure of vertebrate limbs requires some further explanation than the fitting of means to ends.

The extreme teleological view, that similarity and difference in structure is to be explained solely in terms of similarity and difference in function, is usually attributed to Cuvier, but this is not entirely correct. Outram (1986) emphasizes the distinction between the teleology attributed to Aristotle and that of Immanuel Kant. Aristotle's teleology implied that all features of an organism were designed to some specific end. Kant, who (like Hume) criticized this view, held that apparent design existed as a condition of existence rather than representing an intention. The Aristotelian view implies a designer: the Kantian does not necessarily do so. Cuvier, who prided himself on being a cautious empirical scientist, was a follower of Kant. But he was also concerned with the functional integrity of the animals he studied — an animal must be organized so that all its functions are compatible. This represents his law of the "conditions of existence," which leads to his law of the "correlation of parts." The dentition and other attributes of a carnivore are never found with the cloven hoof of a herbivore.

Aristotelian teleology, the doctrine of final causes, is better represented by the English tradition of natural theology, by John Ray's (1691) *The Wisdom of God Manifested in the Works of the Creation*, William Paley's (1802) *Natural Theology*, and the eight Bridgewater treatises "on the Power Wisdom and Goodness of God as manifested in the Creation" (F. H. Egerton, eighth Earl of Bridgewater, cited in Panchen, 1992, p. 255). Publication of these treatises was paid for from a bequest left by Bridgewater.

In his published lectures and his theoretical work that we have considered so far, Owen seems little concerned with natural theology (Sloan, in Owen, 1992, p. 72); to him it was Kantian teleology that was the incomplete philosophy of nature, needing supplementation by the morphology represented by homology and the archetype. It was Darwin (and perhaps Wallace) who saw that a positivist natural history could be achieved only by replacing Aristotelian teleology as represented by natural theology, by evolution and natural selection. However, Rupke (1992, and personal communication) has pointed out that in some of his primary anatomical work, notably his *Memoir on the Pearly Nautilus* (Owen, 1832), *Memoir on the Megatherium* (Owen, 1860b), and his famous reconstruction of the skeleton of *Dinornis*, the moa (in Owen, 1879), Owen was under some pressure from those colleagues who constituted the Oxford and Cambridge establishment to adopt the stance of natural theology. Owen saw these men, such as Buckland, Conybeare, and Sedgwick (Rupke, 1993) as necessary allies if he was to achieve his aim of a National Museum of Natural History.

The last possible explanation of homology suggested here is that of essentialism, also attributed to Aristotle. In taxonomy essentialism is often conflated with Platonic idealism as "typology," notably by Mayr (1969, 1982, 1987, 1988). But the two, whether correctly attributed to Aristotle and Plato, respectively, are different concepts. I attempted to distinguish them in talking about the reality of species and higher taxa (Panchen, 1992, pp. 109-117, 337-341).

In the case of Platonic idealism taxa owe their reality to the real existence of an archetype (or *Bauplan*); in the case of essentialism to the belief that every taxon is characterized by an *essence*. Taking Aristotle's example, "man" (i.e., *Homo sapiens*), is defined as a "rational animal." "Animal" is the *genos*, "rational" the *diafora*, and "man" the *eidos*, one of the subordinate taxa into which "animal" is divided. This example represents the method of logical division, the basis on which Linnaeus, and less successful predecessors, based their taxonomy (Cain, 1958). But in Aristotle's system the *genos* (Latin *genus*) was any class of entities whose constituent subordinate classes — each an *eidos* (Latin *species*) — were being investigated. With Linnaeus genus and species became definitive taxonomic categories. But in both cases the stance of *essentialism* is that the genus owes its reality to the fact that has an immutable definition, similarly the species, whose definition consists of that of the genus, plus the *diafora* (Latin *differentia*).

But how can this confer reality on the concept of homology, i.e., be an explanation of homology? It is because the *differentia* consists of one or more homologies.

According to Aristotle all humans share the characteristic of being "rational," thus rationality is a homologous feature of all humans. If the class of all human beings is a real entity, because all members of that class share the same essence, then "rational" is part of that essence. Similarly all tetrapods are vertebrates with four dactyl limbs (bearing digits) — excepting cases of agreed secondary loss. Thus the wing of a bird and the forelimb of a mammal are homologous *because* they represent part of the diagnosis of Tetrapoda. Although *diagnosis* is the term used in taxonomy, the correct term from an essentialist viewpoint is *definition*. (Ghiselin, 1987, Panchen, 1992, pp. 343-344). This essentialist position seems to me to be the stance adopted by transformed cladists, to whom classification is (correctly) logically prior to phylogeny (Panchen, 1992, pp. 343-344).

V. CONCLUSIONS: RICHARD OWEN AND HOMOLOGY

In this introductory chapter I have been concerned mainly to discuss the concept of homology as it appeared to Richard Owen during that period from the late 1830s to the early 1850s, when his principal theoretical work in comparative anatomy was done, stimulated by the annual courses of lectures that he was obliged to deliver before the Royal College of Surgeons. But my introductory section, based on de Beer's (1971b) undergraduate reader, was written to make two particular points. First, although de Beer discussed some of the triumphs of comparative anatomy, achieved by means of the concept of homology, he was also concerned to demonstrate the fallibility of various criteria for homology, taking examples of what he regarded as undoubted examples of the phenomenon, forelimbs and occipital arches of several vertebrates, the alimentary canal in major vertebrate groups, the eye lens in two closely related frog species, eusociality in ants, bees, and wasps, plus spiral cleavage in various invertebrate phyla. The fallible and failing criteria were, respectively, derivation from the same body segment, development from the same region of the embryo, similarity in embryonic induction and either the primitive or the general occurrence in a number of taxa of features taken to be homologous among those taxa. To those criteria I added an example from the polymorphic butterfly *Papilio dardanus* to illustrate that homologous features need not be produced by the same genotype, a further point made by de Beer.

Thus, I began by talking about criteria of homology to take us back to Owen and other early comparative anatomists. We saw that Owen was quite cavalier in his approach to the embryological criterion (and also was quite happy to homologize, for instance, dermal with endochondral bones, thus rejecting similarity as well as embryonic derivation as a criterion). The only criterion of homology to which Owen seems to have been loyal was Geoffroy's principle of connections. But, consideration of the fallible criteria leads to another conclusion of some interest, already hinted at. In Owen's time, after the taxonomy of Linnaeus, based mainly on external characters, but before the Darwinian revolution, data for inferring animal relationships and for constructing classifications were those of comparative anatomy, achieved by dissection and close observation. Homology was a concept in comparative anatomy; if embryological criteria conflicted with those of adult anatomy so much the worse for embryology. This stance is not surprising when one looks at the early nineteenth century: practitioners like Cuvier, Geoffroy and Owen dominated comparative anatomy, and comparative anatomy dominated biology. But move forward 150 years and we find de Beer taking it for granted that homology is a purely anatomical concept. He does not talk about conflict between homologies determined by anatomical criteria and by embryological criteria, but about the failure of the embryological (and other) criteria. At about the time that de Beer was writing his "reader," biochemistry and molecular biology were beginning to use concepts of homology that are so very different that it is worth wondering whether the same term, *homology*, should be applied to them, but even today homology to most taxonomists implies anatomical comparison of adult organisms.

My second reason for introducing de Beer's "reader" was a heuristic one. Homology is a subject of endless fascination to biologists, but there is something about the apparently endless stream of papers and articles on the subject that suggests that the concept itself is intractable and unsatisfactory and/or that there is a lot of muddled thinking going on. I believe it would help if, as in the previous section, definitions of homology, criteria for testing proposals of homology, and explanations of the phenomenon of homology, were clearly separated. In a stimulating discussion of homology from a cladistic point of view that has become something of a classic, Patterson (1982) records four categories of definitions — classical, evolutionary, phenetic, and cladistic, together with more informal utilitarian ones. But the evolutionary definitions and two of the four cladistic ones conflate definition and explanation. Evolution, or rather phylogeny, is the explanation of homology, as it is of the phenomenon of natural classification, itself reconstructed from the hierarchy of homologies. It should not be part of the definition. However unsatisfying, homology should be defined simply as "structural [and positional] similarity" rather than as "any structural similarity due to common ancestry" (Boyden, 1973, p. 82). Or in Owen's words, "HOMOLOGUE. The same organ in different animals under every variety of form and function."

ACKNOWLEDGMENTS

The substance of this chapter was delivered as a lecture at a meeting organized in April 1992 by the Society for the History of Natural History to mark the centenary of Richard Owen's death. But I must also acknowledge a special debt of gratitude to Dr. Nicolaas Rupke for sending me a manuscript copy of his paper on the archetype long before its publication, and for the interest he has shown in this work.

I am indebted to the Robinson Library, University of Newcastle upon Tyne, for special permission to borrow works of Richard Owen published before 1850, which are not normally allowed to be removed from the library. The Natural History Society of Northumbria allowed photography of one of their most precious books, a first edition of Belon (1555). I am also very grateful to Ms Gina Douglas, Librarian at the Linnean Society of London, who copied the table from MacLeay (1821) (our Figure 7) and helped track down a difficult reference to Geoffroy's work. Photographs for the figures were taken by the Audiovisual Centre, University of Newcastle upon Tyne, and the manuscript was typed by Mrs. Yvonne Humble.

REFERENCES

Agassiz, J. L. R. (1857). "Essay on Classification. Contributions to the Natural History of the United States, vol. 1, part 1." Little, Brown, Boston.

Appel, T. A. (1987). "The Cuvier-Geoffroy Debate: French Biology in the Decades before Darwin." Oxford Univ. Press, New York and Oxford.

Barry, M. (1837). Further observations on the unity of structure in the animal kingdom, and on congenital abnormalities, including 'hermaphrodites'; with some remarks on embryology, as facilitating animal nomenclature, classification, and the study of comparative anatomy. *Edinb. New Philos. J.* **22**, 345-364.

Belon, P. (1555). "L'Histoire de la Nature des Oyseaux." Guillaume Cavellat, Paris.

Boyden, A. (1973). "Perspectives in Zoology." Pergamon Press, Oxford and New York.

Cain, A. J. (1958). Logic and memory in Linnaeus's system of taxonomy. *Proc. Linn. Soc. London* **169**, 144-163.

Carpenter, W. B. (1841). "Principles of General and Comparative Physiology," 2nd ed. Churchill, London.

Carus, C. G. (1828). "Von den Ur-Theilen des Knochen - und Schalengerüstes." Fleischer, Leipzig.

Chambers, R. (1844). "Vestiges of the Natural History of Creation," 1st ed. Churchill, London.

Clarke, C. A., and Sheppard, P. M. (1960). The evolution of dominance under disruptive selection. *Heredity* **14**, 73-87.

Cuvier, G. (1812). Sur un nouveau rapprochement à établir entre les classes qui composent le règne animal. *Ann. Mus. Hist.Nat.* **19**, 73-84.

Cuvier, G. (1817). "Le Règne Animal Distribué d'après son Organisation," 1st ed., 4 vols. Fortin, Paris.

Darwin, C. R. (1859). "On the Origin of Species by Means of Natural Selection, or the Preservation of Favoured Races in the Struggle for Life" 1st ed. John Murray, London.

de Beer, G. R. (1971a). "Some General Biological Principles Illustrated by the Evolution of Man," Oxford Biol. Readers No. 1. Oxford Univ. Press, London.

de Beer, G. R. (1971b). "Homology: an Unsolved Problem," Oxford Biol. Readers No. 11. Oxford Univ. Press, London.

de Beer, G. R. (1972). "Adaptation," Oxford Biol. Readers No. 22. Oxford Univ. Press, London.

de Beer, G. R. (1975). "The Evolution of Flying and Flightless Birds," Oxford Biol. Readers No. 68. Oxford Univ. Press, London.

Desmond, A. (1989). "The Politics of Evolution: Morphology, Medicine, and Reform in Radical London." Univ. of Chicago Press, Chicago.

Ford, E. B. (1975). "Ecological Genetics," 4th ed. Chapman & Hall, London

Geoffroy Saint-Hilaire, E. (1818). "Philosophie Anatomique, vol. 1, Des Organes Respiratoires sous le Rapport de la Détermination et de l'Identité de leurs Pièces Osseuses." J. B. Baillière, Paris.

Geoffroy Saint-Hilaire, E. (1825). Mémoire sur la structure et les usages de l'apparelle olfactif dans les poissons, suivi de considerations sur olfaction des animaux qui odorent dans l'air. *Ann. Sci. Nat.* **6**. 322-352.

Ghiselin, M. T. (1987). Species concepts, individuality, and objectivity. *Biol. Philos.* **2**, 127-143.

Goethe, J. W. von (1807). Bildung und Umbildung organischer Naturen [new introduction] *In* "Versuch die Metamorphose der Pflanzen zu erklären" (reprint). Carl Wilhelm Ettinger, Gotha.

Goethe, J. W. von (1817-1824). "Zur Natur-Wissenschaft überhaupt, besonders zur Morphologie," 2 vols. J. G. Cotta, Stuttgart & Tübingen.

Gould, S. J. (1977). "Ontogeny and Phylogeny." Harvard Univ. Press, Cambridge MA.

Lankester, E R. (1870). On the use of the term Homology in modern Zoology, and the distinction between homogenetic and homoplastic agreements. *Ann. Mag. Nat. Hist.* [4] **6**, 34-43.

Lewes, G. H. (1864). "The Life of Goethe," 2nd ed. Smith, Elder & Co., London

Lovejoy, A. O. (1936). "The Great Chain of Being". Harvard Univ. Press, Cambridge, MA.

MacLeay, W. S. (1821). "*Horae Entomologicae*: or Essays on the Annulose Animals, Vol. I, Part II. Containing an Attempt to Ascertain the Rank and Situation which the Celebrated Insect, *Scarabaeus sacer*, Holds Among Organised Beings." S. Bagster, London.

Maclise, J. (1846). On the nomenclature of anatomy (addressed to Professors Owen and Grant). *Lancet*, March 14, 298-301.

Maclise, J. (1847). "Comparative Osteology: Being Morphological Studies to Demonstrate the Archetype Skeleton of Vertebrate Animals." Taylor & Walton, London.

Mayr, E. (1969). "Principles of Systematic Zoology." McGraw-Hill, New York.

Mayr, E. (1982). "The Growth of Biological Thought: Diversity, Evolution, and Inheritance." Harvard Univ. Press, Cambridge, MA.

Mayr, E. (1987). The ontological status of species: Scientific progress and philosophical terminology. *Biol. Philos.* **2**, 145-166.

Mayr, E. (1988). A response to David Kitts. *Biol. Philos.* **3**, 97-98.

Milne-Edwards, H. (1844). Considérations sur quelques principes relatifs à la classification naturelle des animaux. *Ann. Sci. Nat., Zool. Biol. Anim.* [3] **1**, 65-99.

Oldroyd, D. (1986). "The Arch of Knowledge: An Introductory Study of the History of the Philosophy and Methodology of Science." Methuen, New York and London.

Outram, D. (1986). Uncertain legislator: Georges Cuvier's laws of nature in their intellectual context. J. Hist. Biol. **19**, 323-368.

Owen, R. (1832). "Memoir on the Pearly Nautilus (*Nautilus pompilius* Linn.), with Illustrations of its External Form and Internal Structure". Royal College of Surgeons, London.

Owen, R. (1838). A description of Viscount Cole's specimen of *Plesiosaurus macrocephalus* (Coneybeare). *Proc. Geol. Soc. London* **2**, 663-666.

Owen, R. (1840). A description of a specimen of the *Plesiosaurus macrocephalus*, Conybeare, in the collection of Viscount Cole, M.P., D.C.L., F.G.S., etc. *Trans. Geol. Soc. London* [2] **5**, 515-535.

Owen, R. (1843). "Lectures on the Comparative Anatomy and
    Physiology of the Invertebrate Animals, Delivered at the
    Royal College of Surgeons, in 1843." Longman, Brown,
    Green, and Longmans, London.
Owen, R. (1846). "Lectures on the Comparative Anatomy and
    Physiology of the Vertebrate Animals, Delivered at the
    Royal College of Surgeons of England, in 1844 and 1846.
    Part I. Fishes". Longman, Brown, Green, and Longmans,
    London.
Owen, R. (1847). Report on the archetype and homologies
    of the vertebrate skeleton. *Meet. Br. Assoc. Adv. Sci., Rep.*
    16, 169-340.
Owen, R. (1848). "On the Archetype and Homologies of the
    Vertebrate Skeleton". John van Voorst, London.
Owen, R. (1849). "On the Nature of Limbs. A Discourse
    Delivered on Friday, February 9, at an Evening Meeting of
    the Royal Institution of Great Britain." John van Voorst,
    London.
Owen, R. (1851). Lyell — on life and its successive develop-
    ment. *Q. Rev.* **89**, 412-451.
Owen, R. (1855). "Lectures on the Comparative Anatomy and
    Physiology of the Invertebrate Animals, Delivered at the
    Royal College of Surgeons," 2nd ed. Longman, Brown,
    Green, and Longmans, London.
Owen, R. (1860a). Darwin on the origin of species.
    Edinburgh Rev. **111**, 483-532.
Owen, R. (1860b). "Memoir on the Megatherium or Giant
    Ground-sloth of America (*Megatherium americanum*,
    Cuv.)." Taylor & Francis, London.
Owen, R. (1866). "On the Anatomy of Vertebrates," Vols. I
    and II. Longmans, Green, London.
Owen, R. (1868). "On the Anatomy of Vertebrates," Vol. III.
    Longmans, Green, London.
Owen, R. (1879). "Memoirs on the Extinct Wingless Birds of
    New Zealand, with an Appendix on Those of England,
    Australia, Newfoundland, Mauritius and Rodriguez," 2
    vols. John van Voorst, London.
Owen, R. (1992). "The Hunterian Lectures in Comparative
    Anatomy May-June, 1837 (Edited, and with an Intro-
    ductory Essay and Commentary, by Philip Reid Sloan)".
    Univ. of Chicago Press, Chicago.

Paley, W. (1802). "Natural Theology: Or Evidences of the Existence and Attributes of the Deity Collected from the Appearances of Nature". J. Faulder, London.

Panchen, A.L. (1992). "Classification, Evolution and the Nature of Biology". Cambridge Univ. Press, Cambridge and New York.

Panchen, A. L. (1993). "Evolution" (Mind Matters series) Bristol Classical Press, London.

Patterson, C. (1982). Morphological characters and homology. In "Problems of Phylogenetic Reconstruction" (K. A. Joysey and A. E. Friday, eds.), pp. 21-74. Academic Press for the Systematics Association, London.

Ray, J. (1691). "The Wisdom of God Manifested in the Works of Creation: Being the Substance of Some Common Places Delivered in the Chappel of Trinity College in Cambridge". Samuel Smith, London.

Richards, E. (1987). A question of property rights: Richard Owen's evolutionism reassessed. Br. J. Hist. Sci. **20**, 129-171.

Rieppel, O. C. (1988). "Fundamentals of Comparative Biology". Birkhäuser, Basel and Boston.

Rupke, N. A. (1992). Richard Owen and the Victorian museum movement. *J. Proc. R. Soc. N. S. W.* **125**, 133-136.

Rupke, N. A. (1993). Richard Owen's vertebrate archetype. *Isis.* **84**, 231-251.

Russell, E. S. (1916). "Form and Function: a Contribution to the History of Animal Morphology". John Murray, London. (page quotations from reprint 1982, Univ. of Chicago Press).

Strickland, H. E. (1846). On the structural relations of organized beings. *Philos. Mag. J. Sci.* [3] **28**, 354-364.

Turner, J. R. G. (1962). Geographical variation and evolution in the males of the butterfly *Papilio dardanus* Brown (Lepidoptera: Papilionidae). *Trans. R. Entomol. Soc. London* **115**, 239-259.

Uzzell, T. M., Gunther, R., and Berger, L. (1977). *Rana ridibunda* and *Rana esculenta*: A leaky hybridogenetic system (Amphibia, Salientia). *Proc. Acad. Nat. Sci. Philadelphia* **128**, 147-171.

Vane-Wright, R. I. and Smith, C. R. (1991). Phylogenetic relationships of three African swallowtail butterflies, *Papilio dardanus, P. phorcas* and *P. constantinus*: a cladistic analysis (Lepidoptera: Papilionidae). *Syst. Entomol.* **16**, 275-291.

von Baer, K. E. (1828). "Über Entwickelungsgeschichte der Thiere: Beobachtung und Reflexion." Bornträger, Königsberg.

2

HOMOLOGY, TOPOLOGY, AND TYPOLOGY: THE HISTORY OF MODERN DEBATES

Olivier Rieppel

Department of Geology
Field Museum of Natural History
Chicago, Illinois 60605-2494

Homology: The Hierarchical Basis of Comparative Biology
Copyright © 1994 by Academic Press, Inc.
All rights of reproduction in any form reserved.

I. INTRODUCTION

When Aristotle realized that dolphins were related not to sharks but to mammals, he must have made this discovery on the basis of what we, today, would call homology. His look at organisms transcended their outward appearance as conditioned by functional demands imposed by their mode of life, in a search for characters that would indicate the organisms' affinities with greater reliability.

However, the modern history of the concept of homology begins not so much with Aristotle's system of nature, but with Belon's (1555) famous illustration of structural correspondences in the skeleton of a bird and of a human (see Figure 8 in the previous chapter). To depict this correspondence, Belon again had to abstract from both form (shape) and function of the compared structures. He pictured bird and human as suspended from the skull, the limbs dangling down — a highly unnatural position and at the same time an artistic trick, forcing the reader to look at a bird skeleton in an unusual way. Once Belon had taught his readers this new "way of seeing," he conceptually cut the bird into pieces. He labelled individual bones with letters, and used the same letter to indicate structural equivalence of bones in the skeleton of the bird and human. He did not look at the skeleton as an integrated whole, but as a composition of parts, and he compared these parts neither in terms of shape, nor in terms of function, but in terms of another criterion of similarity: topology.

Today, homology continues to be a much debated issue in comparative biology. Arguments are about continuity of information (Van Valen, 1982), shared developmental pathways (Roth, 1984, 1988) and shared developmental constraints (Wagner, 1989a,b), congruence and homoplasy (Patterson, 1982), mistakes in the identification of homology (Brady, 1985), molecules and morphology (Patterson, 1987), and finally population genetics and the concept of the structural plan (*Bauplan*; Roth, 1991). This is not the place to summarize current debates about homology (see Rieppel, 1992, and other chapters in this volume); instead, I would like to clarify some of the points that seem to have been responsible for keeping the debate on the road it threatens to follow endlessly.

II. DEFINING HOMOLOGY

Definitions are tricky things. Some think that definitions are, or ought to be, independent of time and space, just as natural laws are, or ought to be. This ideal is inspired by Euclidean geometry: the sum of the angles in a triangle is always and only 180°, regardless of time and space. True, indeed, as long as parallels do not cross, but somebody will eventually point out that parallels do meet in the infinite![1]

A. Similarity Due to Common Descent

The definition of homology, at least in the post-Darwinian era, has been similarity due to common descent; see Chapter 1 and Boyden (1947) for reviews. There were other definitions before Darwin; "essential similarity," or "philosophical similarity," postulated by Etienne Geoffroy Saint-Hilaire (1830), and there are different definitions today; "structural similarity" (Goodwin, 1984; Webster, 1984), or "biological homology" as opposed to "historical" or "phylogenetic" homology (Wagner, 1989a,b). Obviously, different intentions must lie at the heart of different definitions of homology; different causal mechanisms are invoked to explain shared similarity. Definitions do not have to entail causal explanations however, nor do they have to specify operational criteria for the recognition of homology (Bock, 1973). What they have to do is specify in a short yet complete manner the significance of a word or thing.

It has become commonplace to criticize the phylogenetic definition of homology (similarity due to common

[1] Bookstein in Chapter 5 (Section I) deals with precisely this issue in relation to curvature in geometry, cladistics, morphometrics, and homology (Ed.).

ancestry) as circular, or tautological, since knowledge of common ancestry seems to be required for the recognition of homology. That is not true. "On my theory, unity of type is explained by unity of descent," wrote Darwin (1859, p. 206), meaning that his theory of evolution (descent with modification) provides a causal explanation for homology. Homology was (and still is) the concept on which the *Unity of Type* was (and is) built; it disclosed the "hidden affinities" amongst organisms as discovered by Darwin's predecessors in the science of comparative anatomy. When he proclaimed the "hidden affinities" as being real and material, i.e., due to common ancestry, Darwin proposed a natural causal explanation for the "philosophical" or "essential" similarity observed by comparative anatomists of the School of Etienne Geoffroy Saint-Hilaire and those influenced by him, including Owen (Desmond, 1989; and see Chapter 1). As Richards (1992, p. 132) put it: "Darwin historicized and thereby naturalized teleology." The definition of homology as similarity due to common ancestry is, indeed, utterly unproblematic, if it is understood that common ancestry is the causal explanation for shared similarity, which itself has to be discovered by means which are decoupled, or independent of, the theory of evolution (Brady, 1982, 1985).

B. *Biological Homology and Developmental Constraints*

But there is another kind of homology, *biological homology*, some authors contend (Wagner, 1989a,b), which is shared similarity of "individualized developmental units" due to shared developmental constraints, rather than common descent. This argument is perfectly logical, but lacks an empirical basis. The observation of structural invariance is explained by developmental constraints, which in turn are explained by reference to structural invariance. Developmental constraint thus proves to be an empirically empty concept, a redescription of structural invariance, unless some underlying causal mechanism for developmental invariance is specified. For the time being, developmental constraint is but a descriptive concept derived from observations of shared similarity — but why "shared"? The notion of shared similarity requires a causal explanation as to why some similarity (whether structural or developmental) should be shared. Even if causal mechanisms of developmental constraints were known, we still would be left to consider why these should be shared by two or more organisms. If the answer is that developmental constraints are shared because of common descent, homology is no longer solely dependent on causal mechanisms of development, but now invokes the phylogenetic history of the entities compared. Developmental mechanisms may be a proximal cause of similarity, but phylogeny remains the ultimate cause of shared similarity (for a distinction of proximal and ultimate causes, see Mayr, 1982).

If shared developmental pathways, or shared developmental constraints, are not only the proximal cause of homology, but the core of its definition, the concept of shared similarity would appear to be amenable to an empirical, even experimental research program, namely that of developmental biology. At the same time, however, this research program would have to be decoupled from evolutionary history and thus would have to be framed as an ahistorical, or structuralist (Rieppel, 1990), approach to analysis of biodiversity. Evolution, after all, means descent with modification — modification of structures, but also modification of developmental pathways, as well as potential modification of developmental constraints!

It has been widely recognized that homologous structures may originate through different developmental mechanisms (Roth, 1984, 1988; Wagner, 1989a,b; for reviews of the literature see Rieppel, 1992, and Hall, 1992), which has led to the rejection of shared developmental pathways as a means for the identification of homologies. Developmental constraints have been invoked instead (Wagner, 1989a,b) — and from a logical, i.e., ahistorical point of view it is, indeed, possible to claim that all similarity due to shared developmental constraints is homologous. But this then renders the issue of homology a matter of logical deduction from a definition which is independent from time and space and hence has nothing to do with history or phylogeny. While similarity of developmental constraints may cause similarity of structure, no explanation is offered as to why developmental constraints are shared, nor is it even necessary to address the issue as to whether similar developmental con-

straints are shared because of chance, because of independent acquisition, because of hidden affinities determined by
a Creator, or because of common descent.

Let **a** = **b**, and **b** = **c**: if these two premises are accepted,
a = **c** must follow, not as a consequence of causal mechanisms, but as a consequence of logics, independent of time
and space. If **a** is shared similarity, **b** is shared developmental constraints, and **c** is homology, it follows that similarity
due to shared developmental constraints can give rise to a
logical classification of organismic diversity of the basis of
the concept of biological homology (Wagner, 1989a,b). Such
a classification remains independent of time and space,
lacks a natural causal explanation (of shared similarity), and
hence is irrelevant for historical biology. For purely logical
reasons, there can be no historical change of developmental
constraints; there can be no problem of convergence or, in
more general terms, of homoplasy; and there can be no
appeal to common descent. The questions left unanswered
by this structuralist perspective are the natural causal explanation of "order versus disorder" (Patterson, 1982, p.
42), of hierarchy versus homoplasy (Rieppel, 1990).

The modern debate on homology has brought some
central issues relating to the concept of homology back into
focus, issues glossed over all too easily in the heyday of
evolutionary systematics (Mayr, 1969). What does shared
similarity mean? What is similarity in the first place? And if
shared similarity calls for a causal explanation, how is it
possible to recognize similarity independent of its causal
explanation — indeed a necessity if circular reasoning is to
be avoided! And last: what is it, after all, that we compare?
Life, phyla, organisms, structures, organs, development,
nucleotides, characters?

III. RECOGNIZING HOMOLOGY

Phylogeny reconstruction has been wedded to
Popperian falsificationism (Platnick and Gaffney, 1978a,b)
which, in the face of the classical problem of induction, resolved the problem by denying it: induction, in fact, would
simply not exist (Popper, 1972). All scientific investigation

would follow from some hypothesis, and only these hypotheses which generate testable and potentially falsifiable predictions would be truly scientific.

If all justification of scientific knowledge derives from testing and potential refutation of hypotheses, the origin of these hypotheses would appear to be irrelevant to the outcome of scientific investigation. As long as any idea, or intuition, generates testable predictions, scientific progress would continue. It is this Popperian paradigm that has given rise to the notion that homologies are conjectures of similarity (to be explained by common ancestry) "whose source is immaterial to their status" (Patterson, 1982, p. 58). The real problem lies in testing these hypotheses adequately. The test is that of congruence of characters (Patterson, 1982). Homology, after all, is a conjecture of similarity to be explained by common descent. The more characters (conjectured similarities) congruently supporting a hierarchy of relationships, the stronger the evidence for a regularity of character distribution among taxa, suggestive of an underlying cause — which is evolution. What matters is not so much how we get to conjectures of similarities, but that these hypotheses of homology be tested against all other characters known in the search for regularity, i.e., congruent character distribution.

Another way of putting the same insight is to claim that the initial conjecture of similarity is a "primitive concept" for systematics (de Pinna, 1991, p. 377), the primordial ooze from which a significant hypothesis may, or may not, emerge. However, Popperian falsificationism starts with the insight that there is no theory-free observation. If true, there cannot be a primitive concept for systematics, nor can the origin of an hypothesis of homology be immaterial to its status. Second, falsificationism views an hypothesis as the general statement, from which particular statements may be deduced, and in turn tested. The general statement is more inclusive than the particular statement which is to be tested — a point which in the final analysis adds to the problems of falsificationism [as it plunges into an infinite regress (Chalmers, 1982)], but which also clearly demarcates Popperian falsificationism from the test of congruence. Here, it is one character testing the other and *vice versa*, which means that if observation could, indeed, be a primitive concept of systematics (which is impossible), one observation would be allowed to test the other and *vice versa* — and a perfect circle would obtain. Something must be added to the observation of similarity which guides this observation (since theory free observation is nonexistent) but must be justified outside the context of one character testing the other. The conceptual tool added to, or rather guiding observation in the search for homology is *topology*.

Biologists observe and analyze organismic diversity in the attempt to make sense of it, seeking regularity of phenomena, a regularity of shared similarities which, for example, would allow the classification of organismic diversity in a hierarchical system of groups within groups. This is, after all, the way every child starts to master the diversity of the world it experiences: to distinguish the dead from the living, animals from plants, bushes from trees, and so on. A hierarchical system provides the basis for efficient, i.e., economical, information storage and retrieval. It is, of course, perfectly possible to hypothesize any wild kind of shared similarity: it might seem feasible to start grouping all flying objects, all living things that are of a rounded shape, or that share red color. The problem with this approach becomes apparent if no consistent hierarchical system obtains. We could start to group all organisms which share similarities due to similar developmental constraints: again, an interesting system of order in nature might become apparent, logical and consistent within its own premises, but perhaps not very economical and hence not very useful. If making sense of organismic diversity means the search for an efficient and economical system of information storage and retrieval, congruence of characters becomes the criterion of success for (rather than a test of) conjectures of similarity (Brady, 1985). Congruence, in turn, signifies a regularity of character distribution calling for a causal explanation such as evolutionary theory. It may, indeed, be true that the origin hypotheses is immaterial to their scientific status, but it is also true that topological relations of similarity have, historically, led to greater success in the search for a natural (hierarchical) system of groups within groups than have other premises, such as similarity of form and function.

When Geoffroy Saint-Hilaire (1830) defined *analogie* (our homology) as "essential similarity," or "philosophical similarity," he stressed the distinction of his research program in comparison to that of Georges Cuvier (a detailed analysis of the competing research programmes of Cuvier and Geoffroy Saint-Hilaire can be found in Appel, 1987; see also Rieppel, 1988, and references therein). Cuvier was an anachronistic believer in the doctrine of preexistence (Rieppel, 1986, and references therein), which led him to maintain that organisms are determined by the Creator's plan, providing a perfect match between the organismal structure and the environment (biotic and abiotic) for which the respective structure had been created. In that sense, Cuvier was a committed functionalist (Rieppel, 1990): organismic structure would be determined by its function. Geoffroy was far from denying the importance of functional influence on structure. But he sensed that there was something more at issue, something beyond functional constraints, namely laws of structure. Why should we recognize fins of teleosts, salamanders' limbs, and birds' wings, i.e., paired appendages, as part of the evidence for a common structural plan underlying the *embranchement* (a term used by Cuvier for major divisions of the animal kingdom) of vertebrate animals? There is an underlying similarity of these structures, notwithstanding their differences in form and function, which justified this conclusion. Although Cuvier did recognize "embranchements" on the basis of a shared structural plan (*Bauplan*), he did not provide any explanation for the conservation of underlying structural relations. And yet these are readily perceived: paired appendages in a symmetrical arrangement! Number and relative position, the law of symmetry above everything else, and the law of compensation — these were the cornerstones on which Geoffroy Saint-Hilaire built his *Philosophie Zoologique* (1830). If a structure diminishes in importance, volume or otherwise, this would have to be compensated for by the complementary increase of another structural component: if snakes lost limbs, they increased the number of vertebrae.

There is room, in Geoffroy's system, for modification of structures under the influence of the environment, but the underlying structural plan always remains visible due to developmental constraints. Geoffroy contended that structures may increase or decrease in volume, depending on the amount of nutritive material they receive from the blood vascular system: however, since the volume of blood remains constant, diminished growth of one organ must be compensated for by increased growth of another organ, and the topological relations between organs, all interconnected by the blood vascular system, remain invariant. Connectivity is determined by the blood vascular system, which provides the nutritive material necessary for development and growth. Connectivity also determines relative topological position within the structural complex. Laws of development determine invariant topological relations — this is how Geoffroy's *analogie* (our homology) came to mean something else than similarity due to functional necessity; see Chapter 1. It became evidence for the "hidden affinities" of organisms, which Charles Darwin set out to explain by his theory of evolution (descent with modification).

IV. CONNECTIVITY VERSUS TOPOLOGY IN THE SEARCH FOR HOMOLOGY

Topology is based on connectivity and *vice versa* (Rieppel, 1988), yet both concepts depend on an invariance of number and relative position of constituent elements in a biological structure (Rieppel,1990). In the tetrapod limb, the zeugopodium (radius/ulna, tibia/fibula) is identified, in topological terms, as "paired elements located distal to the unpaired proximal element" of the stylopodium (humerus, femur). The dynamics of embryonic development document the bifurcation (for a definition of the term, see Crumly, 1991) of the zeugopodial elements from the stylopodium (Shubin and Alberch, 1986): zeugopodial elements are identified by their connectivity to the stylopodium.

Homology, as well as topology, are concepts related to a research program rooted in atomism (Rieppel, 1986). They require the decomposition of organic structures into their constituent elements, and their comparison not in terms of shape, nor in terms of function, but in terms of connectivity, or topological relations within the structural complex. This, however, is by no means an easy task, since the subdivision of an organic whole into its constituent elements will never be unproblematic. Indeed, comparison of organisms on the basis of homology, recognized by connectivity or topology, will depend on different conceptualizations of embryonic development — which, in fact, reach back to the origins of comparative embryology. If the modern debate on the meaning of homology has highlighted one most important issue, it is certainly the units of comparison. Wagner (1989a) speaks of "individualized developmental units," a choice of words which highlights the necessity that homologues have some individuality imparted on them by ontogeny and phylogeny. The individuality of homologues is usually rooted in topology and genetics (Tabin, 1992): topology provides guidance toward the observational recognition of homology, continuity of information through heredity provides the causal explanation for historical identity and individuality of homologues (Van Valen, 1982; Roth, 1991).

Individuality and historical identity are prerequisites for recognition of constituent elements in a complex organic structure as potential homologues, a requirement which may not be easy to fulfill. Consider the famous example of the ichthyosaur fin, a highly derived structure, difficult to compare to the "archetypical" tetrapod limb. How can individuality and historical (phylogenetic) identity be postulated for every element in the ichthyosaur fin in comparison to the standard tetrapod limb (Goodwin, 1984)? Obviously, the claim that homologues share individuality and historical identity depends on a particular conueceptualization of ontogeny and of the develomental units resulting from it. To clarify these issues, I commented, on an earlier occasion (Rieppel, 1986), on the differences of the atomistic versus epigenetic approaches to ontogeny, and the respective implications of these models for the concept of homology. To do so, I pursued an historical approach which I will use again in this context, although presenting a different episode in the history of comparative embryology.

A. Preformation and Epigenesis

Terminology was blurred in the heated debates characteristic of early phases of comparative embryology, and continues to be confused today. Then and now, the doctrine of "preformation" was and is being pitched against the doctrine of "epigenesis," as if preformation meant the same thing as "preexistence," and again as if the epigenesist William Harvey shared the same model of embryogenesis as the atomist P.-M. L. de Maupertius (see Rieppel, 1986, 1988). There is no need to review details of the historical debate. For understanding the concept of homology a brief review of the essentials will have to do (for a more detailed discussion and references see Rieppel, 1986, 1988).

The doctrine of preformation maintained that the parts of an embryo preexist in the seminal fluid of the male and/or female, and that embryos form by juxtaposition of those preexistent parts. This doctrine adhered to an ancient model of atomism, according to which male and female seminal fluids would contribute "blueprint particles" to the formation of their offspring, which form by juxtapposition of those particles. Accordingly, embryogenesis is reduced to growth by terminal addition, and differentiation does not takes place. Epigenesis, in contrast, maintained that no single part ("individualized developmental unit") of the embryo was preformed and supplied by either male or female seminal fluids, but that developmental units individualize, or emerge *de novo*, by a process of growth, subdivision and differentiation of primordial tissue. Preexistence finally claimed that the embryo preexisted, *en miniature* but in its total *Gestalt*, in the spermatozoid (animalculism) or in the egg (ovism).

B. Harvey and Malpighi

The distinction of atomism versus epigenesis can be well illustrated with reference to the opposing views held by William Harvey (1578-1657) and Marcello Malpighi (1628-1694) respectively.

Without going into the details of their contemporary history (found in Adelmann, 1966), Malpighi was not only an admirer, but also a critic, of Harvey.

> Malpighi was an observer (Adelmann, 1966, Vol. II, p. 820), (he) was no friend of hypotheses not based on observation (Adelmann, 1966, Vol. II, p. 821), (and)in his first dissertation on the development of the chick, Malpighi cites the 'immortal Harvey' (Adelmann, 1966, p. 821). However: Malpighi criticizes Harvey's contention that the early stages of the chick resemble a worm (Adelmann, 1966, Vol. II, p. 824),

which would successively differentiate the characteristics of its class through further development. Indeed, Malpighi was understood to be a preformationist by his contemporaries (Roe, 1981, p. 6; see also Bowler, 1971, pp. 226-227), which meant that the parts which would eventually come together during embryogenesis were preformed, even if they became visible in succession only:

> ... in the egg itself we behold the animal already almost formed ... (Adelmann, 1966, Vol. II, p. 935), (but)... being unable to detect the first origins, we are forced to await the manifestation of the parts as they successively come to view. (Adelmann, 1966, Vol. II, p. 937)

According to Malpighi, these parts, as they successively come to view, do not differentiate from one another through a process of growth, subdivision, and divergent differentiation, as was maintained by Harvey. Instead: "things which must afterwards be fitted into the assemblage may first be viewed as separate elements" (Adelmann, 1966, Vol. II, p. 935). Accordingly, in his description of the seventh day of development of the chick, Malpighi contended that the heart was:

> composed of paired ventricles, contiguous saccules, so to speak, which were joined above to the superposed body of the auricles ... (Adelmann, 1966, Vol. II, p. 967). On the fourteenth day, however, "The heart was composed of the united ventricles, and several arterial tubules like fingers of a hand, which formerly lay at a distance from the heart, were now joined directly to it ...." (Adelmann, 1966, Vol. II, p. 973)

Whatever Malpighi's real agenda, and however he was understood by his contemporaries, this last quote aptly illustrates the idea that during embryogenesis, things (or organs) that initially lie "at a distance" from each other come together: embryonic development, essentially, is a "coming together" of parts, juxtapositional growth of primordial particles, or atoms. The same is true for the brain:

> When the head was incised, the mass of the brain was formed to be more solid, for the vesicles, which had formerly been separate but now were united, produced twin eminences in which ventricles were formed .... (Adelmann, 1966, Vol. II, p. 967)

The "coming together" of parts, themselves preformed or of some other origin, was a far cry from what Harvey had in mind. In pursuit of his research program, Malpighi found it indeed "superfluous to inquire whether the heart is formed before the brain, or the blood before the heart ..." (Adelmann, 1966, Vol. II, p. 867). This outlook was clearly directed against Harvey, who, after ascertaining ("I have many times observed") that "the point and the blood exist first, and that pulsation arrives only afterwards" [Harvey, 1981, p. 98; meaning that blood exists first, the heart makes its appearance, and pulsation of the heart (of the *punctum saliens*) follows], concluded:

> "Now this is most certain, that nothing of the future foetus appears at this time except the streaks of blood and the *punctum saliens* and those blood vessels which are all derived from one trunk (as the trunk itself derived from the *punctum saliens*).... (Harvey, 1981, p. 99)

This is not the "coming together" of parts which previously existed in isolation from one another; but the generation of one part from another:

> ... it is certain that the chick is built by epigenesis, or the addition of parts *budding out* from one another ... (Harvey, 1981, p. 240; emphasis added)

Embryogenesis is not a process of juxtaposition of (preformed) parts, but a process of growth, budding, sub-division, and differentiation. Following Malpighi's atomistic paradigm, each developmental unit would be identified by its geometrical topology as a constituent element in a complex structure. Following Harvey's epigenetic paradigm, constituent elements in a complex structure "evolve" (individualize) one from the other, and are identified by their connectivity, rather than geometrical identity, to developmental precursors. Connectivity, too, is a topological criterion (Rieppel, 1988), but it allows for more plasticity of structure, for relative (rather than geometrically identical) position within a complex whole.

C. Polydactyly and Hyperphalangy

The contrast of an atomistic versus an epigenetic conceptualization of ontogeny (Rieppel, 1986) may be difficult
to understand, but will become immediately apparent in a
discussion of polydactyly and hyperphalangy in the
ichthyosaur fin (Fig. 1). As homology requires individuality
and historical (phylogenetic) identity of constituent elements in a biological structure (Goodwin, 1984), it seems
reasonable to address the question of hyperphalangy in
terms of an atomistic research program. Supernumerary
phalanges in the ichthyosaur fin, identified by number and
topological (geometrical) relations, would have been added
to the generalized phalangeal count of the archetypal tetrapod limb, and therefore would have no homologues in other
tetrapods.

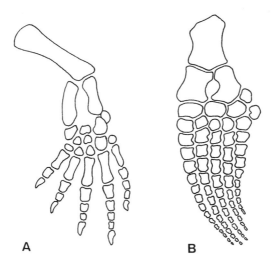

A   B

Fig. 1.   The limb of (A) a generalized tetrapod (reptile)
compared to (B) an ichthyosaur fin (showing the forelimb of
the Triassic genus *Mixosaurus*). [(A) after Kuhn-Schnyder,
1974; not to scale.]

The generalized tetrapod phalangeal count is 2-3-4-5-3(4), where numbers signify the number of bones in each digit, beginning preaxially with digit 1. As the number of phalanges exceeds this count in ichthyosaurs, the atomistic research program favors the view that the proximal, and only the proximal, phalanges in the ichthyosaur fin are homologous (i.e., topologically and historically identical) with phalanges in the generalized tetrapod limb. Supernumerary phalanges (and digits) would be identified as *additions* to what existed before (in phylogenetic history); neomorphs which have no equivalent in the "ancestral" condition. Accordingly, the addition of neomorphs would have transformed the "ancestral" into a "descendant" condition, which is the synapomorphic character state diagnostic of a monophyletic group. This "way of seeing" is in perfect accordance with the atomistic research program, such as expounded by Maupertuis, who investigated polydactyly in the befriended family of the surgeon Jakob Ruhe in Berlin — as an hereditary juxtaposition of supernumerary parts (Rieppel, 1989a). This perspective is also in accordance with the Darwinian paradigm:

> It may be said that natural selection is daily and hourly scrutinizing, throughout the world, every variation, even the slightest; rejecting all that is bad, preserving and adding up all that is good.... (Darwin, 1859, p. 84, emphasis added) [2]

This view contrasts with current understanding of the development of hyperphalangy, which supports an epige-

---

[2] The atomistic research program outlined above (and represented, among others, by Maupertuis) is identified as "recapitulationist" by Richards (1992). Its relation to Darwin's research program may accordingly seem problematical, since his views on embryology were mainly influenced by von Baer's (epigenetic) critique of recapitulationism (Ospovat, 1981). To make von Baerian epigenesis work for his theory, Darwin had to corrupt it by making the "archetype" (i.e., the common early ontogenetic stage in von Baer's terms) an actual adult ancestor (Richards, 1992, p. 123, 125). The von Baerian concept of deviating differentiation thus became a concept of terminal addition (Richards, 1992, p. 152), bringing epigenesis back into line with atomism. But this was by no means what von Baer had in mind; his views were perfectly summarized by E. Geoffroy Saint-Hilaire when the latter pointed out that the adult bird lung is so specialized that it could not possibly have originated by transformation of an adult lizard lung — yet the two types of lung could easily derive from a common early ontogenetic *anlage* (for more details, see Rieppel, 1988).

netic perspective. Discussing embryonic hyperphalangy in crocodiles, Kükenthal (1893; see also Müller and Alberch, 1990) noted that during early stages of development, the digits form contiguous *anlagen*. During subsequent development, these originally contiguous *anlagen* become subdivided in different ways, resulting in different numbers of phalanges. Phalanges "segment" (for a definition of the term, see Crumly, 1991) in a proximodistal sequence, but within an originally contiguous anlage of chondrogenic tissue (confirmed by O. Rieppel, personal observation). In crocodiles, digits may become subdivided into a greater number of cartilaginous phalangeal precursors than will eventually ossify. The atomistic research program would require that some (i.e., the proximal), but not the supernumerary phalangeal precursors are homologous to the phalanges in the generalized tetrapod limb, since hyperphalangy would develop by the *addition* of elements to those already existent. Viewed from the epigenetic perspective, the unit "digit" becomes *subdivided* into a different number of phalanges in different taxa. Different taxa share a more generalized condition of form, the originally contiguous *anlage* of digits. From the more general condition of form differentiates the less general condition, i.e., a different number of phalanges in different taxa. Different taxa become individualized (or diagnosable) by a different number of phalanges in their digits, but the individual phalanges cannot be homologized across the board.

Topology identifies every phalangeal condensation on its geometrical identity; connectivity (Shubin and Alberch, 1986) identifies phalangeal condensations by their differentiation from common primordia, in this example the digits. The digit is a developmental field, which will become subdivided in different ways. Homology is not with the resulting morphological units (phalanges), but with the developmental mechanisms determining the subdivision of the developmental field (Goodwin, 1984). However, digits in themselves result from the subdivision of a developmental field, the limb bud, and this developmental field can again be subdivided in different ways, giving rise to a pentadactyl limb, but also to a limb with six, seven, or eight digits (Coates and Clack, 1990). Again, the digits do not appear to share a positional identity which would allow their homologization (as the first, second, etc.) across the board

(Goodwin, 1984). Instead, there is a more general condition of form (that of a limb bud), from which the less general condition of form (number of digits) develops through differentiation, providing the basis for the individualization of taxa. Not all tetrapods, such as birds, for example (Hinchliffe, 1989), pass through the stage of a pentadactyl limb (see also Hinchliffe and Griffiths, 1983) during ontogeny, but all tetrapods share the limb bud stage.

Connectivity rather than topology has been proposed as a solution for the homologization of individual digits, since it identifies digit IV as part of the primary axis of the limb, from which preaxial digits originate in succession along the digital arch (Shubin and Alberch, 1986). Aside from the fact that the model of a primary axis of limb development running through digit IV does not provide an immediate solution to the problem of digital reduction in theropod dinosaurs [which retain digits I, II, and III (Gauthier, 1986)], connectivity will ultimately have to be rooted in topology. The primary axis of limb development (Burke and Alberch, 1985; Shubin and Alberch, 1986) runs through the stylopodium (humerus, femur) and continues, in accordance with the pattern of connectivity, through the ulna (fibula), ulnare (fibulare), distal carpal (tarsal) 4 into the fourth digit. But the stylopodial element originates as a *de novo* condensation, and cannot be identified other than by its topological position (Rieppel, 1988). The same is true of limb buds which, even if viewed as a developmental field, will have to be identified on topological grounds as paired protuberances from the ventrolateral body wall. In that sense, there exists a complementarity of topology and connectivity, just as there is a complementarity relationship of atomistic mechanisms of transformation, and epigenetic mechanisms of deviation and differentiation.

Most recently, *Hox* gene expression in the limb bud has been used to specify the identity of tetrapod digits (Tabin, 1992). The four vertebrate *Hox* gene clusters were found to subdivide the limb bud into five genetically marked and internested zones along the anteroposterior axis, determining pentadactyly in tetrapods. Polydactyly is explained as duplication or multiplication of digits within any one of these zones. *Hox*-4 gene expression would therefore "not preclude growing more than five digits, rather it prohibits developing more than five morphologically distinct types of digits" (Tabin, 1992, p. 295). Questions to be raised at this point relate not only to the applicability of this developmental model to classical cases of polydactyly as seen in some ichthyosaurs, but also to whether the identity of a structure can, in fact, be tied to the signal inducing its development (G.P. Wagner, personal communication). Nevertheless, recourse to the pattern of expression of the *Hox* gene clusters results in a topological concept, albeit allowing for serial homology of duplicated or multiplicated digits within each individual compartment of the limb bud.

D. *Atomism versus Epigenetics*

Atomistic versus epigenetic models of ontogeny may lead to different hypotheses of homology. The issue can become increasingly complex as the questions raised become increasingly detailed.

Some reptiles, such as turtles, crocodiles, and lepidosaurs, show a modification of the fifth metatarsal, which assumes a very distinctive "hooked" morphology. The bone consists of a slender distal shaft, and a broad proximal head, bent inward and articulating with the calcaneum (fibulare) proximally, and the distal tarsal 4 medially (Fig. 2). Ever since Goodrich's (1916) seminal paper (see also Robinson, 1975), the "hooked fifth metatarsal" has played an important role in reptile classification. To judge from its morphology, it seems reasonable to assume that the proximal head of the hooked fifth metatarsal incorporates distal tarsal 5, an independent ossification in the tarsus of generalized tetrapods. Yet in none of the extant reptiles which exhibit the hooked fifth metatarsal (turtles, crocodiles, and lizards) has the actual fusion of an originally independent distal tarsal 5 with the fifth metatarsal been observed.

Goodrich summarized all logically possible hypotheses for homology of the hooked fifth metatarsal:

> Some believe it to represent the modified fifth distal tarsal, others the 5th metatarsal combined with its tarsal, and yet others that it represents the modified fifth metatarsal only. (Goodrich, 1942, p.264)

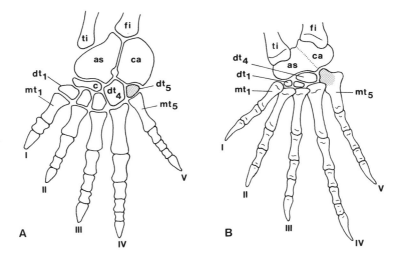

Fig. 2. The "hooked" fifth metatarsal in reptiles, and its relation to distal tarsal 5. (A) *Captorhinus aguti*, a "stem reptile" from the Lower Permian; (B) *Sphenodon punctatus*. Abbreviations: as, Astragalus; ca, calcaneum; dt1-5, digits 1 to 5; c, proximal centrale; fi, fibula; mt1-5, metatarsal 1-5; ti, tibiadistal tarsal 1-5, (after Romer, 1956; not to scale.)

Goodrich believed the last interpretation to be the most plausible one based on "Sewertzoff's excellent account" (*ibid*). Modern literature (Robinson, 1975, Gauthier *et al.*, 1988) agrees that the hooked fifth metatarsal incorporates distal tarsal 5. The latter conclusion is based on an atomistic approach, viewing organisms as a result of "parts coming together," each part characterized by a particular topological (geometrically identical) position. As the proximal head of the hooked fifth metatarsal expands into the space which, on geometrical grounds, should be occupied by the distal tarsal 5, the conclusion must be that the proximal head of the hooked fifth metatarsal fused with the distal tarsal 5 in the phylogenetic past. However, "Sewertzoff's excellent account" (Goodrich, 1942, p. 264) nowhere mentions fusion of an originally separate distal tarsal 5 with the fifth metatarsal. Instead, Sewertzoff (1908, p. 130) emphasized that although distal tarsal 4 dissociates ("segments," in Shubin and Alberch's, 1986, terms, see Crumly, 1991) from metatarsal 4, distal tarsal 5 never segments from metatarsal 5. For this reason, and in contradiction of Goodrich (1916), Sewertzoff (1908) believed the hooked metatarsal 5 to "incorporate" distal tarsal 5, not as a result of fusion, but as a result of nondifferentiation.

In atomistic terms, the fifth distal tarsal is identified, on geometrical grounds, as part of the proximal head of the fifth metatarsal, and the conclusion is that metatarsal 5 incorporates distal tarsal 5. Viewed from the epigenetic perspective, metatarsal 5 incorporates distal tarsal 5 not as a consequence of fusion, but because distal tarsal 5 fails to segment from the fifth metatarsal. Accordingly, distal tarsal 5 cannot have an historical individuality or positional identity in taxa characterized by a hooked 5th metatarsal. Consequently, incorporation, in Sewertzoff's (1908) sense cannot mean a material identity due to observed fusion of originally separate primordia — it must be an ideal identity based on the failure of one component to actually differentiate into its potential components. The potential is "ideal" (a logical rather than a material-historical relation, which is the inverse of the Platonic understanding)[3], the actualized component of the potential is material or "real." Another way of looking at the evidence views topology of the constituent elements of the tarsus to change not because of fusions of originally separate rudiments, but as a consequence of changing generative mechanisms subdividing the developmental field in a different manner (Goodwin, 1984).

Whereas the atomistic view allows for transformation of the descendant from its ancestor by means of fusions to (or obliterations from) what existed before, the epigenetic view allows for deviation of descendants from a shared embryonic condition of form through differentiation. In other words, the atomistic view allows conceptualization of ancestor-descendant relationships, whereas epigenesis results in the individualization of sister groups; the latter become individualized (diagnosable) by divergent ontogenetic differentiation from a common and more generalized condition of form, represented by early ontogenetic stages. It is evident that the atomistic view gives rise to what has been termed

---

[3] Plato thought the ideal condition to be "real," whereas the material object subject to the constraints of time and space he believed to be imperfect appearances of the underlying idos.

Haeckelian recapitulation (Løvtrup, 1978) or Heckelian ancestry (Patterson, 1983), whereas the epigenetic way of seeing supports the von Baerian concept of recapitulation (Løvtrup, 1978) or von Baerian ancestry (Patterson, 1983). Just as there is a conceptual relationship between topology and connectivity, between atomism and epigenesis, there is also one between Haeckelian and von Baerian recapitulation: the first is a special case of the latter (Løvtrup, 1978), Haeckelian recapitulation corresponds to terminal deviation (see also footnote 2).

V. CHANGING IDENTITIES AND TAXIC HOMOLOGY

Ontogenetic repatterning (Wake and Roth, 1989) refers to ontogenetic mechanisms changing the identities (in topological terms) of constituent elements in a complex structure during development. A simple example for ontogenetic repatterning is the change of identity and individuality of constituent elements of a biological structure when cartilage is replaced by bone. Let this biological structure be the tetrapod tarsus.

A. The Tetrapod Tarsus

In the generalized tetrapod condition, represented by the fossil anthracosaurian amphibian *Proterogyrinus* (Holmes, 1984), the tarsus comprises a multitude of elements — intermedium, fibulare, centralia 1 through 4, and five distal tarsals. In amniotes, the putative descendants of anthracosaurian amphibians, the number of tarsal elements is much reduced. In particular, only two bones ossify in the proximal tarsus, these being the astragalus [believed to represent the product of fusion of the intermedium, tibiale, and proximal centrale (centralia 4 and 3)], and the calcaneum (believed to represent the fibulare). Homologies indicated within brackets correspond to the atomistic approach: ancestor-descendant relationships are explained on the basis of additions, or fusions, of constituent elements in a biological structure. This explanation, however, appears to be insufficient for the reptile astragalus. Whatever the cartilaginous precursors of the astragalus, they all fuse into a single cartilage within which the astragalus of reptiles ossifies from a single ossification center (Rieppel, 1993a). The cartilaginous precursors lose their individuality and identity as they fuse. An astragalus, ossifying from a single ossification center, no longer reflects the multiple cartilaginous precursors supposed to participate in its formation. The atomistic approach to animal organization fails: there is no actual fusion of separate ossifications (intermedium, tibiale, and proximal centrale) to form the astragalus at any stage of ontogeny, or phylogeny of the Reptilia. Reptile embryos show multiple proximal tarsal cartilages, comparable to the proximal tarsal cartilages, and the corresponding proximal tarsal ossifications of anthracosaurian amphibians (but see Burke and Alberch, 1985; Shubin and Alberch, 1986). However, during subsequent development, reptiles deviate from this more general condition of form: cartilaginous precursors of proximal tarsal elements fuse and thereby lose their topological identity and individuality, and multiple elements are replaced by a single ossification, the astragalus.

Reptiles cannot be derived from anthracosaurian amphibians by fusion of proximal tarsal ossifications (*contra* Peabody, 1951), because such fusion never takes place. Instead, reptiles and anthracosaurian amphibians share a more generalized condition of form, the existence of several proximal tarsal cartilages in the embryo. Reptiles deviate from this more generalized condition of form more drastically than anthracosaurian amphibians. In the latter group, each cartilaginous precursor in the proximal tarsus presumably ossified separately, whereas in reptiles, ontogenetic repatterning results in replacement of several cartilaginous precursors by a single ossification. It is theoretically possible to homologize the cartilaginous precursors in the tarsus of reptiles with the cartilages in the tarsus of their putative ancestors — which represents the more general condition of form. It would even seem possible to homologize the cartilaginous precursors in the reptile tarsus with ossifications in the adult tarsus of their putative ancestors, since the latter did not deviate from the more general condition of form, every cartilaginous precursor giving rise to an individualized ossification. It is, however, impossible to homologize the reptile astragalus with the cartilaginous precursors in the reptile tarsus, or with the cartilaginous precursors and the corresponding ossifications in the putative ancestor. This is because the cartilaginous precursors lose their individuality and topological identity as they fuse prior to ossification of the astragalus from a single ossification center.

Chameleons are highly derived arboreal squamates with zygodactylous hands and feet. Modifications in the tarsus have resulted in the development of a ball and socket joint involving one proximal and one distal element. Whereas the distal element corresponds to distal tarsal 4, the proximal element is the only one to ossify within the proximal tarsal cartilage which forms by fusion of multiple cartilaginous precursors in the proximal tarsus. This single proximal element has traditionally been interpreted as the product of fusion of astragalus and calcaneum (e.g. Romer, 1956, Fig. 190), an atomistic interpretation. However, developmental studies show that only one element ossifies, and no fusion of two originally separate ossifications ever takes place (Rieppel, 1993b). Again, ontogenetic deviation and the corresponding failure of the atomistic approach preclude an easy answer to the question of whether the single proximal tarsal ossification in the chameleon tarsus is homologous to the astragalus or to the calcaneum of other reptiles (see Rieppel, 1993b, for further details).

Cartilaginous precursors, however, need not necessarily lose their identity as they fuse. In the carpus of the turtle *Chelydra serpentina*, for example, the originally separate precursors of two centralia (centralia 3 and 4 according to Burke and Alberch, 1985) fuse prior to ossification. Once ossification begins, both originally separate cartilaginous rudiments are represented by separate ossification centers in the combined cartilaginous precursor. Again in *Chelydra* (and in some other, but not all, turtles: see Rieppel (1993b) for details), the proximal centrale fusing into the proximal tarsal cartilage is represented by a separate ossification center, independent from astragalus and calcaneum. In the leatherback turtle (*Dermochelys coriacea*), the intermedium fuses with the proximal centrale in the carpus, while distal carpal 5 fuses with distal carpal 4. During ossification, each of the originally separate cartilaginous rudiments is represented by a separate ossification center (Völker, 1913). All this means is that simple fusion needs not invalidate the topological identity of cartilaginous precursors — they still may be represented by separate ossification centers in the may be represented by separate ossification centers in the combined cartilage. This is not the case, however, for the various cartilaginous precursors of the astragalus.

In conclusion, the astragalus is homologous throughout reptiles (probably throughout amniotes), but as a proximal tarsal *ossification* it has no homologue outside the reptiles (probably the amniotes). This means that the astragalus, homologous throughout the Reptilia (Amniota), cannot be derived from the *adult* ancestral condition by transformation based on fusion or obliteration of parts. There is no adult ancestral character state, which can be thought to have been transformed into the adult descendant character state in an atomistic fashion. Instead, the putative ancestor and its descendant (the sister groups) share a common early onto- genetic stage (multiple cartilaginous precursors in the proximal tarsus), from which the descendant deviates to a greater degree than the putative ancestor as cartilages are replaced by bone. The astragalus, therefore, is a *taxic homology* (Patterson, 1982) of the Reptilia (Amniota). As the less general condition of form, it is homologous throughout the Reptilia (Amniota), and hence diagnostic of the Reptilia (Amniota) as a monophyletic taxon, but it has no ancestral condition in the adult putative ancestor, the anthracosaurian amphibians, from which it could be derived by means of "terminal deviation." Viewed from the epigenetic perspec- tive, taxic homology is synonymous with *synapomorphy* (Patterson, 1982): a taxic homology is a character diagnostic of a monophyletic taxon. This contrasts with transforma- tional homology, whereby an "ancestral" character state is believed to be transformed into a "descendant" character stage by "terminal deviation" (terminal addition).[4]

---

[4] In a more general sense, taxic and transformational homology become complementary terms within the hierarchy of monophyletic taxa. Just as every synapomorphy turns into symplesiomorphy al lower levels of inclusiveness, so can a taxic homology provide the basis for transfor- mational hypotheses at included levels of the hierarchy (for further details see Patterson, 1982; Rieppel, 1988). It is recognition of the taxic homology of the tetrapod limb throughout Tetrapoda which allows the hypothesis that the wing of *Archaeopteryx* derives from the theropod hand; it is the taxic homology of paired appendages throughout Gnathostomata which allows the hypothesis that the sarcopterygian fin transformed into the ichthyostegid limb.

VI. TRANSFORMATIONAL AND TAXIC APPROACHES

Evolution means change. Originally, the term was used to denote ontogenetic "unfolding" of preexistent structures (Bowler, 1975; Richards, 1992). With Darwin it came to mean phylogenetic diversification of life on earth. However, under the atomistic research program, all of this diversification ultimately would be reduced to the addition, deletion, or transformation ("terminal deviation") of adult ancestral parts as one generation gives rise to the next. As Darwin's "bulldog," Thomas H. Huxley, so aptly remarked in his Croonian lecture of 1858, when he tried to explain the complexity of the vertebrate skull:

> ... a community of plan is discernible amidst the manifold diversities of organic structure ... there is nothing really aberrant in nature ... an apparently new and isolated structure will prove ... to be only a *modification of something which existed before* ....(Huxley, 1859, p. 382; emphasis added)

What "existed before" is the ancestral condition of form. In transformational terms, any monophyletic taxon, including the ancestor and all of its descendants, is diagnosed by the "terminal addition" of an evolutionary innovation to what already "existed before." The evolutionary innovation "added" may be the deletion of parts — an observation which justifies the distinction of "absence" from "loss" of a character. The evolutionary innovation is the synapomorphy supporting the monophyly of the derived taxon. Since the synapomorphic ("descendant") character state is a transformation of the plesiomorphic ("ancestral") character state, homology must encompass both plesiomorphic and apomorphic character states.

The transformational approach has culminated in the idea of an archetypal tetrapod limb, from which all existing tetrapod limbs could be derived by addition, deletion, fusion, or terminal modification of constituent elements. The Haeckelian theory of recapitulation (Løvtrup, 1978) has motivated the search for archetypal limb geometry in developing limb buds of extant tetrapods. There must be some ancestral condition of form, which would have been transformed, under the influence of natural selection, into a descendant condition of form. Throughout this process of transformation, the constituent elements of the tetrapod limb, all essentially present in the archetypal condition, would preserve their topological, historical, and ultimately genetic identity — the basis for their homology.

This atomistic research program failed, at least in the case of tetrapod limbs (Hinchliffe and Griffiths, 1983): bird wings never pass through the stages of salamander, frog, or lizard forelimbs. Bird wings are always and only bird wings — but they share with frogs, salamanders, and other tetrapods a common early ontogenetic stage of limb development: the more general condition of form. From this more general condition, the less general condition of form develops by deviation and differentiation (von Baer, 1828), in terms of field theory, by a reorganization of the developmental field (Goodwin, 1984). The bird wing exists always, and only, as a bird wing: it bears the characteristics of its taxon from its first deviation from the common early ontogenetic stage of limb development (Hinchliffe and Griffiths, 1983; Hinchliffe, 1989).

Deviation from common early ontogenetic stages provides the basis for taxic homology, or synapomorphy, but the common early ontogenetic stage of limb development is not the adult ancestral condition of form (see also footnote 1). Nor is it a bird wing, a salamander leg, or a lizard limb. This more generalized condition of form represents "*the* tetrapod limb", but not in actuality (i.e. in the differentiated stage), only in potential. "*The* tetrapod limb" does not actually exist in nature, only this or that particular limb, in this or that particular group of tetrapods (taxon), adapted to these or those particular conditions of life. The archetypal limb is an ideal construct, abstracted from generalities of structure apparent in the developing tetrapod limb — generalities which are not actual or actualized in extant tetrapods, but which are potential in the common early ontogenetic stages of the limb bud. The limb bud entails, potentially, all of the particulars of tetrapod limbs which will eventually differentiate from it, and in that sense it represents all those potentialities, but it does not actualize all those particulars in any extant or fossil taxon. Consequently, the archetypal limb is not an actually observed structure; it is the ideal concept of a *Bauplan*, not observable, nevertheless rooted in empiricism, in the analysis of developmental mechanisms — it is an abstraction.

Viewed from the von Baerian perspective of differentiation, homology is no longer an issue of material and historical identity of parts accumulated and transformed during phylogenetic history, i.e., of character transformation from a plesiomorphic to an apomorphic character state. Instead, it is a matter of actualization of potential conditions of form through ontogeny. It is the differentiation of divergent conditions of form from a more generalized condition which results in the "individuation" (a von Baerian term) of (monophyletic) taxa. Homology is not a transformation of an ancestral to a descendant character state. Instead, homology is a unique and deviant condition of form, diagnostic of the particular taxon in question.

VII. HOMOLOGY, TOPOLOGY, AND INDIVIDUALITY

To von Baer (1828), development meant individualization. Starting from shared ontogenetic primordia, divergent differentiation would result in the actualization of individual traits of organization. In modern terms, the von Baerian processes of deviation and differentiation are the causal basis for the individualization of developmental units (Wagner, 1989a,b); shared individualized developmental units are homologies, or synapomorphies, diagnostic of monophyletic groups. "The tetrapod limb" is such a character (homology), diagnostic of the Tetrapoda; "the bird wing" is a subordinated character, differentiated from "the tetrapod limb" and diagnostic of the Aves, a subgroup of the Tetrapoda.

If (taxic) homologies, or synapomorphies, are individualized developmental units, do they individuate the taxa for which they are diagnostic? And if so, are these individual taxa units of evolution in a Darwinian world?

In order to take part in a Darwinian process of evolution through variation and natural selection, any evolutionary unit must be able to act as a replicator as well as an interactor, indeed as an individual (Hull, 1988). Viewed from the Darwinian perspective, it is not ontogenetic differentiation which provides a criterion of individualization of taxa, but rather their potential to act as replicators and interactors in an historical process subject to the constraints of time and space. If "individual" means the same thing from a von

Baerian as from a Darwinian perspective, it must follow that monophyletic taxa are the empirical units of evolution (discovered on cladistic principles), individualized by the relation of (taxic) homology and hence able to perform as replicators and interactors, i.e. as ancestors and descendants, in a Darwinian process of transformation. Tetrapoda unquestionably is a monophyletic taxon, individualized by "the tetrapod limb" (and many more features), as Aves is individualized by "the bird wing." Also, "the bird wing" unquestionably differentiates from an early ontogenetic stage characterized as "the tetrapod limb" — but can Tetrapoda be ancestral to Aves? Obviously, the ancestor of the Aves must be a tetrapod animal; but "the Tetrapoda" cannot, as an entity, have given rise to "the Aves," just as "the tetrapod limb" never was the adult ancestral structure which transformed into a bird wing. Obviously, the concept of an "individual" does not mean the same thing if based on a von Baerian conceptualization of ontogenetic deviation versus a Darwinian conceptualization of a selective process among replicators and interactors (see also footnote 2).

The level of generality at which a character, or taxic homology, is diagnostic of a monophyletic group must be determined with reference to the subordinated hierarchy of taxa. With respect to tetrapod taxa, "the tetrapod limb" characterizes a more general condition of form from which a less general condition of form such as "the bird wing" differentiates, or is actualized, by ontogenetic deviation. With respect to nontetrapod vertebrates, "the tetrapod limb" represents a less general condition of form, differentiated from "paired appendages." "Paired appendages" is a (taxic) homology of Vertebrata, not an actual adult ancestral character state which has transformed, in a material and historical sense, into "the tetrapod limb". The sarcopterygian fin may be hypothesized to have historically transformed into an ichthyostegid limb (see footnote 4), the wing of *Archaeopteryx* may be viewed as a transformation of the theropod forelimb, but within the subordinated hierarchy of taxa, "the bird wing" is again an abstract and more general condition of form, from which the wings of ostriches, penguins and pelicans develop.

If "the tetrapod limb" is an abstraction, a more general condition of form holding the potential for the actualization of a whole variety of tetrapod limbs through ontogeny, then

"Tetrapoda" must likewise be an abstract generalization, rather than an individual acting as a replicator and interactor in a Darwinian process of evolution (Hull, 1988). Nevertheless, and even if "Tetrapoda" is an abstraction — it is a monophyletic taxon, as is "Vertebrata," a more inclusive taxon characterized by "paired appendages," and "Aves," a less inclusive taxon characterized by "the bird wing." Taxic homology specifies an *inclusive* hierarchy of taxa which cannot provide the basis for an *exclusive* sequence of actual ancestors and descendants in an evolving world (Rieppel, 1988; see also above for the distinction of Haeckelian and von Baerian concepts of ancestry). The problem is that inclusive taxa always only stand for, or represent a relatively (i.e., relative to the derived taxon) generalized condition of form, and by that token are not real, i.e., actual replicators and interactors, but only exist as a potential, or as a logical relation (as is the case for "the tetrapod limb"). This is why the (von Baerian) process of differentiation results in a branching order of life (Ospovat, 1981; Richards, 1992), terminal taxa sitting at the end of diverging trajectories of ontogenetic differentiation; but the hierarchy specifying the relative interrelationships between those terminal taxa, implying the specification of several subordinate levels of monophyly, or of inclusive taxa (of nodes from where those ontogenetic trajectories diverge), cannot be a hierarchy of actual individuals, but must be a hierarchy of potential conditions of form or a hierarchy of types (Rieppel, 1985).

Webster (1984) distinguished the "nominal" from the "real" essence of form. One might say that the characters diagnostic of a hierarchy of inclusive taxa (taxic homologies or synapomorphies: Patterson, 1982) represent the nominal essence of these taxa, justifying their being named but with no claim to justification of their actual existence. As "nominal" essence, the homologies of inclusive taxa justify the naming of these taxa, but these characters are not a material bond tying these taxa together; they are abstractions representing taxa. It is developmental mechanisms, i.e., causal relations underlying the differentiation of characters, which tie monophyletic groups together. Shared developmental mechanisms, or constraints, are the "real" essence of monophyletic taxa (Webster, 1984).

VIII. THE HIERARCHY OF TYPES

To view shared developmental trajectories and/or constraints as proximal causes of homology does not mean that they provide clues to the recognition of homology (see above), if homology is defined as similarity to be explained by common descent. In an evolving world, everything may change, including developmental trajectories and/or constraints, geometry, topology, and connectivity. Again, homology cannot be treated as an observational fact, but only as an hypothesis of similarity based on topological relations and with potential phylogenetic information content to be tested by congruence.

What this means is that we will never know, with any certainty, about the homology of characters. We can only hypothesize that two or more characters are homologous. But support for this hypothesis does not derive from the analysis of these particular characters *per se* (in terms of topology, connectivity, or development), but from the relation of these particular characters to all other characters known. Yet even if congruence justifies the acceptance of two or more particular characters as homologues, the addition of new taxa, or of new characters, to the analysis may radically change the picture, and support the interpretation of formerly accepted homologies as homoplasies. Homology cannot be genealogical continuity in any material sense — it can only stand for or represent hypothesized historical identity. Homology is always only an hypothesis of similarity due to common ancestry, to be tested by congruence, i.e., in relation to all other characters known. As such, every hypothesis of homology may turn into a homoplasy on further analysis, just as every homoplasy holds the potential to turn into an hypothesis of homology on further analysis. As such, homology is a logical relation, not a relation of material identity. The empirical basis of homology is rooted in the observation of topological equivalence; its corroboration comes from a regularity of character distribution (i.e., congruence) suggestive of an underlying cause: evolution.

Geoffroy Saint-Hilaire (1830) realized that there was more to the reconstruction of the order of life than Cuvier's insistence on perfect adaptation (Ospovat, 1981), functional anatomy, and the principle of the correlation of parts derived therefrom. There was, he insisted, a "philosophical," "transcendental," or simply an abstract level of structural comparison transcending the immediate appearance of the individual organism in relation to its environment and revealing those "hidden affinities" which Geoffroy called *analogies* and which Darwin explained by common descent.

Recognition of Geoffroy's *analogies*, corresponding to the modern notion of homology, involves a process of abstraction from form and function of the constituent elements of a biological structure, and their comparison in terms of topological relations instead. Topology and connectivity guide observation in the hypothetical identification of homology. As explained above, connectivity is reduced to topological criteria in the last analysis. Topology, therefore, is the ultimate operational clue to homology (Riedl, 1975), a first principle which, indeed, cannot be justified any further, but which justifies itself, if it generates a regularity of phenomena, i.e., a regularity (congruence) of character distribution. But whereas topology is a *sine qua non* for the hypothesis of homology, it is not also proof or verification of homology. Characters satisfying the criterion of topology may show incongruent distribution. The test of any hypothesis of homology is neither topology, nor the developmental pathways or developmental constraints adduced to explain topological equivalence, but congruence. Inasmuch as topology, or connectivity, allow the identification of hypothetical homologies, and so as far as these homologies do, indeed, show congruent distribution among taxa, topology is justified as a guide in the search for homology — and at the same time the question of typology becomes one of topology.

IX. CONCLUSION

It has been argued above, along two different and independent lines, that homology cannot be historical and material identity, but represents an abstract or logical relation instead.

The first line or argument relates to the von Baerian notion of ontogenetic divergence, which roots the notion of homology not in the actualized morphology of terminal taxa, but in the potential expressed by shared ontogenetic primordia.

The second line of argument relates to the fact that any conjecture of homology may turn into a homoplasy on the addition of taxa or characters to the analysis, and *vice versa*. The notion of homology therefore can only stand for, or represent, material and historical identity. Cladistic tools of analysis identify logical relations of similarity, defining (or diagnostic of) taxa which themselves are not material entities (in the sense of individual interactors and replicators in an evolving world), but *ideal* entities (abstract or logical constructs) — i.e., types (Rieppel, 1985). However, "ideal" does not mean any wild idea, but rather the outcome of an empirical research program, searching for congruence of topological correspondence. What could be the underlying cause of such congruence? What mechanisms cause topological correspondence, or conservation of connectivity, as shape and function of the structures in question change? Conversely, what are the mechanisms which cause homoplasy of topological equivalence, of developmental trajectories or of developmental constraints (Rieppel, 1989b)? The problem is not that types are ideal constructs. The problem is to find what the types of biological classifications stand for, what they represent. Knowledge of mechanisms which conserve topological relations (even in a relative sense, as expressed by connectivity) may provide the answer, in which case typology becomes a problem of topology, and types become a group of organisms tied together by shared mechanisms preserving relative topological relations of constituent elements in a biological structure.

ACKNOWLEDGMENTS

I thank Brian K. Hall, Charles R. Crumly, Neil Shubin, and Günter P. Wagner who kindly read an earlier draft of this chapter, offering much helpful advice and criticism.

REFERENCES

Adelmann, H. B. (1966). "Marcello Malpighi and the Evolution of Embryology," Vols. I-IV. Cornell Univ. Press, Ithaca, NY.

Appel, T. (1987). "The Cuvier-Geoffroy Debate. French Biology in the Decades before Darwin." Oxford Univ. Press, New York and Oxford.

Belon, P. (1555). "L'Histoire de la Nature des Oyseaux." Guillaume Cavellat, Paris.

Bock, W. J. (1973). Philosophical foundations of classical evolutionary classification. *Syst. Zool.* **22**, 375-392.

Bowler, P. J. (1971). Preformation and prexistence in the seventeenth century. *J. Hist. Biol.* **4:** 221-244.

Bowler, P. J. (1975.) The changing meaning of 'Evolution'. *J. Hist. Ideas* **4**, 95-114.

Boyden, A. (1947). Homology and analogy. A critical review of the meanings and implications of these concepts in biology. *Am. Midl. Nat.* **37**, 548-669.

Brady, R. H. (1982). Dogma and doubt. *Biol. J. Linn. Soc.,* **17**, 79-96.

Brady, R. H. (1985). On the independence of systematics. *Cladistics* **1**, 113-126.

Burke, A. C., and Alberch, P. (1985). The development and homologies of the chelonian carpus and tarsus. *J. Morphol.* **186**, 119-131.

Chalmers, A. F. (1982). "What is This Thing Called Science?," 2nd ed. Univ. of Queensland Press, St. Lucia.

Coates, M. I., and Clack, J. A. (1990). Polydactyly in the earliest known tetrapods. *Nature (London)* **347**, 66-69.

Crumly, C. R., (1991). Morphogenesis and pattern formation. *In* "Magill's Survey of Science: Life Science" (F. N. Magill and L. L. Mays Hoopes, eds.), pp. 1783-1789. Salem Press, Pasadena, CA.

Darwin, C. (1859) "On the Origin of Species." John Murray, London.

de Pinna. M. C. C. (1991). Concepts and tests of homology in the cladistic paradigm. *Cladistics* **7**, 367-394.

Desmond, A. (1989). "The Politics of Evolution." Univ. of Chicago Press, Chicago.

Gauthier, J. (1986). Saurischian monophyly and the origin of birds. *Mem. Calif. Acad. Sci.* **8**, 1-55.

Gauthier, J., Estes, R. and de Queiroz, K. (1988). A phylogenetic analysis of Lepidosauromorpha. *In* "Phylogenetic Relationships of the Lizard Families" (R. Estes and G. Pregill, eds.), pp. 15-98. Stanford Univ. Press, Stanford, CA.

Geoffroy Saint-Hilaire, E. (1830). "Principes de Philosophie Zoologique, discutés en Mars 1830, au Sein de l'Académie Royale des Sciences." Pichon et Didier, Paris.

Goodrich, E. S. (1916). On the classification of the Reptilia. *Proc. Roy. Soc. London, Ser. B* **89**, 261-276.

Goodrich, E. S. (1942). The hind foot in *Youngina* and fifth metatarsal in Reptilia. *J. Anat.* **76**, 308-312.

Goodwin, B. C. (1984). Changing from an evolutionary to a generative paradigm in biology. *In* "Evolutionary Theory. Paths into the Future" (J. W. Pollard, ed.), pp. 99-120. Wiley, Chichester.

Hall, B. K. (1992). "Evolutionary Developmental Biology." Chapman & Hall, London and New York.

Harvey, W. (1981). "Disputations Touching the Generation of Animals." (Translated with introduction and notes by G. Whitteridge). Blackwell, London.

Hinchliffe, J. R. (1989). An evolutionary perspective of the developmental mechanisms underlying the patterning of the limb skeleton in birds and other tetrapods. *Geobios, mém. spec.* **12**, 217-225.

Hinchliffe, J. R. and Griffiths, P. J. (1983). The prechondrogenic patterns in tetrapod limb development and their phylogenetic significance. *In* "Development and Evolution" (B. C. Goodwin, N. Holder, and C. C. Wylie, eds.), pp. 99-121. Cambridge Univ. Press, Cambridge.

Holmes, R. (1984). The Carboniferous amphibian *Proterogyrinus scheelei* Romer, and the early evolution of tetrapods. *Trans. R. Soc.* London, Ser. B **306**, 431-527.

Hull, D. L. (1988). "Science as a Process." The Univ. of Chicago Press, Chicago.

Huxley, T. H. (1859). The Croonian Lecture — on the theory of the vertebrate skull. *Proc. Zool. Soc. London* **9**, 381-457.

Kuhn-Schnyder, E. (1974). Die Triasfauna der Tessiner Kalkalpen. *Neujahrsbl.-Naturforsch. Ges. Zürich,* **1974**, 1-119.

Kükenthal, W. (1893). Zur Entwicklung des Handskelettes des Krokodils. *Gegenbaurs Morphol. Jahrb.* **19**, 42-55.

Løvtrup, S. (1978). On von Baerian and Haeckelian recapitulation. *Syst. Zool.* **27**, 348-352.

Mayr, E. (1969). "Principles of Systematic Zoology." McGraw Hill, New York.

Mayr, E. (1982). "The Growth of Biological Thought." Harvard Univ. Press (Belknap), Cambridge, MA.

Müller, G. B., and Alberch, P. (1990). Ontogeny of the limb skeleton in *Alligator mississippiensis*: Developmental invariance and change in the evolution of archosaur limbs. *J. Morphol.* **203**, 151-164.

Ospovat, D. (1981). "The Development of Darwin's Theory: Natural History, Natural Theology, and Natural Selection, 1838-1859." Cambridge Univ. Press, Cambridge.

Patterson, C. (1982). Morphological characters and homology. *In* "Problems of Phylogenetic Reconstruction" (K. A. Joysey and A. E. Friday, eds.), pp. 21-74. Academic Press, London.

Patterson, C. (1983). How does phylogeny differ from ontogeny? *In* "Development and Evolution" (Goodwin, B.C., N. Holder, and C.C. Wylie, eds.), pp. 1-31. Cambridge Univ. Press, Cambridge.

Patterson, C., ed. (1987). "Molecules and Morphology in Evolution: Conflict or Compromise?" Cambridge Univ. Press, Cambridge.

Peabody, F. E. (1951). The origin of the astragalus of reptiles. *Evolution (Lawrence, Kans.)* **5**, 339-344.

Platnik, N. I., and Gaffney, E. (1978a). Evolutionary biology: a Popperian perspective. *Syst. Zool.* **27**, 137-141.

Platnik, N. I., and Gaffney, E. (1978b). Systematics and the Popperian paradigm. *Syst. Zool.* **27**, 381-383.

Popper, K. R. (1972). "Objective Knowledge." Oxford Univ. Press, Oxford.

Richards, R. J. (1992). "The Meaning of Evolution." The Univ. of Chicago Press, Chicago.

Riedl, R. (1975). "Die Ordnung des Lebendingen." Paul Parey, Hamburg.

Rieppel, O. (1985). Ontogeny and the hierarchy of types. *Cladistics* **1**, 234-246.

Rieppel, O. (1986). Atomism, epigenesis, preformation and pre existence: A clarification of terms and consequences. *Biol. J. Linn. Soc.* **28**, 331-341.

Rieppel, O. (1988). "Fundamentals of Comparative Biology." Birkhäuser, Basel and Boston.

Rieppel, O. (1989a). "Unterwegs zum Anfang. Geschichte und Konsequenzen der Evolutionstheorie." Artemis Verlag, Zürich and München.

Rieppel, O. (1989b.) Character incongruence: Noise or data? *Abh. Naturwiss. Ver. Hamburg*, (N.S.) **28**, 53-62.

Rieppel, O. (1990). Structuralism, functionalism, and the four Aristotelian causes. *J. Hist. Biol.* **23**, 291-320.

Rieppel, O. (1992). Homology and logical fallacy. *J. Evol. Biol.* **5**, 701-715.

Rieppel, O. (1993a). Studies on skeleton formation in reptiles. IV. The homology of the reptilian (amniote) astragalus revisited. *J. Verteb. Paleontol.* **13** (in press).

Rieppel, O. (1993b). Studies on skeleton formation in reptiles. II. *Chamaeleo hoehnelii* (Reptilia, Chamaeleoninae); with comments on the homology of carpal and tarsal bones. *Herpetologica* **49** (in press).

Robinson, P. L. (1975). The functions of the hooked fifth metatarsal in lepidosaurian reptiles. *Colloq. Int. C. N. R. S.* **218**, 461-483.

Roe, S. A. (1981). "Matter, Life and Generation. Eigtheenth-century embryology and the Haller-Wolff debate." Cambridge Univ. Press, Cambridge.

Romer, A. S. (1956). "The Osteology of the Reptiles." Univ. of Chicago Press, Chicago.

Roth, V. L. (1984). On homology. *Biol. J. Linn. Soc.* **22**, 13-29.

Roth, V. L. (1988). The biological basis of homology. *In* "Ontogeny and Systematics" (C. J. Humphries, ed.), pp. 1-26. Columbia Univ. Press, New York.

Roth, V. L. (1991). Homology and hierarchies: Problems solved and unresolved. *J. Evol. Biol.* **4**, 167-194.

Sewertzoff, A. N. (1908). Studien über die Entwicklung der Muskeln, Nerven und des Skeletts der Extremitäten der niederen Tetrapoda. *Bull. Soc.Imp. Nat. Moscou*, [N.S.] **21**, 1-430.

Shubin, N. H. and Alberch, P. (1986). A morphogenetic approach to the origin and basic organization of the tetrapod limb. *Evol. Biol.* **20**, 319-387.

Tabin, C. J. (1992.) Why we have (only) five fingers per hand: *Hox* genes and the evolution of paired limbs. *Development (Cambridge, UK)* **116**, 289-296.

Van Valen, L. (1982). Homology and causes. *J. Morphol.* **173**, 305-312.

Völker, H. (1913). Über das Stamm-, Gliedmassen- und Hautskelet von *Dermochelys coriacea* L. *Zool. Jahrb. Abt. Anat. Ontog. Tiere* **33**, 431-552.

von Baer, K. E. (1828). "Über Entwickelungsgeschichte der Thiere. Beobachtung und Reflexion," Vol.1. Bornträger, Königsberg.

Wagner, G. P. (1989a). The biological homology concept. *Annu. Rev. Ecol. Syst.* **20**, 51-69.

Wagner, G. P. (1989b). The origin of morphological characters and the biological basis of homology. *Evolution (Lawrence, Kans.)* **43**, 1157-1171.

Wake, D. B., and Roth, G. (1989). The linkage between ontogeny and phylogeny in the evolution of complex systems. *In* "Complex Organismal Functions: Integration and Evolution in Vertebrates" (D. B. Wake and G. Roth, eds.), pp. 361-377. Wiley, Chichester.

Webster, G. (1984). The relations of natural forms. *In* "Beyond Neo-Darwinism" (M.-W. Ho and P. T. Saunders, eds.), pp. 193-217. Academic Press, London.

3

HOMOLOGY AND SYSTEMATICS

G. Nelson

Department of Herpetology and Ichthyology
American Museum of Natural History
New York, New York 10024
and
School of Botany
The University of Melbourne
Parkville, Victoria 3052
Australia

Homology: The Hierarchical Basis of Comparative Biology
Copyright © 1994 by Academic Press, Inc.
All rights of reproduction in any form reserved.

La vérité ... réside dans le système de leurs
différences et de leurs communes propriétés.
                              C. Lévi-Strauss (1962)

Homology, homology,
there's nothing like homology,
There never was a law of such
deceitfulness and suavity.
                              Anonymous graffiti

I. INTRODUCTION

Systematics is about taxa and homologies, or in plain
language, about organisms and what science knows of their
similarities and differences. As a human activity, systematics
has a long history, which teaches that systematics works to
some degree, and that science learns about taxa and
homologies. Why and how systematics works are variously
conceived in one or another theory, most of which assent to
the meaning of the epigraph above, when applied to organisms: "the truth resides in the system of their differences
and similarities."

Recent theories, such as evolutionary systematics, phenetics, and phylogenetic systematics explain how and why,
and on occasion how not and why not. Theories also elaborate goals, methods, and standards for evaluation of results.
At best, the elaborations make systematics more intelligible
and more efficient. Better intelligibility and efficiency are
worthy objectives, but to aspire to them is easy and to

achieve them is difficult. Such is evident also in the recent history of systematics, with its conflicting discussions of basic principles such as homology, as well as conflicting claims of achievement.

Theories attract supporters and detractors, and time changes their relative abundance. Theories once popular become less so and fall into oblivion, sometimes to be resurrected with increased vigor. New theories emerge with few if any enthusiasts. New theories are seen as subversive and alien to proper understanding, but new theories become popular and with time even old, outmoded, and forgotten. Perhaps cladistics is a new theory, but here it is treated as a critique, or criticism, of systematics, in the belief that more criticism rather than less is appropriate. Cast as a new theory, cladistics is prone to distortion, for some elements of cladistics are not new (see, e.g., Donoghue and Kadereit, 1992), and some are incompletely analyzed and, therefore, of ambiguous status. An initial encounter with cladistics and its jargon is sometimes like the discovery, made late in life by M. Jourdain, that his ordinary speech of over 40 years had all along been prose:

> Par ma foi! il y a plus de quarante ans que je dis de la prose sans que j'en susse rien, et je vous suis le plus obligé du monde de m'avoir appris cela (Molière's *Le Bourgeois Gentilhomme*, Act II, Scene V)

II. CLADISTICS

The cladistic critique of systematics began decades ago, nominally with the publication of Hennig's *Grundzüge* (Hennig, 1950) and its translation into English (Hennig, 1966). To the critique Brundin (1966) added emphasis on biogeography. New dimensions arose from analysis of molecular data and development of computer-implemented algorithms. The critique now has a life of its own as a collection of outlooks, methods, and philosophical underpinnings. History of the critique is best reflected by the Willi Hennig Society, founded in 1980, and in its quarterly journal *Cladistics*, now in its ninth volume following two early volumes of *Advances in Cladistics* (Funk and Brooks, 1981; Platnick and Funk, 1983). Portions of history, distorted by the author's theory of scientific change, are in a lengthy book by Hull (1988, also 1989), relevant review of which seems limited to Farris and Platnick (1989). Other historical comment (Donoghue, 1990) is superficial (Nelson and Patterson, 1993), or reflects various points of view (Dupuis, 1979; Rosen *et al.*, 1979; Schafersman, 1985; Tassy, 1991; Craw, 1992).

A. Homology as Synapomorphy

Early in the critique, homology was equated with (or replaced by) Hennig's concept of shared derived character indicative of phylogenetic relationship (*synapomorphy* in the cladistic jargon). The equation persists in recent reviews (Patterson, 1982, 1988; Janvier, 1984; Stevens, 1984; Brooks and McLennan, 1991; Panchen, 1992; de Pinna, 1991; Rieppel, 1992; Wenzel, 1992b), which consider molecular, morphological, ecological, and behavioral data, and which should be consulted for relevant literature; dissenting comment (de Queiroz, 1985) is problematic (de Pinna, 1991). The equation means that homology is illuminated, if not revealed, by cladistic analysis, which seeks the tree, or cladogram, that best fits available data. Data fit to (optimized on) such a tree specify particular values for nodes (branchings). Values at a node are interpreted as synapomorphies or homologies, and as characters of an ancestral taxon (*character* and *homologue*, as used throughout, refer to a part of an organism, even an organism of a hypothetical species, not to a definition; *synapomorphy* and *homology* refer to the phylogenetic relationship between parts, or homologues, of different organisms).

B. Data and Nodes

For example, birds have feathers, absent from other forms of life. If "feathers," as a datum, is fitted to the usual tree showing relationships of recent vertebrates, it best fits only one node (Fig. 1). The relationship between feathers is a synapomorphy of birds. Feathers are assumed to have appeared for the first time in the taxon ancestral to birds, and to have been passed to descendant taxa through mechanisms of heredity.

The criterion of best fit is construed as evolutionary parsimony, which means that hypothetical evolutionary events (e.g., appearance of feathers) need not be multiplied beyond necessity. The criterion accords with Aristotle's advice to generalize to the extent possible. Rather than repeatedly to associate the datum "feathers" with every bird specimen, species, or other avian taxon (Fig. 2), it is enough once to associate "feathers" with birds in general.

Rather than to overgeneralize the datum, such that it associates with organisms other than birds (Fig. 3), it is enough to associate "feathers" only with birds. Overgeneralization requires, if accuracy be maintained, that other groups of organisms be qualified as not having feathers, and the force, or meaning, of the generalization diminishes with each qualification (*homoplasy* in cladistic jargon): for example (Fig. 3), all amniotes have feathers except crocodiles, lizards, snakes, and mammals. The sense of evolutionary parsimony is the same as above: hypothetical evolutionary events (e.g., disappearance of feathers) need not be multiplied beyond necessity.

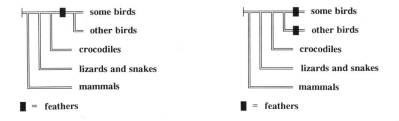

Fig. 1. (left). The datum "feathers" fitted to one node of cladogram of some recent vertebrates.

Fig. 2. (right). The datum "feathers" misfitted to two nodes of cladogram (cf. Fig. 1).

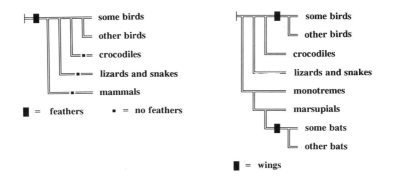

Fig. 3. (left). The datum "feathers" misfitted to one node of cladogram (cf. Figs. 1 and 2).

Fig. 4. (right). The datum "wings" fitted to two nodes of cladogram of some recent vertebrates.

For counterexample, birds and bats have wings, absent from other recent vertebrate forms of life. Fit to the usual tree, "wings" associates with two nodes — one leading to birds, one leading to bats (Fig. 4). The relationship among wings, therefore, is not one but two synapomorphies, or two homologies: one of birds and another of bats.

"Wings" is not best associated with the node leading both to birds and bats (Fig. 5). At that node the datum is overgeneralized, for a number of groups of other organisms must be qualified as having no wings: all amniotes have wings except crocodiles, lizards, etc. The sense of evolutionary parsimony is the same as above (disappearance of wings). In any case, the ancestral taxon leading to birds and bats is assumed to have had no wings, and wings of birds and bats are not homologous but homoplastic (parallel or convergent).

Were relationships of vertebrates seen otherwise, then synapomorphies, homologies, and ancestral characters might be differently construed. Were bats and birds considered as close relatives (Fig. 6), then "wings" would associate with a single node, bat wings and bird wings may be seen as

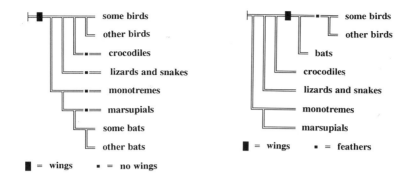

Fig. 5. (left). The datum "wings" misfitted to one node of cladogram (cf. Fig. 4).

Fig. 6. (right). The data "wings" and "feathers," each fitted to a different node of one possible cladogram of some recent vertebrates (cf. Figs. 1-5).

homologous, and the ancestral taxon common to birds and bats may be assumed to have had wings. de Pinna (1991, p. 375) refers to bat and bird wings, stating that, despite their many anatomical differences, there would be no "fierce opposition" to their homology. Van Valen (1982, p. 305) notes that for many accepted homologies "no resemblance need be apparent" between them, and cites as an example the homology of middle ear bones of mammals and jaws of fishes. He comments that further "examples abound in molecular evolution," a point made earlier by Ghiselin (1969b).

C. Retrospective

In retrospect, the early equation of homology and synapomorphy seems neither problematic nor critical. It was nevertheless viewed as problematic by persons who believed that homologies and ancestral characters (and taxa) are, at least sometimes, directly observable: when an observation is construed to conform to a definition of homology, or when some taxon (or character), usually known from fossil specimens, is construed as ancestral to another.

Such empirical claims have diminished, if not disappeared, from the primary literature of systematics, and their dwindling occurred in reaction to the critique. Homologies and ancestral characters are now generally seen as derived from data fit to a particular tree. Data from fossil and recent life are treated in the same way (e.g., Schoch, 1986; Forey, 1992). Some commentators consider data to be homology-free prior to analysis, with homologies emerging as a result (Sober, 1988). Other commentators consider data to be conjectures of homology, which are later corroborated or refuted by analysis (de Pinna, 1991). In either case, data are seen, naturally enough, as of less meaning and significance before analysis than after. Even after analysis homologies are conceded to be provisional, subject to change if required by new data or better methods of analysis.

Traditional notions of homology are essential similarity, structural correspondence, and similarity due to common ancestry. As considered above the cladistic critique offers nothing contrary to these notions. The critique provides a method to specify particular details of (assign values to) notions such as essence, correspondence, and common ancestry. Because any datum fits a single node on some tree, a particular tree serves as a standard. In general, cladists prefer the tree that best fits all data at hand; hence conjectures of homology are corroborated, or refuted, by that tree of best fit.

How to find the tree that best fits a sample of, or even all available, data will not be considered here, except to note that the task becomes technically more difficult as data, especially the number of taxa, increase. The task falls among the class of problems called NP-complete (e.g., Penny and Hendy, 1985; Kitching, 1992), which means basically that the best fitting tree(s) cannot be directly calculated and that all possible trees must be examined to ensure an exact solution (namely that there are no trees equally or better fitting than the one or more already found). Even for a modest sample of data concerning five or six taxa, to find the best fitting tree may be too time consuming for hand (noncomputer) solution, because of the great number of possible trees (Felsenstein, 1978). With development of efficient programs for personal computers (Platnick, 1989b; Sanderson, 1990), their use has become standard, but existing programs do not always deliver exact solutions for complex data when taxa exceed a dozen or so. Nevertheless, Farris and Platnick (1989, p. 308, also Platnick, 1989a), lamenting that Hull (1988) fails to mention any of the "real progress" in systematics during the last 20 years, epitomize that progress as discovery of three noble truths: phenetics is false, parsimony is necessary, and exact solutions to parsimony problems are sometimes possible.

If the critique has accomplished anything then surely it has proven, although not perhaps to everyone's satisfaction, that biological data relevant to taxa are hierarchically structured, that this structure is not subjective or artifactual but is an objective property of the living world, and that the structure reflects the history of life insofar as that history may be apprehended through scientific investigation of similarities and differences of organisms. Elements of this structure, which are entirely relational, have been known under a variety of names, *homology* and *taxon* being two of them. In this view the information content both of homology and taxon are one and the same particular hierarchy — life's hierarchy, exemplified in the title of a recent Nobel Symposium (Fernholm *et al.*, 1989). In that same symposium is a contrasting, and oddly anachronistic, view:

> If the history of the phenetic school of taxonomy has anything to teach us, it is that indefinitely many patterns exist out there in nature depending on a variety of considerations including the sort of clustering techniques one uses. (Hull, 1989, p. 8)

Homology is phylogenetic relationship between parts of different organisms, as indicated for example by the tree (cladogram) relating the organisms themselves. Taxon and homology are the same phylogenetic relationship, as seen either between organisms (taxon) or between their parts (homology). The fully resolved tree of life, reconstructed from biological data, would specify all phylogenetic relationships (taxa) of organisms and all phylogenetic relationships (homologies) of their parts. How particular organisms are comprehended in terms of taxa is the province of taxonomic specialities able, for example, to determine that, according to current understanding, a particular bird soaring overhead is a black-bellied storm petrel, *Fregetta tropica* (family Oceanitidae), and not a greater frigate-bird *Fregata minor* (Fregatidae).

How the biology of organisms is comprehended — of what parts organisms and their activities are composed — also is the province of particular disciplines with specialized lore, such as anatomy, physiology, behavior, ecology, and molecular biology, each of which uses criteria of homology of parts relevant to that discipline. Criteria of homology are useful, even necessary, to organize data, but are not decisive in resolving conflict among data during the search for the tree that best fits all data. To interpret such a tree once found, criteria are again useful, even necessary, to specify homologies and homoplasies, either those inherent in data already fitted to a tree, or in other data that will be. A tree may depict, for example, birds and crocodiles as related more closely to one another than to frogs. The tree need not specify what particular parts of birds and crocodiles are related more closely than to parts of frogs, even though the tree implies that there are such parts.

III. CONTENTIONS

A. *True Criteria*

Some commentators sense that the cladistic view of homology as synapomorphy is a radical change for the worse from the traditional and true meaning of Darwinism (Szalay and Bock, 1991, is a late examples). They construe homology as logically prior to cladistic analysis. They defend certain criteria of homology, as if homology were a property capable of direct and exact measurement by careful attention to true criteria of a privileged discipline. They argue, in effect, that some data are better than others, and that data may be, and should be, evaluated directly, independently of cladistic analysis. Even after such evaluation, so far as known, all samples of data of reasonable quantity, when analyzed cladistically, display both homology (data that fit one node) and homoplasy (data that fit two or more nodes in the above sense) which can be specified only with cladistic analysis and only with a particular tree as a standard of reference. In these respects there is no difference between classes of biological data, such as molecular and morphological (Patterson 1987a,b; Fernholm *et al.*, 1989; Hillis and Moritz, 1990; Miyamoto and Cracraft, 1991; Donoghue and Sanderson, 1992).

Current measures of the mixture of homology and homoplasy are the consistency index and the related retention index (Farris, 1989b). The consistency index is the ratio of ideal fit to actual fit, as measured in evolutionary steps (nodes), for a given sample of data fitted to a particular tree. Either index for real data, however carefully collected and evaluated prior to analysis, typically is significantly less than 1.0 (the maximum) and frequently about 0.5, indicating that data frequently are about half as good as they might ideally be. There might exist a correlation between care in collecting and organizing data and either index, but there is no evidence to support such a claim, however intuitively reasonable it might appear, or however hopefully one may wish that it were true. Even if there were a significant correlation, there is no reason to expect that homoplasy would disappear no matter how carefully data were collected and organized. The conclusions seem inescapable that systematics works not because its data tend toward perfection, but because its methods are generally efficient in analysis of data that are always significantly less than perfect; and that recent progress in systematics stems not from better data but from better methods and more data of roughly the same quality.

B. Golden Age

These conclusions are at odds with the expectation, seemingly perennial, that an approaching golden age of technology will produce better or even perfect data. This wishful expectation, however understandable in the context of a discipline, history invariably finds naive. According to Blackwelder (1977, p. 115) "What is new in these fields, is the sudden availability of types of comparative data not previously available.... New types of data are potentially of great importance, but they do not replace other types except in problem cases." According to Grant (1992, p. 690), "New kinds of evidence in plant taxonomy have often been heralded when first introduced as sources of decisive criteria for resolving difficult taxonomic problems, but the early hopes have not held up in practice."

However naive, the expectation has emanated for decades from, among other sources, molecular biology. An early manifestation in the 1960s was the belief that as one approaches DNA one necessarily approaches the truth (E. Margoliash, personal communication; also the "factual phylogenetic relationships" of Margoliash and Fitch, 1969). With regard to homology, Gould (1985, p. 25) broadcast that "We finally have a method [DNA hybridization] that can sort homology from analogy.... Perhaps we should mount the parapets and shout: 'The problem of phylogeny has been solved.'" And again (1986, p. 68) that "Molecular phylogenies work not because DNA is 'better,' more real, or more basic than morphology, but simply because the items of a DNA program are sufficiently numerous and independent to ensure that degrees of simple matching [DNA hybridization] accurately measure homology." DNA hybridization subsequently proved notoriously problematic (e.g., Cracraft and Mindell, 1989; Sarich *et al.*, 1989; Mindell, 1992), as did the entire episode of phenetics in systematics, confirming the early view of Ross (1964, p. 108) that phenetics would prove "an excursion into futility."

Phenetics, with its varied comment on homology (e.g., Sneath and Sokal, 1973) is not considered here except to note its overall similarity with another episode of the past (Knight, 1981, p. 23): "In the search for the natural system the stakes are high; one may go quite wrong and find that one has wasted most of one's working life. This happened, for example, to those chemists in the nineteenth century who determined atomic weights with ever higher accuracy, believing that they were finding fundamental constants of nature that would enable the elements to be classified unambiguously" (exact atomic weights are perturbed by the mix of different isotopes in a sample; and exact measures of similarity, by the mix of symplesiomorphy, synapomorphy, and homoplasy). No pheneticist has yet confessed to wasted effort; only Simpson implied as much (Hull, 1988, p. 169) for reasons arising more from paleontology and Darwinism than from phenetics (Nelson, 1989a).

Goodman (1989, p. 43) claims that premolecular biology never provided a practical means to separate homology from analogy (and by homology he explicitly refers to synapomorphy, not to phenetics and its notion of overall similarity). He claims that because of its revolutionary advances, the new age of molecular biology can now reveal "synapomorphy at the genetic level" and realize "Darwin's dream" — namely, to discover the true genealogical history of life.

Goodman's claims are overstated. The "practical means" of cladistics are the same regardless of the nature of the data to which the means are applied. That these same means have developed to some degree independently among traditional systematists on the one hand, and molecular systematists on the other, indicates real and significant progress toward solution only of one and the same problem, which is variously seen as history, method, or both. Cladists of this period might agree with Lévi-Strauss (1962, p. 347; 1966, p. 262) that "history is tied neither to man nor to any particular object. It consists wholly in its method." They might agree, too, with Danaher's (1992) observation, made in another context, that beyond systematics and biogeography "there is no easy way ... to find present meaning in past events" of biological evolution.

C. Homology in a Straitjacket

Other commentators saw the cladistic view as too narrowly constrained (by "the straight-jacket [*sic*] of Hennigian phylogenetic systematics" of Fryer, 1978), omitting resemblances of scientific value as seen from traditional, or Darwinian, points of view. Among others Roth (1988, p. 4) asserts that:

> I see no point in narrowing the definition of homology, a word of broader connotation, so drastically .... To insist that 'homology' be used interchangeably with 'synapomorphy' beclouds the issue .... To equate homology with synapomorphy is analogous to equating Primates with *Homo sapiens*.

Such comment creates the impression that there is disagreement over mere definition of words, such as homology, with one faction (cladists) insisting that everyone else conform to its idiosyncratic usage. In reality disagreement stems from different perceptions about the nature of taxa and their study, both past and present. Disagreement is more substantial than displayed or implied by anticladist comment, which typically attempts to trivialize disagreement (and to portray cladists as perversely narrow minded) rather than to explore, or to extend the basis of, the disagreement, which was most acute over symplesiomorphy and paraphyly.

D. Symplesiomorphy and Paraphyly

Shared primitive characters in cladistic jargon are termed *symplesiomorphy*. Commentators reckoned symplesiomorphy (in contrast to *synapomorphy*) as homology that cladists improperly ignore. This reckoning, perhaps now obsolete, seems a mere mistake, as cladists claim when they point out that symplesiomorphy, too, is synapomorphy at a higher taxonomic level. The sense of the cladist claim is easy to grasp through an example, such as the occurrence of limbs (arms and legs) in most nonbat and nonbird land vertebrates. Cladists perceive that limbs are a character of land vertebrates including birds and bats (Fig. 7). That frogs, crocodiles, humans, and other organisms have arms and no wings is false (symplesiomorphy) if it means that birds and bats have wings but no arms. Cladists view wings, too, as arms and two peculiar and different sorts of them.

Taxa based on symplesiomorphy cladists term paraphyletic groups, construed as composed of some but not all descendants of an ancestral taxon, such as tetrapods that are not birds or mammals (herptiles), amniotes that are not birds or mammals (reptiles), anthropoids that are not humans (apes). Anticladist Darwinites claim that paraphyletic groups are as real as monophyletic groups, and that cladists, in distinguishing between them and deeming the difference as that between unreal and real, or nonexistence and existence, argue merely over definition or classification (see below).

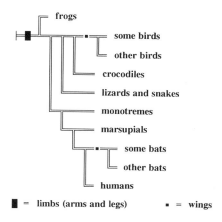

frogs

some birds

other birds

crocodiles

lizards and snakes

monotremes

marsupials

some bats

other bats

humans

■ = limbs (arms and legs)   ▪ = wings

Fig. 7. The data "limbs" and "wings," each fitted to a different node of cladogram of some recent vertebrates.

Prior to the critique, and frequently even after it began, definitions of homology did not differentiate between synapomorphy and symplesiomorphy (e.g., the definitions listed in Patterson, 1982, pp. 1-4, as classical, evolutionary, phenetic, and cladistic). The claim that traditional homology embraces and sees value in symplesiomorphy (and paraphyly) seemingly overlooks the apparently complete lack of explicit reference to that value in older literature, whereas statements expressing that of synapomorphy, some explicit, have been found (e.g., Nelson and Platnick, 1981, pp. 324-328). Claims that traditional homology embraces symplesiomorphy (and paraphyly), and that the traditional notion is more general and more meaningful than synapomorphy, are overstated at best and false at worst.

E. Latent Homology

Another class of resemblance allegedly omitted from the cladistic critique is that sometimes termed parallelism (inside parallelism of Brundin 1976, and underlying synapomorphy of Saether 1979; review by Cranston and Humphries, 1988).

This class of resemblance has a long history in systematics (de Beer, 1971), usually discussed as orthogenesis. If
considered an example of parallelism in this view, then
wings of birds and bats may be interpreted as related
because of common ancestry. The ancestral taxon might
have had no wings but would have had some wing-related
attribute (or gene), favoring the subsequent and repeated
development of wings (or a gene switching on or off). In
cladistic analysis, inside parallelism is problematic if no
related attribute (gene) can be recognized, as is commonly
the case. If an attribute can be recognized then it can be
treated as an independent datum with a significance the
same as that of an inside parallelism, which then becomes
superfluous. Farris and Platnick (1989, p. 299) regard inside
parallelism as "the only trace of idealism that anyone could
accurately identify in purportedly cladistic literature." They
characterize concern for inside parallelism as the wish to
alter ugly data to fit beautiful hypotheses.

F. Homoplasy

Homology and synapomorphy aside, *homoplasy*, even if
it, too, is derivative of a particular tree, appeals, and in
recent years has become a focus for the study of adaptation
(e.g., Harvey and Pagel, 1991; Gittleman and Luh, 1992).
The idea here is that biological phenomena are explicable
either as the effects of phylogeny (homologies, synapomorphies) or as the effects of adaptation (homoplasies). It is
curious that in little more than 100 years the phenomenon
of adaptation, which for Darwin pervaded biological data and
explained biological phenomena, could become so severely
narrowed. And curious, too, that the narrowing has named
itself "the comparative method," as if cladists really were
"that ill-behaved, radical fringe that is safely ignored and
best not invited to tea" (Wenzel, 1992a, p. 97 — a theme
propagated most notably, and with characteristic virtuosity,
by Gould in many commentaries over many years). Relevant
review of "the comparative method" seems limited to
Siebert and Humphries (1991) and Carpenter (1992).

Characteristically, cladists view adaptation not in terms
of homoplasy, but in terms of synapomorphy (e.g., the
exchange between Coddington, 1988, 1990, and de Pinna

and Salles, 1990). It may not escape the reader that homoplasy is (sometimes more than one) synapomorphy at lower taxonomic levels (see above), or that the problem of homology is effectively solved when data are associated with the proper nodes of the proper tree of taxa — as if, in the end, all characters were homologous (see below).

G. Idealistic Morphology

Among homoplasy (and homology, too) falls a "great diversity of structural relationships and transformations" (Sattler, 1984, p. 392; and see Chapter 13) that one or another author finds interesting for whatever reason, and that sometimes are referred to as homologies of one sort or another (Wagner, 1989; Minelli and Peruffo, 1991; Roth, 1991; reviews by Rieppel, 1992 and Donoghue, 1992). Sattler (1984, p. 391) considers that for them "a distinction of different kinds" seems appropriate, and that "a synthesis of different concepts may be possible through the notion of a semi-quantitative or quantitative homology concept." Perhaps to their disadvantage, such concerns, at times deliberate, lose touch with systematics (Smith, 1987), and ally with the tradition of idealistic morphology represented, for example, by D'Arcy Thompson (1860-1948) and his modern acolytes (e.g., Rohlf and Bookstein, 1990): "the philomorphs of Cambridge, the world [Michigan] and beyond, where D'Arcy Thompson must lie in the bosom of Abraham" (Gould, 1977, p. v).

For Donoghue (1992, p. 179) such concerns, even homology as synapomorphy, revolve around choice of a definition of homology, "a means of forcing other scientists to pay close attention to whatever one thinks is most important," and of determining "whose agenda will attract the most attention." The cladistic critique has had as its objective to discover empirically what these concepts mean, how they interrelate, and how science might progress as a result of such discovery. Homology as synapomorphy has already helped to illuminate systematics and, insofar as systematics is basic, biology as a whole, a change having nothing to do with definitions as such. One can hardly imagine that systematics would ever revert to a precladistic state, no matter how words, homology included, might come to be defined in future times.

Idealistic morphology sees homology as meaningful regardless of any other considerations. Its homologies are not seen as conjectures refutable by cladistic analysis. If morphology becomes idealistic when, and because, it denies cladistic refutation, then it is easily recognized by the standards of today even if not by the standards of times past.

Idealistic morphology is the domain also of certain nineteenth century figures relevant to the history of homology concepts, such as Oken, Goethe, Geoffroy Saint-Hilaire, Owen, and Louis Agassiz, and of abstract (and perhaps irrefutable) transformation of organic form, such as Goethe's series of leaves of plants, Oken's and Owen's visualization of the skull as a series of vertebrae, Owen's graphic depiction of the vertebrate archetype (see Chapter 1), Agassiz's threefold parallelism in the mind of God, etc.

Idealistic morphology is sometimes used as a term of abuse, implying kinship with modes of thought of the Ancient World, or the Middle Ages, rather than with those of modern science. For reasons of his own, Hull (1988, 1989) argues that some cladists are idealistic morphologists in this abusive sense, even that Dwight Davis and Rainer Zangerl (more "idealistic morphologists"), who produced the English translation of Hennig (1966), "may well have transformed Hennig's views no matter how unintentionally" (Hull, 1989, p. 5). Hull sees (1989, p. 5) their effort as "equivalent to an anti-evolutionist translating Darwin into German." His argument seems aimed only to provoke controversy among systematists, of which he is self-appointed chronicler and interpreter, as if systematists really were fruitflies (his metaphor) that could be stirred into activity by a verbal probe into their bottle. Hull offers no evidence that Davis and Zangerl transformed any view. One may say with equal justification that Davis and Zangerl may have transformed nothing. Hull's view is that "According to ideal morphologists, there is something out there in nature properly termed *the* pattern which all perceptually normal people consistently recognize" (1989, p. 4). Applied to idealistic morphologists of the nineteenth, or even twentieth, century, his characterization is absurd because it exactly describes what philosophers (Hull perhaps excepted) recognize as naive realism. To state that no one before Hull ever had such a vision would go too far, for that would confound vision and mere wordplay. Besides, according to Dullemeijer (1980, p. 175; cf. Young, 1993), "Hennig's system only says what idealistic comparative anatomy has done for ages" — a contention that Farris and Platnick (1989, p. 299) find equally "puerile."

H. Paraphyly and Darwinism

Modern systematics is sometimes portrayed as embedded in a Darwinian paradigm. Recent advocates of this viewpoint disagree on many levels (e.g., de Queiroz, 1988; Mayr and Ashlock, 1991). The Darwinian paradigm, at least in the context of systematics, has been said to be false, because its modern advocates claim to be Darwinites good and true, and explicitly embrace symplesiomorphy and paraphyly (Nelson and Platnick, 1984; Nelson, 1989a). Anticladist Darwinites typically attempt to trivialize discussion by placing it in a purely formalistic context, concerned with definition, classification, and the like. "It is now fully evident that the proposal to disqualify paraphyletic groups from recognition in classification is not only impractical and destructive but scientifically untenable" (Mayr and Ashlock, 1991, p. 225).

As mentioned above, cladists address the nature of biology, not merely arcane terminology and classification, although these topics receive more than their due attention (e.g., de Queiroz and Gauthier, 1992). That discussion of paraphyly and plesiomorphy should have this poignant meaning may seem strange. Stranger still is Hull's argument, remarkable even in the annals of anticladist comment, that "if cladistic terminology becomes standard, then subsequent generations will assume that cladists won even if the system that eventually becomes standard turns out to be indistinguishable from the system that Hennig set out to refute" (1988, p. 141 — Hull's "idealistic morphology"). In reply, Farris and Platnick (1989, p. 299) repeat that the issue concerns groups such as invertebrates, apterygotes, and reptiles, paraphyletic groups without empirical support. Specifically the issue is not terminology but whether such groups are ultimately rejected from classification for reason of their nonexistence as biological phenomena. Patterson (1982, p. 59) mentions the particular example of a "real" paraphyletic group offered by Van Valen (1978, p. 289), namely insectivores, said to be adaptively unified because they are always small and terrestrial, feed on invertebrates captured one at a time, have inflexible learning, etc. Patterson invites readers to achieve greater success than he in distinguishing insectivores, on the basis of their adaptive unity, from "opposums, caenolestids, notoryctids, lizards, toads, pitcher plants," etc.

I. Taxon and Homology as Relationship

Literature is troublesome when it confounds homology, its possible causes, its manifestations, and their possible causes. Cladistic literature is no exception. To a cladist, feathers are a synapomorphy, or homology, of birds, as if feathers themselves constitute, comprise, or compose a particular homology that is part of the matter constituting, comprising, or composing birds. It should be clear, however, that homology, in general and in particular, is a relationship between homologues (e.g., feathers of different birds) and not simply the homologues themselves.

There is an evident resonance between homology and taxon. In the systematic literature a taxon almost universally is conceived as a collection, or group, of material organisms, whether living or dead, fossil or recent, even those yet unborn. Similarly, it should be clear that a taxon, in general and in particular, is a relationship between organisms (e.g., different birds) and not simply the organisms themselves. In sum, it should be clear that *taxon relates whole organisms and homology relates their parts*, otherwise taxon and homology have exactly the same significance: the whole (taxon) is what parts (homologies) jointly comprise (Nelson, 1989b).

J. Taxa as Individuals

From the viewpoint of taxa as relationships, the extended discussion (e.g., Ruse, 1989) about taxa as individuals vs. taxa as classes seems pointless. Both alternative views (taxa as individuals, taxa as classes) construe taxa as collective groups, as if taxa were literally composed of material organisms (de Queiroz and Donoghue, 1988, among other authors). The discussion was begun by Ghiselin and fostered by Hull in several papers that received some attention and favorable comment. For example, Wiley (1981, p. 74) states that to understand taxa one must first understand differences between philosophical notions of class and individual. Crediting Hennig (1966) with insight into these differences, he states that Ghiselin (1966b, 1969a, 1974, 1980) "independently resolved the ontological status of individuals and classes as they relate to biological taxa." Wiley confesses that these ideas, of "profound implications for systematics," influenced and guided his thinking, and recommends the "excellent discussion" of Hull (1976).

There has been little comment on these matters from the critique, but what little there is indicates underlying, and complicating, factors. Farris and Platnick (1989, p. 310) find Hull's book "despicable" for a variety of reasons, not least of which is that Hull's account is compromised. In their view, the book is partisan, allied with a particular faction of biologists ("entropists"), namely Brooks and Wiley (Carpenter, 1987, Figs. 1-3). Entropists on their part support Hull's notion of taxa as individuals (Hull's "pet hypermodern species concept"), meaning that a taxon is like an organism, e.g., a person. Farris and Platnick view Hull's conclusions as determined not by evidence but by the same factors that determined the alliance — namely (Farris, 1989a, p. 108), the "opportunity to conduct a sociological experiment ... [to] assess the degree to which an utterly baseless analogy can become established as a 'revolutionary theory' by sheer force of propaganda." Farris (1989a, p. 103) mentions, for example, Hull's extraordinary appraisal of *Evolution as Entropy* (Brooks and Wiley, 1986, 1988, published in the book series of which Hull is editor), as "the greatest advance in evolutionary biology since the introduction of Mendelian genetics." No reply to Farris and Platnick has yet appeared.

In the imaginary world where taxa really are like people, with organisms (instead of organs) as their parts, homologies bear no direct relation to taxa. By design of that world, homologies are not parts of taxa, and taxa need have no homologies (as if organisms were related and their parts were not). Interested readers may explore the significance of such possibilities, bearing in mind perhaps that the resonance between taxon and homology, however conceived, is important to practice as well as to theory of systematics, and on this point there has been until recently universal agreement.

K. Causes

Literature dealing with homology and taxon is fraught with definitions that invoke *causes*: similarity due to common ancestry, correspondence due to continuity of information, relationship due to common descent, etc. All sorts of causes (common ancestry, continuity, descent) are invoked to explain relationships (homologies, taxa) taken to be already discovered, or potentially discoverable, through empirical investigation. To a cladist the causes all refer to by-products of cladistic analysis, derivatives of a particular tree and a particular analysis, that in themselves are without empirical content. All they refer to in an objective sense are the particular causes, whatever they might be (for they are largely if not totally unknown), of one or more nodes of life's hierarchy.

Choose a node, any node will do (e.g., the node relating humans and chimpanzees among themselves more closely than to gorillas, as is sometimes shown in published trees of primates). Then make two lists of causes of that node in particular, causes that are known to have been operative (because there is evidence of them), as well as causes that might have been operative (because there is no evidence against them). Compare the lists and determine which is the longer.

Discussion of definitions is complex because of other dimensions of causation, for example the cause of a particular conjecture. Definitions that refer to resemblance, similarity, and correspondence, for example, construe homology as those phenomena whereas resemblance, similarity, and correspondence are really causes of particular conjectures of homology, not homology itself. A particular conjecture always involves difference as well as resemblance among homologues, dissimilarity as well as similarity, noncorrespondence as well as correspondence in all possible proportion (difference, dissimilarity, and noncorrespon- dence, too, are causes of particular conjectures). Other dimensions of causation refer to particular criteria of homology (e.g., common embryonic origin) and particular homologues (e.g., eye lenses) that have many causes oper- ative in embryonic development, all of which may be deemed relevant, if one is so inclined, to a thorough dis- cussion of homology. Still, causes of homologues are neither causes of homology nor homology itself. Even causes of homology are not homology.

de Pinna (1991, pp. 372-373) describes the first of these dimensions (cause of a particular conjecture) under the name "primary homology." For de Pinna each proposi- tion of homology passes through two stages, its generation (primary homology) and its legitimation (secondary homol- ogy). Citing the observational and theoretical components of Jardine (1967, 1970), the topographical and phylogenetic homologies of Rieppel (1988), the preliminary and final testing of Kluge and Strauss (1985), and the homology and homogeny of Lankester (1870), de Pinna argues that the two stages have been long recognized under these different names, and that they are perceived in a particularly clear way by cladistics. He finds that all criteria of primary homol- ogy reduce to topographical correspondence (Riedl, 1975, 1979), which curiously accords both with the "unity of plan" of Etienne Geoffroy Saint-Hilaire (1771-1844) and with modern molecular biology. de Pinna observes that despite "legitimate and insightful efforts" propositions of primary homology are not objective (self-legitimizing); hence the second stage (cladistic analysis). He agrees with Patterson (1982, p. 58) that "hypotheses of homology are conjectures whose source is immaterial to their status." If so, criteria of homology, however useful or reasonable they may appear or even actually be, are neither decisive nor relevant in the court of last resort, namely the tree of best fit to all available data.

It seems not to have occurred to Patterson, or to any other nonmolecular cladist for that matter, to announce a general solution to the problem of homology, even though that solution might never be improved (at any rate it is the same solution found, more or less independently, by molecular biologists (e.g., Mindell, 1991). Perhaps it did not seem a real solution because its results are always tentative, and relative rather than absolute, in contrast to the extravagant claims, mentioned above, of perfect knowledge of better, if not perfect, data (those of Gould and Goodman). Perhaps it seemed a solution already achieved long ago.

As seen by cladists, a particular homology is a function, or derivative, of a particular tree, but it and the tree from which it is derived are also inferred possibly to be objective knowledge and to reflect the true nature of things. After all, a tree might truly depict the evolutionary relationships of the organisms thereon (if a tree were not a true depiction what indeed would be its possible significance?). Similarly, a homology might truly depict the evolutionary relationships of the parts of organisms to which the data refer.

L. Taxa as Ancestors

Early in the critique it was realized that there was a problem (paraphyly) with the traditional notion of evolutionary relationship, meaning that taxa, construed as groups of organisms, stand in relation as ancestor and descendant, a notion sometimes construed as nothing more than a metaphor based on human reproduction. Gilmour (1961, pp. 35-36) condemned the use of terms such as *ancestor, parent, phylogenetic tree, descent,* and *relationship,* because they seemed not to apply to taxa in general. He recommended that this borrowed terminology be abandoned as well as the idea that "one taxon originates from another in the same way as a child from its mother." Gilmour's criticism is even more appropriate to today's cladistic jargon, which has added terms and notions such as *mother-, daughter-,* and *sister-species* and *sister-groups.*

Familiar examples of the traditional notion are statements such as: fishes evolved into amphibians, amphibians evolved into reptiles, reptiles evolved into birds and mammals, and so on. An early assessment suggested a solution — that ancestral taxa be conceived as hypothetical species, one for each node of the tree, and that data optimized at a node be construed as the characters of a particular hypothetical species, revealing ancestral species insofar as they might ever become known. Even with Hennig's added burden of metaphor (mother-, daughter-, sister-), this solution proved effective to a degree and became standard practice early in the cladistic literature, despite protests from paleontologists who found it subversive of their authority — the "spectre haunting paleontology" of Campbell (1975, p. 87). According to Patterson (1987b, pp. 8-9), cladists' style of analysis was seen by paleontologists as a threat or insult because cladists do not treat species, recent or fossil, as ancestors and descendants of one another, but place them at the tips of branches of trees (not at nodes of trees) as if they were all descendants. He recalls that "by about 1960 paleontology had achieved such a hold on phylogeny reconstruction that there was a commonplace belief that if a group had no fossil record its phylogeny was totally unknown and unknowable." Patterson views cladistic methods as appropriate for reconstructing the history of groups regardless of their material content — whether they refer only to fossils, only to living organisms, or to both — a view now standard. He considers this view — "non-terminal branches of the evolutionary tree as hypothesis" — as one cause of the dispute over cladistics. He sees it of greatest significance, however, because its "methods, results, and style" are the same as those of systematists working with molecular data and permit comparison that is both easy and universal.

Later in the critique it was found that the problem (paraphyly), rather than being solved (with reference only to ancestral species that are hypothetical), had been merely reduced to the lowest possible taxonomic level (species) and exposed as without empirical content. In retrospect the effort seems one of minimization of a problem rather than its solution (Nelson, 1989a). Cladists might not disagree among themselves on whether ancestral species have empirical content, but they do disagree about species and their significance for evolutionary models (e.g., de Queiroz and Donoghue, 1988; Nixon and Wheeler, 1990; Vrana and Wheeler, 1992). Anticladist Darwinites typically claim that the problem, whether framed in terms of species or of taxa in general, is entirely about semantics and arbitrary rules. Comment from Maynard Smith (1985, p. 38, 1988, p. 159;), for example (other examples in Nelson, 1989a), refers to the "rule" that every taxon must be monophyletic (as if there were a choice). If so, he reasons that the notion that there is no such thing as an ancestral taxon "is obviously true" and "self-evident". It is an argument about what we should call

things, and not about what the world is or was like." He
misses the point of the cladists' claim regarding taxa in
general and ancestral taxa in particular, which is not merely
a theoretical claim but first and foremost an empirical one,
based on painstaking study, namely that taxa do not exist as
biological phenomena related as ancestor and descendant.

To claim that something does not exist may be dis-
missed as a universal negation not susceptible of proof in
the ordinary sense. Accordingly, one may dispute the value
of the claim with reference to the discoveries of paleontologists, during the centuries of their activity. Many discoveries were initially touted as ancestral taxa of one sort or
another. It is to these purported ancestral taxa that the
negative claim actually pertains, for example taxa of horses
and their relatives (Prothero and Schoch, 1989). The claim
means that subsequent study invariably shows that taxa once
touted as ancestral are really not ancestral, for a variety of
empirical reasons widely discussed in the early cladistic
literature (e.g., Englelmann and Wiley, 1977). To learn from
empirical study that there are no ancestral taxa proves
sometimes a troubling experience akin to loss of religious
faith.

M. Monophyly

In the cladistic view the only taxa deemed real are
termed *monophyletic*, which is an old term given, or acknowledged to have, precise meaning by the critique. A
taxon is generally conceived as a group of organisms (or
taxa) with an ancestral taxon (species) unique to itself, or as
a group composed of all descendants of one ancestral
species. Cladists contrast monophyletic groups, which they
deem as real, with paraphyletic and polyphyletic groups,
which they deem as unreal. With the usual tree of vertebrates as a standard, an example of a paraphyletic group is
Reptilia (including crocodiles, lizards, snakes, etc., but not
birds). An example of a polyphyletic group would be one
including birds and bats but excluding reptiles and other
mammals. Early in the critique it was perceived that monophyly relates to synapomorphy, paraphyly to symplesiomorphy, and polyphyly to convergence, or in other words that
the characters of a monophyletic group are synapomorphies,

those of a paraphyletic group are symplesiomorphies, and those of a polyphyletic group are convergences. Patterson (1982) proposes that homology (synapomorphy) be defined not in terms of similarity, correspondence, or ancestry but in terms of monophyly: "homology is the relation which characterizes monophyletic groups [real taxa]."

N. Characters as Ancestors

Groups of characters themselves were conceived as if they, too, were related by descent with modification, or in the cladistic jargon as parts of a *transformation series*. Thus wings, arms, and fins were construed as parts of a transformation series, such that wings evolved from arms, which evolved from fins. "Fins" (excluding arms and wings) is symplesiomorphy (of fishes), but fins (including arms and wings) is synapomorphy (of gnathostomes); "arms" (excluding wings) is symplesiomorphy (of nonvolant tetrapods), but arms (including wings) is synapomorphy (of tetrapods); finally bird and bat wings have no meaning other than wings, which are (two) synapomorphies (of birds and bats). Characters seen as part of the same transformation series were claimed by early cladists to be homologous, and it is now evident that the claim is defective, for it construes the transformation series as "fins"-"arms"-wings, that is as including symplesiomorphy. If the series is construed as fins-arms-wings then the meaning of the claim is trivial. The series is merely a more complicated way of saying that wings are arms and that arms are fins (wings = arms = fins) — in other words, characters considered homologous (part of the same transformation series) are homologous.

It is curious, but not surprising in retrospect, that cladists abandoned the traditional notion of relationship of taxa (one group of organisms ancestral to another) but retained the same notion for relationship of characters (one group of characters ancestral to another) — as if whole organisms were related in a way different from their parts. Cladists routinely reject statements such as birds evolved from reptiles, and accept statements such as bird wings evolved from arms, when the two sorts of statements, to the extent that they mean anything at all, mean exactly the same thing: such as, the relationship among birds (and some of their parts) is one and the same node of life's hierarchy. Put in other words, if birds are a monophyletic group of organisms, then bird wings are a monophyletic group of organs.

Cladists do not write of mother-, daughter-, and sister-characters (homologues). The burden of metaphor aside, cladists might yet do so to advantage, and treat characters of organisms (as they do taxa) as terminal entities, and, for example, remedy afflictions of transformation series analysis (e.g., Mickevich and Lipscomb, 1991; Lipscomb, 1992). One recalls Patterson's remark (see above) that "This view — nonterminal branches of the evolutionary tree as hypothesis — was one cause of the dispute over cladistics." Perhaps cladists are fated to repeat the dispute, this time over characters rather than taxa, as if this science after all moves gradually, but one pace at a time.

O. Minimum of Three

A statement of relationship, if explicitly informative, contains at least three terms. Cladists universally agree that an explicit statement of relationship among taxa requires at least three taxa, in the form, for example, that taxa A and B are related among themselves more closely than either is to taxon C. To state that a parrot is related to an ostrich (or even that all birds are related) conveys no explicit information of relationship, in contrast to the statement that a parrot and an ostrich are related more closely to each other than either is to a lizard. An episode notorious in the history of cladistics began with the observation that "a lungfish is more closely related to a cow than to a salmon" (Halstead, 1978; Schafersman, 1985).

Similarly to state that feathers of a parrot and of an ostrich are homologous (or even that all bird feathers are homologous) conveys no explicit meaning. What is required for explicit meaning is a third term, such as in the statement that bird feathers (parrot, ostrich) are homologous relative to lizard scales or, perhaps, to foot bones of lizards. Fortunately, or not as the case may be judged, the usual homology statement contains an implicit third term, such as bird feathers are homologous (relative to some unspecified

ʹⱯrt of other organisms). That systematics could actually work as efficiently as it does with statements so imprecise may seem remarkable (see below).

Patterson (1982, p. 34) notes in his review that Bock (1969, 1974) and Ghiselin (1966a) "insisted that statements of homology are meaningless unless accompanied by a 'conditional phrase.'" Bock's (1974) model phrase is "the wing of birds and the wing of bats are homologous as the forelimb of tetrapods." And Ghiselin's is "the wings of eagles are homologous as wings to those of hawks, and as derivatives of fore-limbs to those of bats." Such conditional phrases do not supply a third term, but merely affirm that wings are arms (Bock) or, oddly, that wings are not arms but are derivatives of them (Ghiselin).

P. Binary and Multistate Data

Cladists conventionally organize data in the form of binary and multistate characters. Feathers (Table I, Character 1: 0 = absence, 1 = presence) may be construed as a binary character, present in birds and absent from other forms of life. Wings may also be construed as a binary character (Table I, Character 2). Wings, arms, and fins may be construed as a multistate character (Table I, Character 3: 0 = absence; 1 = fins; 2 = arms; 3 = wings), or as a series of binary characters (Table I, Character 3a = fins, 3b = arms, 3c = wings). Conversion from multistate to binary characters (additive binary coding) is construed as exact (but see below).

Viewed as a homology statement and as a binary character, "feathers" includes a third term, represented by the "0," or absence from organisms other than birds (Table I, Character 1). Although it is true in a conventional sense that feathers are absent from nonbirds, the absence need not be construed to mean that there really is no homologue of feathers in these organisms. In a strict sense the "0" means only that no homologue is specified. Were one specified, for example the body scales of lizards, snakes, and crocodiles, then the character may be construed as a multistate character (Table I, Character 4). Were the homology extended

further, for example to the body scales of fishes, then the
multistate character would become even more complex
(Table I, Character 5).

Table I. Matrix Representation of Some Data[a]

Taxa	Character							
	1	2	3	3a	3b	3c	4	5
Amphioxus	0	0	0	0	0	0	0	0
Fishes	0	0	1	1	0	0	0	1
Frogs	0	0	2	1	1	0	0	0
Some birds	1	1	3	1	1	1	2	3
Other birds	1	1	3	1	1	1	2	3
Crocodiles	0	0	2	1	1	0	1	2
Lizards/snakes	0	0	2	1	1	0	1	2
Monotremes	0	0	2	1	1	0	0	0
Marsupials	0	0	2	1	1	0	0	0
Some bats	0	1	3	1	1	1	0	0
Other bats	0	1	3	1	1	1	0	0
Humans	0	0	2	1	1	0	0	0

[a] See text for details

Greater complexity in a multistate character means that
the character might include more of the real structure of
life's hierarchy. There is no theoretical limit to the complexity of a multistate character, except that of life's
hierarchy itself. One may suppose that as knowledge
augments, data would increasingly be organized as multistate characters, themselves increasing in complexity.
There is a limit in representing a multistate character in
the usual data matrix, and the limit, determined by complexity of branching, is reached with as few as four states
(Wiley et al., 1991, pp. 38-39, and Brooks and McLennan,
1991, pp. 55-58, discuss the technicalities involved in the
same hypothetical example). Distortion of data in the conventional matrix, through use of additive coding and other
devices required by current programs, is evident through

three-item analysis (e.g., Nelson and Platnick, 1991; Nelson and Ladiges, 1993a,b), relevant here for its concept of homology as the simplest possible statement of relationship (011 or, alternatively, 001; the former should be read as zero, one, one — not the integer eleven; the latter as zero, zero, one — not the digit one). In this view, homology literally is 011, meaning exactly that of three things two are more closely related to each other than either is to the third. Formally, the statement involves both similarity (the similarity of the two ones) and difference (the difference between the two ones and the zero), but these formalities should not be confused with the similarity and difference always displayed by actual homologues — the real parts, of which the putative relationship is represented by ones in any particular case. Three-item analysis, a recent proposal based on the notion that taxon and homology are the same relationship, dispenses with the formalisms of binary and multistate characters (Nelson, 1993).

Q. Orthology and Paralogy

Progress in analysis of molecular data has followed a path similar to that of nonmolecular systematics. Parallels with cladistics, as well as perspectives unique to molecular biology, are evident (Patterson, 1988, should be consulted). Nearly 25 years ago Fitch (1970, p. 113) asserted that there are really two classes of homology: orthology and paralogy (for older paralogy see Hunter, 1964; Inglis, 1966; Kaplan, 1984). *Paralogy* is based on the idea of gene duplication, such that different copies of the same gene come to occur in the genome of a species, where they are members of one gene family. Fitch offered the example of alpha and beta hemoglobins, which he considered to have arisen by means of gene duplication from an early hemoglobin that was neither alpha nor beta. Alpha and beta hemoglobins occur in gnathostomes generally. Fitch proposed that the homology between alpha hemoglobins, of say mouse and human, is *orthology*, whereas the homology between alpha and beta hemoglobins of the same organism, or even different organisms, is paralogy. For Fitch "phylogenies require orthologous, not paralogous, genes."

Fitch's terms, and criterion of homology, have become standard in molecular systematics. Moritz and Hillis (1990, p. 9), for example, state that "for problems of species phylogeny, the sequences examined must be orthologous," as if, again, the real meaning of data (homologies) can be ascertained prior to, and independently of, cladistic analysis (see above and Chapter 10). Molecular homology is complicated by the phenomenon, or model as the case may be (Sanderson and Doyle, 1992), of concerted evolution, which operates to convert the material basis of paralogy into that of orthology (e.g., Doyle et al., 1992, p. 228): "The problem of mixing orthologous and paralogous comparisons would appear to be overcome when concerted evolution operates to 'homogenize' the members of a multigene family."

Patterson (1988, p. 611; also de Pinna, 1991), in his review of homology in classical and molecular biology, asserted that orthology really is equivalent to homology of classical systematics, and that paralogy is equivalent to serial homology and homonomy. *Homonomy* refers, for example, to the numerous feathers on a single bird, with the implication that they are all homologous with each other (so also for hair and red blood cells). *Serial homology* refers also to numerous parts, usually repeated along the body axis, such as the individual bones of the backbone, as well as their associated nerves and blood vessels, with the implication that all are homologous with each other. It should be clear that paralogy, in the sense of homonomy or serial homology, is a relationship, created by ontogeny, between parts of a single organism. In this sense paralogy has no direct relevance to synapomorphy, or homology as considered above, which is the relationship, created by descent, between different organisms, or parts of different organisms. de Beer (1971, p. 9) considers serial homology a "misnomer," because he considers homology to refer to "tracing organs in different organisms to their representatives in a common ancestor," and serial homology to refer to "organs repeated along the anteroposterior axis of one and the same organism."

Paralogy evidently has meaning beyond homonomy and serial homology, if paralogy refers to the relationship between homologues of homologies construed as different, and particularly those homologies, associated with the same node, that provide necessary corroboration for a tree well supported by data. Mammals, for example, are said to be amniotes with hair and mammary glands — two among many mammalian synapomorphies associated with the node leading to mammals.

What is the relationship, if any, between hair and mammary glands? Homology (orthology) between these features would be universally denied, as if there were no relationship at all. Both, however, are epidermal derivatives, and each has homologues in other vertebrates and in other organisms as well (feathers and scales are possible homologues of hair, and sundry epidermal glands are possible homologues of mammary glands). Ultimately it would seem that hair and mammary glands are historically related, through other structures that are themselves homologous (orthologous), in a manner perhaps only analogous to gene duplication, but related nonetheless.

If this relationship (between hair and gland) be construed as a subclass of homology (paralogy), then it follows that all characters are homologous. The only problems are determining what multistate characters data actually comprise, and with what nodes (of a taxon tree) data are truly associated. These may reasonably be seen as tasks difficult to accomplish. Once conceptualized as discovery of putative homology (orthology and paralogy) and its cladistic analysis, they are easily seen as tasks which systematics, from its beginning, has adopted as its own, and reiterated from time to time in one or another context (e.g., Goodman, 1989, p. 60): "The immense task of systematics, therefore, in its new age of exploration is not only to work out the detailed history of each family of genes but also the history of the ramifying branches of millions of species in the tree of life."

Discovery of gene families has led to study of relationships between genes of the same family and to trees showing relationships between genes (Fig. 8). It is always emphasized that a tree of genes is not the same as a tree of organisms (or taxa), yet there is an evident relation between the two types of trees. Gene trees are the equivalent of character (state) trees wherein characters (states) are placed at terminal positions (see above). It is evident that each node of a gene tree has an exact equivalent in one or more node of an ordinary tree (of taxa), and that different nodes of a gene tree may, but need not, have the same equivalent node in an ordinary tree. It is evident also that

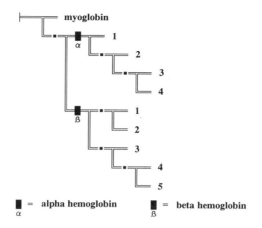

Fig. 8. Gene tree of alpha and beta hemoglobins, show-ing members of each gene subfamily (four members of alpha, five members of beta hemoglobin). Divergence of myoglobin and hemoglobin is estimated at 600-800 million years ago; that of alpha and beta hemoglobins, at 450-500 million years ago. [Data after Li and Graur (1991:152, Fig. 8).]

the relationships expressed in the gene tree, even though they are termed paralogous, are the exact equivalent of classical homology (the relations between genes on a gene tree might just as well be termed orthologous).

It should be clear that the exact meaning of orthology and paralogy, in a systematic sense, is not apparent until the information (nodes) of a gene tree is considered in relation to an ordinary tree of taxa, i.e., after cladistic analysis. Were all ten globins (of Fig. 8), for example, confined to verte-brates, found in all species of vertebrates and in no other organisms, then all nodes of the gene tree would associate with only one node of the tree of taxa — that node leading to all vertebrates and to no other organisms. In that circum-stance the gene tree would be, from a systematic standpoint, minimally informative, but maximally corrob-orative, of relationships among taxa. In contrast, were each node of the gene tree to associate with a different node of the tree of taxa, then the gene tree would be maximally informative, but minimally corroborative, of any particular relationship among taxa.

R. Truisms

If homology pervades biological data then all of biology relates to homology. All manner of biological investigations may be viewed as contributing to yet unsolved problems of systematics. One might be tempted to say even that "nothing in biology makes sense except in the light of systematics," but Dobzhansky already said it (inexplicably, he used the word "evolution" instead of "systematics"). Of truisms there seems no end and little use.

IV. BIOGEOGRAPHY

Homology is not usually discussed in the context of bio-geography, which relates to systematics because historical relationships among organisms display geographic patterns of perennial interest. Platnick and Nelson (1989, p. 412) note that Craw (1983, 1984, 1988) repeatedly pointed out the need for a "biogeographic equivalent of homology" (1983, p. 437), supplied in his view by Croizat's panbiogeography, but fatally absent from other approaches to the study of life's geography. Platnick and Nelson suggest that in cladistic biogeography homologies are congruent distributions of taxa, as in systematics homologies (corroborating paralogies, in the above sense) are congruent characters of taxa. Ladiges et al. (1992) refer to biogeographic distributions that are logical equivalents of paralogy of molecular biology (also Page, 1993)
It would go too far here to review recent developments in biogeography, which to some extent parallel and interconnect with those of systematics (and cladistics in particular). In brief, geographic patterns are viewed, alternatively, as a derivative of an ordinary (cladistic) tree (e.g., Nelson and Ladiges, 1991, 1992), or as primary phenomena

in their own right, as if biogeography were independent of systematics, operating with, or with the hope of, a principle of homology of its own.

Relevant to the cladistic critique of biogeography, the beginning of which is nominally marked by Brundin's *Transantarctic Relationships* (1966), is the life's work of Croizat (bibliography in Craw and Gibbs, 1984).

Craw and his colleagues in New Zealand attempted, with some success, to revive Croizat's views, more or less in their original form, and to develop them, independently of cladistics (Craw and Gibbs, 1984; Craw and Sermonti, 1988; Matthews, 1990; cf. Platnick and Nelson, 1989; Mayden, 1992). The development is relevant to homology in terms of geographic relationship: in what does geographic relation-ship consist and how is it evidenced? Interested readers may explore the large literature on these questions (Nelson, 1978, 1983 are not the worst starting points). Historical perspective, supplied by Croizat and ignored by Craw *et al.* in their effort to establish Croizat's panbiogeography as the dominant paradigm, reveals that the British botanist John Christopher Willis (1868-1958) is particularly relevant

> I have not copied Willis, nor has he borrowed from me, but on the ground that we have both paid living nature perhaps better than passing attention we have inevitably reached in many respects conclusions to match. It is not impossible that my own work is essentially original, and that it may as such finally impress the majority of its readers and users. However, were this work to be 'placed' somewhere in the stream of our times anyhow, it would surely not fall with Darwinism in the least (and even less with Matthewism, Mayrism, Simpsonism, Darlingtonism, Goodism, etc.) but find more or less distal place in the sphere of Willisism. (Croizat, 1958, p. 116)

Willis is known for a constellation of ideas centered on the aptly named concept of age and area (Willis,1922; also discussed in Willis, 1940, 1949). In very clear prose he explained how age of a taxon may be inferred from its geo-graphic distribution and internal taxonomic differentiation (in both cases the larger, the older), how ancestral taxa may be inferred from relative ages (the ancestral taxon is the oldest), and how a center of origin may be inferred from the geography of differentiation of the ancestral taxon (area of greatest differentiation indicates center of origin). For Willis time is an independent criterion of homology, the great common denominator, and ordering principle of relationship not only taxonomic but geographic as well. For Croizat (1964), time was but part of the trinity, space-time-form.

Willis is best known today for his "hollow curves," showing few large taxa (and many small), few taxa with large geographic distributions (and many with small), few abundant taxa (and many rare), etc. (Willis construed the few as older than, and therefore as ancestors of, the many). He early was convinced that hollow curves are real phenomena requiring explanation, a theme repeated in various contexts in more recent literature (about 150 citations listed in Science Citation Index since 1945: e.g., Anderson, 1974; Brown and Maurer, 1989; Marzluff and Dial, 1991).

From a cladistic viewpoint, even a cursory appraisal of Willis's curves, as well as age and area (and related notions), demonstrates that their purported meaning is artifact derived from paraphyly and from taxa related as ancestor and descendant. How successfully Croizat, perched in the distal space of Willisism, may have avoided these pitfalls is not yet clear (cf. Seberg, 1986). What is clear is that Croizat had a unique perception of the possible imprint left by past events on the geography of life today.

V. CONCLUSIONS

Taxa and homologies are phylogenetic relationships between, respectively, organisms and their parts. Taxa and homologies are not directly revealed but are products of a method of analysis. Criteria of homology nevertheless are useful, even necessary, to organize data as putative homologies, which are either corroborated or refuted by the cladogram that best fits all data.

A cladogram differs from other types of phylogenetic trees in placing all organisms, both fossil and recent, in terminal positions, implying that ancestral taxa are artifacts. Cladistics may possibly be improved if parts of organisms were treated in the same fashion in character (state) trees, with the implication that ancestral characters, too, are artifacts.

Paralogy, a class of homology defined by molecular biology, evidently refers to different synapomorphies (homologies) of the same taxon (node), necessary for corroboration of a particular tree.

Efforts to construe homology as independent of systematics are in the tradition either of idealistic morphology, which seems eternal, or of the phenetic episode of biology, which seems closed at least for the time being in systematics.

ACKNOWLEDGMENTS

Review by P. Y. Ladiges, J. Carpenter, M. Carvalho, M. de Pinna, C. Patterson, and M. Crisp and the Canberra Cooper's and Cladistics group, discussion with A. N. Drinnan, and help with literature by M. J. Donoghue and C. J. Humphries, are gratefully acknowledged.

REFERENCES

Anderson, S. (1974). Patterns of faunal evolution. *Q.Rev Biol.* **49**, 311-332.

Blackwelder, R. E. (1977). Twenty five years of taxonomy. *Syst. Zool.* **25**, 107-137.

Bock, W. J. (1969). Comparative morphology in systematics. *In* "Systematic Biology: Proceedings of an International Conference Conducted at the University of Michigan, Ann Arbor, Michigan, June 14-16, 1967, Sponsored by the National Research Council" (Anon., ed.), pp. 441-448. National Academy of Science, Washington, DC.

Bock, W. J. (1974). Philosophical foundations of classical evolutionary classification. *Syst. Zool.* **22**, 375-392.

Brooks, D. R., and McLennan, D. A. (1991). "Phylogeny, Ecology, and Behavior." Univ. of Chicago Press, Chicago and London.

Brooks, D. R., and Wiley, E. O. (1986). "Evolution as Entropy: Toward a Unified Theory of Biology." Univ. of Chicago Press, Chicago and London.

Brooks, D. R., and Wiley, E. O (1988). "Evolution as Entropy: Toward a Unified Theory of Biology," 2nd ed. Univ. of 'Chicago Press, Chicago and London.

Brown, J. H., and Maurer, B. A. (1989). Macroecology: The division of food and space among species and continents. *Science* **243**, 1145-1150.

Brundin, L. (1966). Transantarctic relationships and their significance, as evidenced by chironomid midges. *K. Sven. Vetenskapsakad. Handl.*, Fjärde Ser. **11**,1-472.

Brundin, L. (1976). A Neocomian chironomid and Podonominae-Aphroteniinae (Diptera) in the light of phylogenetics and biogeography. *Zool. Scr.* **5**, 139-160.

Campbell, K. S. (1975). Cladism and phacopid trilobites. *Alcheringa* **1**, 87-96.

Carpenter, J. M. (1987). Cladistics of cladists. *Cladistics* **3**, 363-375.

Carpenter, J. M. (1992). Comparing methods. *Cladistics* **8**, 191-196.

Coddington, J. A. (1988). Cladistic tests of adaptational hypotheses. *Cladistics* **4**, 1-20.

Coddington, J. A. (1990). Bridges between evolutionary pattern and process. *Cladistics* **6**, 379-386.

Cracraft, J., and Mindell, D. P. (1989). The early history of modern birds: A comparison of molecular and morphological evidence. *In* "The Hierarchy of Life: Molecules and Morphology in Phylogenetic Analysis" (B. Fernholm, K. Bremer, and H. Jörnvall, eds.), Proc. Nobel Symp. 70, pp. 389-403. Excerpta Medica (Elsevier), Amsterdam.

Cranston, P. S., and Humphries, C. J. (1988). Cladistics and computers: A chironomid conundrum. *Cladistics* **4**, 72-92.

Craw, R. C. (1983). Panbiogeography and vicariance cladistics: Are they truly different? *Syst. Zool.* **32**, 431-437.

Craw, R. C. (1984). Leon Croizat's biogeographic work: A personal appreciation. *Tuatara* **27**, 8-13.

Craw, R. C. (1988). Panbiogeography. *In* "Analytical Biogeography: An Integrated Approach to the Study of Animal and Plant Distributions" (A. Myers and P. Giller, eds.), pp. 405-435. Chapman & Hall, London.

Craw, R. (1992). Margins of cladistics: Identity, difference and place in the emergence of phylogenetic systematics, 1864-1975. *In* "Trees of Life: Essays on the Philosophy of Biology" (P. Griffiths, ed.), pp. 65-107. Kluwer, Dordrecht, The Netherlands.

Craw, R. C., and Gibbs, G. W., eds. (1984). "Croizat's Panbio-geography and Principia Botanica: Search for a novel synthesis." *Tuatara*, **27**, 1-75.

Craw, R. C., and Sermonti, G., eds. (1988). Special issue on panbiogeography: Space-time-form. *Riv. Biol.-Biol. Forum*, **81**, 457-615.

Croizat, L. (1958). An essay on the biogeographic thinking of J. C. Willis. *Arch. Bot. Biogeogr. Ital.* **34**, 90-116.

Croizat, L. (1964). "Space, Time, Form: The Biological Synthesis." Published by the author, Caracas.

Danaher, P. (1992). The bounty of bad language. *Aust. Campus Rev. Week.* (Dec. 17-23), 2, 14.

de Beer, G. R. (1971). "Homology, an Unsolved Problem," Oxford Biol. Readers, No. 11. Oxford Univ. Press, London.

de Pinna, M. C. C. (1991). Concepts and tests of homology in the cladistic paradigm. *Cladistics* **7**, 367-394.

de Pinna, M. C. C., and Salles, L. O. (1990). Cladistic tests of adaptational hypotheses: A reply to Coddington. *Cladistics* **6**, 373-377.

de Queiroz, K. (1985). The ontogenetic method for determining character polarity and its relevance to phylogenetic systematics. *Syst. Zool.* **34**, 280-299.

de Queiroz, K. (1988). Systematics and the Darwinian revolution. *Philos. Sci.* **55**, 238-259.

de Queiroz, K., and Donoghue, M. (1988). Phylogenetic systematics and the species problem. *Cladistics* **4**, 317-338.

de Queiroz, K., and Gauthier, J. (1992). Phylogenetic taxonomy. *Annu. Rev. Ecol. Syst.* **23**, 449-480.

Donoghue, M. (1990). Sociology, selection and success: A critique of David Hull's analysis of science and systematics. *Biol. Philos.* **5**, 459-472.

Donoghue, M. (1992). Homology. *In* "Keywords in Evolutionary Biology" (E. F. Keller and E. A. Lloyd, eds.), pp. 170-179. Harvard Univ. Press, Cambridge, MA, and London.

Donoghue, M. J., and Kadereit, J. W. (1992). Walter Zimmermann and the growth of phylogenetic theory. *Syst. Biol.* **41**, 74-85.

Donoghue, M. J., and Sanderson, M. J. (1992). The suitability of molecular and morphological evidence in reconstructing plant phylogeny. *In* "Molecular Systematics of Plants" (P. S. Soltis, D. E. Soltis, and J. J. Doyle, eds.), pp 340-368. Chapman & Hall, New York and London.

Doyle, J. J., Lavin, M., and Bruneau, A. (1992). Contributions of molecular data to papilionoid legume systematics. *In* "Molecular Systematics of Plants" (P. S. Soltis, D. E. Soltis, and J. J. Doyle, eds.), pp. 223-251. Chapman & Hall, New York and London.

Dullemeijer, P. (1980). Functional morphology and evolution. *Acta Biotheor.* **29**, 151-250.

Dupuis, C. (1979). La systématique phylogenetique de W. Hennig (historique, discussion, choix de references). *Cah. Nat.* **34**, 1-69.

Engelmann, G. F., and Wiley, E. O. (1977). The place of ancestor-descendant relationships in phylogenetic reconstruction. *Syst. Zool.* **26**, 1-11.

Farris, J. S. (1989a). Entropy and fruit flies. *Cladistics* **5**,103-108.

Farris, J. S. (1989b). The retention index and the rescaled consistency index. *Cladistics* **5**, 417-419.

Farris, J. S., and Platnick, N. I. (1989). Lord of the flies: The systematist as study animal. *Cladistics* **5**, 295-310.

Felsenstein, J. (1978). The number of evolutionary trees. *Syst. Zool.* **27**, 27-33.

Fernholm, B., Bremer, K., and Jörnvall, H., eds. (1989). "The Hierarchy of Life: Molecules and Morphology in Phylogenetic Analysis," Proc. Nobel Symp. 70. Excerpta Medica (Elsevier), Amsterdam.

Fitch, W. M. (1970). Distinguishing homologous from analogous proteins. *Syst. Zool.* **19**, 99-113.

Forey, P. L. (1992). Fossils and cladistic analysis. *In* "Cladistics: A Practical Course in Systematics" (P. L. Forey, C. J. Humphries, I. A. Kitching, R. W. Scotland, D. J. Siebert, and D. M. Williams, eds.), pp. 124-136. Oxford Univ. Press, Oxford.

Fryer, G. (1978). Arthropod morphology and evolution. *Nature (London)* **273**, 172-173.

Funk, V. A., and Brooks, D. I., eds. (1981). "Advances in Cladistics: Proceedings of the First Meeting of the Willi Hennig Society." New York Botanical Garden, Bronx.

Ghiselin, M. T. (1966a). An application of the theory of definitions to taxonomic principles. *Syst. Zool.* **15**, 127-130.

Ghiselin, M. T. (1966b). On psychologism in the logic of taxonomic controversies. *Syst. Zool.* **15**, 207-215.

Ghiselin, M. T. (1969a). "The Triumph of the Darwinian Method." Univ. of California Press, Berkeley.

Ghiselin, M. T. (1969b). The distinction between similarity and homology. *Syst. Zool.* **18**, 148-149.

Ghiselin, M. T. (1974). A radical solution to the species problem. *Syst. Zool.* **23**, 536-544.

Ghiselin, M. T. (1980). Natural kinds and literary accomplishments. *Mich. Q. Rev.* **19**, 73-88.

Gilmour, J. S. L. (1961). Taxonomy. *In* "Contemporary Botanical Thought" (A. M. MacLeod and L. S. Cobley, eds.), pp. 27-45. Oliver & Boyd, Edinburgh and London, and Quadrangle Books, Chicago.

Gittleman, J. L., and Luh, H.-K. (1992). On comparing comparative methodology. *Annu. Rev. Ecol. Syst.* **23**, 383-404.

Goodman, M. (1989). Emerging alliance of phylogenetic systematics and molecular biology: A new age of exploration. *In* "The Hierarchy of Life: Molecules and Morphology in Phylogenetic Analysis" (B. Fernholm, K. Bremer, and H. Jörnvall, eds.), Proc. Nobel Symp. 70, pp. 43-61. Excerpta Medica (Elsevier), Amsterdam.

Gould, S. J. (1977). "Ontogeny and Phylogeny." Harvard Univ. Press, Cambridge, MA, and London.

Gould, S. J. (1985). A clock of evolution: We finally have a method for sorting out homologies from "subtle as subtle can be" analogies. *Nat. Hist.* **94** (4), 12-25.

Gould, S. J. (1986). Evolution and the triumph of homology, or why history matters. *Am. Sci.* **74**, 60-69.

Grant, V. (1992). Systematics and phylogeny of the *Ipomopsis aggregata* group (Polemoniaceae): Traditional and molecular approaches. *Syst. Bot.* **17**, 683-691.

Halstead, L. B. (1978). The cladistic revolution — Can it make the grade? *Nature (London)* **276**, 759-760.

Harvey, P. H., and Pagel, M. D. (1991). "The Comparative Method in Evolutionary Biology." Oxford Univ. Press, Oxford.

Hennig, W. (1950). "Grundzüge einer Theorie der phylogenetischen Systematik." Deutscher Verlag, Berlin.

Hennig, W. (1966). "Phylogenetic systematics." Univ. of Illinois Press, Urbana (reprinted in 1979).

Hillis, D. M., and Moritz, C., eds. (1990). "Molecular Systematics." Sinauer Associates, Sunderland, MA.

Hull, D. L. (1976). Are species really individuals? *Syst. Zool.* **25**, 174-191.

Hull, D. L. (1988). "Science as a Process: An Evolutionary Account of the Social and Conceptual Development of Science." Univ. of Chicago Press, Chicago.

Hull, D. L. (1989). The evolution of phylogenetic systematics. *In* "The Hierarchy of Life: Molecules and Morphology in Phylogenetic Analysis" (B. Fernholm, K. Bremer, and H. Jörnvall, eds.), Proc. Nobel Symp. 70, pp. 3-15. Excerpta Medica (Elsevier), Amsterdam.

Hunter, I. J. (1964). Paralogy, a concept complementary to homology and analogy. *Nature (London)* **204**, 604.

Inglis, W. G. (1966). The observational basis of homology. *Syst. Zool.* **15**, 219-228.

Janvier, P. (1984). Cladistics: Theory, purpose, and evolutionary implications. *In* "Evolutionary Theory: Paths into the Future" (J. W. Pollard, ed.), pp. 39-75. Wiley, Chichester.

Jardine, N. (1967). The concept of homology in biology. *Brit. J. Philos. Sci.* **18**, 125-139.

Jardine, N. (1970). The observational and theoretical components of homology: A study based on the morphology of the dermal skull-roofs of rhipidistian fishes. *Biol. J. Linn. Soc.* **1**, 327-361.

Kaplan, D. R. (1984). The concept of homology and its central role in the elucidation of plant systematic relationships. *In* "Cladistics: Perspectives on the Reconstruction of Evolutionary History. Papers Presented at a Workshop on the Theory and Application of Cladistic Methodology, March 22-28, University of California, Berkeley" (T. Duncan and T. F. Stuessy, eds.), pp. 51-70. Columbia Univ. Press, New York.

Kitching, I. A. (1992). Tree building techniques. *In*: "Cladistics: A Practical Course in Systematics" (P. L. Forey, C. J. Humphries, I. A. Kitching, R. W. Scotland, D. J. Siebert, and D. M. Williams, eds.), pp. 44-71. Oxford Univ. Press, Oxford.

Kluge, A. G., and Strauss, R. E. (1985). Ontogeny and systematics. *Annu. Rev. Ecol. Syst.* **16**, 247-268.

Knight, D. (1981). "Ordering the World: A History of Classifying Man." Burnett Books, London.

Ladiges, P. Y., Prober, S. M., and Nelson, G. (1992). Cladistic and biogeographic analysis of the "blue ash" eucalypts. *Cladistics* **8**, 103-124.

Lankester, E. R. (1870). On the use of the term homology in modern zoology, and the distinction between homogenetic and homoplastic agreements. *Ann. Mag. Nat. Hist.* [4] **6**, 34-43.

Lévi-Strauss, C. (1962). "La Pensée Sauvage." Librairie Plon, Paris.

Lévi-Strauss, C. (1966). "The Savage Mind." Weidenfeld & Nicholson, London.

Li, W.-H., and Graur, D. (1991). "Fundamentals of Molecular Evolution." Sinauer Associates, Sunderland, MA.

Lipscomb, D. L. (1992). Parsimony, homology and the analysis of multistate characters. *Cladistics* **8**, 45-65.

Margoliash, E., and Fitch, W. M. (1969). Discussion. *In* "Systematic Biology: Proceedings of an International Conference Conducted at the University of Michigan, Ann Arbor, Michigan, June 14-16, 1967, Sponsored by the National Research Council" (Anon., ed.), pp. 357-360. National Academy of Science, Washington, D. C.

Marzluff, J. M., and Dial, K. P. (1991). Life history correlates of taxonomic diversity. *Ecology* **72**, 428-439.

Matthews, C., ed. (1990). "Panbiogeography Special Issue." *N. Z. J. Zool.* Vol.**16**. DSIR Publishing, Wellington.

Mayden, R. L. (1992). The wilderness of panbiogeography: A synthesis or space, time, and form? *Syst. Zool.* **40**, 503-519.

Maynard Smith, J. (1985). Do we need a new evolutionary paradigm? *New Sci.*, March 14, 38-39 (reprinted in Maynard Smith, 1988).

Maynard Smith, J. (1988). "Did Darwin Get It Right? Essays on Games, Sex and Evolution." Chapman & Hall, New York.

Mayr, E., and Ashlock, P. D. (1991). "Principles of Systematic Zoology." McGraw-Hill, New York.

Mickevich, M. F., and Lipscomb, D. L. (1991). Parsimony and the choice between different transformations for the same character set. *Cladistics* **7**, 111-139.

Mindell, D. P. (1991). Similarity and congruence as criteria for molecular homology. *Mol. Biol. Evol.* **8**, 897-900.

Mindell, D. P. (1992). DNA-DNA hybridization and avian phylogeny. *Syst. Biol.* **41**, 126-134.

Minelli, A., and Peruffo, B. (1991). Developmental pathways, homology and homonomy in metameric animals. *J. Evol. Biol.* **3**, 429-445.

Miyamoto, M., and Cracraft, J., eds. (1991). "Phylogenetic Analysis of DNA Sequencing." Oxford Univ. Press, New York and Oxford.

Moritz, C., and Hillis, D. M. (1990). Molecular systematics: Context and controversies. *In* "Molecular systematics" (D. M. Hillis and C. Moritz, eds.), pp. 1-10. Sinauer Associates, Sunderland, MA.

Nelson, G. (1978). From Candolle to Croizat: Comments on the history of biogeography. *J. Hist. Biol.* **11**, 269-305.

Nelson, G. (1983). Vicariance and cladistics: Historical perspectives with implications for the future. *In* "Evolution, Time and Space: The Emergence of the Biosphere" (R. W. Sims, J. H. Price, and P. E. S. Whalley, eds.), pp. 469-492. Academic Press, London.

Nelson, G. (1989a). Species and taxa: Speciation and evolution. *In* "Speciation and Its Consequences" (D. Otte and J. Endler, eds.), pp. 60-81. Sinauer Associates, Sunderland, MA.

Nelson, G. (1989b). Cladistics and evolutionary models. *Cladistics* **5**, 275-289.

Nelson, G. (1993). Reply to Harvey. *Cladistics* **8**, 355-360.

Nelson, G., and Ladiges, P. Y. (1991). Standard assumptions for biogeographic analysis. *Aust. Syst. Bot.* **4**, 41-58; amendment: **5**, 247 (1992).

Nelson, G., and Ladiges, P. Y. (1992). Three-area statements: Standard assumptions for biogeographic analysis. *Syst. Zool.* **40**, 470-485.

Nelson, G., and Ladiges, P. Y. (1993a). "TAS and TAX: MSDOS Programs for Cladistics," vers. 3.1. Published by the authors, New York and Melbourne.

Nelson, G., and Ladiges, P. Y. (1993b). Information content and fractional weight of three-item statements. *Syst. Biol.* **41**, 490-494.

Nelson, G., and Patterson, C. (1993). Cladistics, sociology and success: A comment on Donoghue's critique of David Hull. *Biol. Philos.* (in press).

Nelson, G., and Platnick, N. I. (1981). "Systematics and Biogeography: Cladistics and Vicariance." Columbia Univ. Press, New York.

Nelson, G., and Platnick, N. I. (1984). Systematics and Evolution. *In* "Beyond Neodarwinism" (M.-W. Ho and P. Saunders, eds.), pp. 143-158. Academic Press, London.

146 G. Nelson

Nelson, G. and Platnick, N. I. (1991). Three-taxon statements: A more precise use of parsimony? *Cladistics* **7**, 351-366.

Nixon, K. C., and Wheeler, Q. D. (1990). An amplification of the phylogenetic species concept. *Cladistics* **6**, 211-223.

Page, R. D. M. (1993). Genes, organisms, and areas: The problem of multiple lineages. *Syst. Biol.* **42**, 77-84.

Panchen, A. L. (1992). "Classification, Evolution, and the Nature of Biology." Cambridge Univ. Press, Cambridge and New York.

Patterson, C. (1982). Morphological characters and homology. *In* "Problems of Phylogenetic Reconstruction" (K. A. Joysey and A. E. Friday, eds.), pp. 21-74. Syst. Assoc. Spec. Vol. No. **25**, pp. 21-74. Academic Press, London.

Patterson, C., ed. (1987a). "Molecules and Morphology in Evolution: Conflict or Compromise?" Cambridge Univ. Press, Cambridge.

Patterson, C. (1987b). Introduction. *In* "Molecules and Morphology in Evolution: Conflict or Compromise?" (C. Patterson, ed.), pp. 1-22. Cambridge Univ. Press, Cambridge.

Patterson, C. (1988). Homology in classical and molecular biology. *Mol. Biol. Evol.* **5**, 603-625.

Penny, D., and Hendy, M. D. (1985). The use of tree comparison metrics. *Syst. Zool.* **34**, 75-82.

Platnick, N. I. (1989a). Cladistics and phylogenetic analysis today. *In* "The Hierarchy of Life: Molecules and Morphology in Phylogenetic Analysis" (B. Fernholm, K. Bremer, and H. Jörnvall, eds.), Proc. Nobel Symp. 70, pp. 17-24. Excerpta Medica (Elsevier), Amsterdam.

Platnick, N. I. (1989b). An empirical comparison of microcomputer parsimony programs. II. *Cladistics* **5**, 145-161.

Platnick, N. I., and Funk, V. A., eds. (1983). "Advances in Cladistics, Proceedings of the Second Meeting of the Willi Hennig Society." Columbia Univ. Press, New York.

Platnick, N. I., and Nelson, G. (1989). Spanning-tree biogeography: Shortcut, detour, or dead-end? *Syst. Zool.* **37**, 410-419.

Prothero, D. R., and Schoch, R. M., eds. (1989). "The Evolution of Perrisodactyls." Oxford Univ. Press, New York and Oxford.

Riedl, R. (1975). "Die Ordnung des Lebendigen: Systembedingungen der Evolution." Paul Parey, Hamburg

Riedl, R. (1979). "Order in Living Organisms." Wiley, Chichester.

Rieppel, O. (1988). "Fundamentals of Comparative Biology." Birkhäuser, Basel and Boston.

Rieppel, O. (1992). Homology and logical fallacy. *J. Evol. Biol.* **5**, 701-715.

Rohlf, F. J., and Bookstein, F. L., eds. (1990). "Proceedings of the Michigan Morphometrics Workshop." Spec. Publ. **2** University of Michigan, Mus. Zool., Ann Arbor.

Rosen, D. E., Nelson, G., and Patterson, C. (1979). Foreword. *In* "Phylogenetic Systematics" (W. Hennig), pp. vii-xiii. Univ of Illinois Press, Urbana.

Ross, H. L. (1964). Principles of numerical taxonomy. *Syst. Zool.* **13**, 106-108.

Roth, V. L. (1988). The biological basis of homology. *In* "Ontogeny and Systematics" (C. J. Humphries, ed.), pp. 1-26. Columbia Univ. Press, New York.

Roth, V. L. (1991). Homology and hierarchies: Problems solved and unresolved. *J. Evol. Biol.* **4**, 167-194.

Ruse, M., ed. (1989). "What the Philosophy of Biology is: Essays Dedicated to David Hull." Kluwer Academic Publishers, Dordrecht, The Netherlands.

Saether, O. (1979). Underlying synapomorphies and ana-genetic analysis. *Zool. Scr.* **8**, 305-312.

Sanderson, M. J. (1990). Flexible phylogeny reconstruction: A review of phylogenetic inference packages using parsimony. *Syst. Zool.* **39**, 414-420.

Sanderson, M. J., and Doyle, J. J. (1992). Reconstruction of organismal and gene phylogenies from data on multigene families: Concerted evolution, homoplasy, and confidence. *Syst. Biol.* **41**, 4-17.

Sarich, V. M., Schmid, C. W., and Marks, J. (1989). DNA hybridization as a guide to phylogenies: A critical analysis. *Cladistics* **5**, 3-32.

Sattler, R. (1984). Homology — a continuing challenge. *Syst. Bot.* **9**, 382-394.

Schafersman, S. D. (1985). Anatomy of a controversy: Halstead vs. the British Museum (Natural History). *In* "What Darwin Began: Modern Darwinian and Non-Darwinian Perspectives on Evolution" (L. R. Godfrey, ed.), pp. 186-219. Allyn & Bacon, Boston.

Schoch, R. R. (1986). "Phylogeny Reconstruction in Paleontology." Van Nostrand-Reinhold, New York.

Seberg, O. (1986). A critique of the theory and methods of panbiogeography. *Syst. Zool.* **35**, 369-380.

Siebert, D. J., and Humphries, C. J. (1991). Speaking comparatively. *Nature (London)* **353**, 615-616.

Smith, M. L. (1987). Morphometrics in evolutionary biology. *Cladistics* **3**, 97-99.

Sneath, P. H. A., and Sokal, R. R. (1973). "Numerical Taxonomy: The Principles and Practice of Numerical Classification." Freeman, San Francisco.

Sober, E. (1988). "Reconstructing the Past: Parsimony, Evolution, and Inference." Massachusetts Institute of Technology, Cambridge.

Stevens, P. F. (1984). Homology and phylogeny: Morphology and systematics. *Syst. Bot.* **9**, 395-409.

Szalay, F. S., and Bock, W. J. (1991). Evolutionary theory and systematics: Relationships between process and patterns. *Z. Zool. Syst. Evolutions-forsch.* **29**, 1-39.

Tassy, P. (1991). "L'Arbre à Remonter le Temps: Les Rencontres de la Systématique et de l'Evolution." C. Bourgeois, Paris.

Van Valen, L. (1978). Why not to be a cladist. *Evol. Theory* **3**, 285-299.

Van Valen, L. (1982). Homology and causes. *J. Morphol.* **173**, 305-312.

Vrana, P., and Wheeler, W. (1992). Individual organisms as terminal entities: Laying the species problem to rest. *Cladistics* **8**, 67-72.

Wagner, G. P. (1989). The biological homology concept. *Annu. Rev. Ecol. Syst.* **20**, 51-69.

Wenzel, J. W. (1992a). What do you get if you cross a cladist with a behavioral ecologist? *Cladistics* **8**, 97-102.

Wenzel, J. W. (1992b). Behavioral homology and phylogeny. *Annu. Rev. Ecol. Syst.* **23**, 361-381.

Wiley, E. O. (1981). "Phylogenetics: The Theory and Practice of Phylogenetic Systematics." Wiley, New York.

Wiley, E. O., Siegel-Causey, D., Brooks, D. R., and Funk, V. A. (1991). The compleat cladist: A primer of phylogenetic procedure. *Univ. Kans Mus. Nat. Hist., Spec. Publ.* **19**, i-ix+1-158.

Willis, J. C. (1922). "Age and Area." Cambridge Univ Press, Cambridge.

Willis, J. C. (1940). "The Course of Evolution." Cambridge Univ. Press, Cambridge.

Willis, J. C. (1949). The birth and spread of plants. *Boissiera* **8**, xi+ 1-561.

Young, B. A. (1993). On the necessity of an archetypal concept in morphology: With special reference to the concepts of "structure" and "homology." *Biol. Philos.* **8**, 225-248.

4

HOMOLOGY, FORM, AND FUNCTION

George V. Lauder

School of Biological Sciences
University of California
Irvine, CA 92717

I. INTRODUCTION

As part of our introduction to biology as undergraduates, most of us were taught the distinction between homology and analogy. Homologies are commonly presented as structures sharing common ancestry, while analogies are presented as similar structures that do not share common ancestry. Homologies may be either similar in structure (e.g., the femur of a cat and a dog) or quite different in appearance. On the other hand, the similarity among analogous characters is explained as being due to common functional or environmental demands imposed on those structures. Insect wings and bird wings are the textbook exemplar of choice. Two important issues arise from this classical explication of homology/analogy and pervade current literature discussions of these concepts.

First, analysis of the concept of homology has been overwhelmingly dominated by an emphasis on structure or form. Most definitions of homology use the term *structure*, involve only discussions of structures, and most authors restrict themselves to structural examples. Some authors have even claimed that the concept of homology applies only to structures and cannot be applied to organismal functions or other traits. More prevalent is the suggestion that without an analysis of structural underpinnings, meaningful analyses of character homology cannot take place. Under this latter view, nonstructural organismal traits such as behaviors cannot be examined exclusive of structures, hence the study of nonstructural characters is logically subservient to the analysis of structural homology. Indeed, the classical notion that function is studied for the purpose of defining analogies rather than homologies reflects the historical bias that has relegated the study of nonstructural characters to a role in identifying nonhomologous traits.

In this chapter I argue that restricting the definition of homology to structures or basing homology of nonstructural characters on an analysis of morphology is unnecessary if an explicitly phylogenetic definition of homology is used. In addition, recognition of the hierarchical nature of organismal design in which many different kinds of traits may be considered as homologues — e.g., structures, functions, behaviors, developmental pathways, and patterns of gene expression —

opens up new and exciting questions about the evolution of form and function.

Second, if the study of character function is a guide to analogy, what is the key to identifying homology? In this chapter I argue that literature on homology is pervaded by a search for the "locus" of homology in the organism: what aspect of organismal design is best for reliably identifying homologous characters? Suggestions for reliable indicators of homology include common developmental patterns, gene sequences, the structure of the nervous system, physiological patterns, and connections among structures themselves. Much of the literature on homology is devoted to arguing in favor of one of these "indicators" of homology: e.g., homologous characters can be recognized by similarity in development, connection, and/or underlying genetic structure.

I discuss below my alternative view: that the search for a locus of homology is fundamentally misguided, and that no matter how similar two organismal traits are (in development, structure, and function) they might still be nonhomologous. In contrast to the approach of recognizing homologues through an analysis of one particular class of data, I advocate a *phylogenetic approach* in which homologous traits are recognized *a posteriori* as a consequence of a global phylogenetic analysis of many characters of all kinds. A phylogenetic approach to homology gives primacy to no one class of data and allows novel questions to be asked about historical patterns of character covariation.

II. HOMOLOGY AND STRUCTURE

The notion that homologous traits of organisms are primarily structural in nature has a long pedigree in biology, founded in debates on the correspondence among parts of organisms that characterized the first half of the nineteenth century (Appel, 1987). Although discussions of archetypes, the correlation of parts within a species, and correspondences of structure across species had strong pre-Darwinian foundations (Russell, 1916; and see Chapter 1), the essential framework for analysis of the relationships among organismal traits carried over into the post-Darwinian era. Many authors, including Darwin, have recognized that traits other than structures in two or more species might be considered to share common ancestry and thus be homologous (e.g., Ghiselin, 1976; Mayr, 1969), but a pervasive theme in the literature on homology is primacy of structural data. Since characters such as behavioral traits and physiological functions can be difficult to link to specific structures, questions have been raised as to the relevance of the concepts of homology and analogy to the analysis of function and behavior.

Table I summarizes a few of the many recent statements supporting a structurally based view of homology. Even Richard Owen's oft-quoted definition of homology as "the same *organ* in different animals under every variety of form and function" (my emphasis) carries with it an implicit structural bias: homologous attributes are organs, structural entities within the organism. In part, the focus on structure in Darwin and Owen's time may have been because information on organismal form constituted the primary data available and thus a natural focus of attention. The rise of experimental and physiological studies of organisms [of increasing interest in Berlin and France during the nineteenth century (Allen, 1975; Coleman, 1977)] occurred primarily in medical schools and had not yet begun to generate comparative data that might have raised questions about the correspondence of function across taxa. Similarly, comparative studies of behavior were not widespread, and few data were available in the last half of the nineteenth century on the distributions of behavioral characters among species. Paleontological discoveries, often a source of new taxa at the time (Desmond, 1982), provided little information on function or behavior.

The homology literature of the last 20 years has maintained the focus on structure (see Table I), at the same time as our growing ability to analyze organismal function has played an increasing role in generating functional characters in a diversity of organisms. In fact, an interesting distinction has developed in the literature between the study of structure and homology on the one hand, and the study of function and analogy on the other. This dichotomy leads to confusion in cases where it is desirable to infer function from structure in taxa in which function has not (or cannot) be studied experimentally.

Paleontologists, for example, often infer function in extinct taxa by reference to living species. Raup and Stanley, in their text on paleontology, ask us to:

Table I. Selected Statements from the Literature Illustrating the View That the Concept of Homology Applies Only to Structural Characters or to the Structural Basis of Nonstructural Traits such as Behaviors or Functions.

Reference	Comment on structural basis of homology
1. Atz (1970, p. 68)	"The extent to which behavior can be homologized is directly correlated with the degree to which it can be conceived or abstracted in morphological terms."
2. Wilson et al. (1973, p. 633)	"Homology is defined as correspondence between two structures which is due to inheritance from a common ancestor."
3. Hodos (1976, pp. 165, 161)	"The concept of behavioral homology is totally dependent on the concept of structural homology..." and "The issue is the degree to which behavior can be related to specific structural entities."
4. Riedl (1978, p. 33)	"Homologies ... are structural similarities which force us to suppose that any differences are explicable by divergence from identical origin."
5. Goodwin (1984, p. 101)	"Homology is an equivalence relation on a set of forms which share a common structural plan...."
6. Carroll (1988, p. 6)	"Homology refers to the fundamental similarity of individual structures...."
7. Wagner (1989, p. 51)	"Only morphological equivalence in terms of relative position, structure, and connections with nerves and blood vessels counts."
8. Hall (1992, p. 57)	"The author's present position is that the term homology should be limited to structures and not used for developmental processes at all ...."

consider first the case in which fossil and Recent taxa that bear similar structures are closely related and the structures are judged to have had a common origin. The fossil structure may then, by homology (having the same origin), be judged to serve the same function as does the Recent structure. (Raup and Stanley, 1971, p. 166.)

This statement reflects a common view (criticized elsewhere; Lauder, 1993) that functions and structures are tightly linked and that if two taxa share a common structure, those structures may be inferred to have shared a common function as well. Within this framework, homology of structures implies homology of function.

But it is common to discover a fossil possessing a novel structure not found in immediate relatives. How then do we infer the function of that structure? One way is to use analogous structures in Recent taxa (where function may be studied directly) unrelated to the fossil taxon of interest (Raup and Stanley, 1971). Thus, if two unrelated taxa (one fossil and one recent) each possess a trait that is similar in structure and that is analogous (that is, has evolved convergently), then inference of function in the fossil taxon will depend on the assumption that analogous structures possess analogous functions. We are then using analogy in structure to infer analogy in function, again assuming that structure and function tend to show concordant historical patterns, an assumption that is often false (Lauder, 1993).

III. SEARCHING FOR THE LOCUS OF HOMOLOGY

The view that structure is a primary vehicle for the recognition of homology and also the primary reference point for the analysis of historical patterns in other types of characters (functions, behaviors, and developmental pathways) is but one aspect of a more general issue that permeates the homology literature. I term this issue the search for the "locus" of information about homology: the search for (or use of) a specific class of data that contains the information necessary to make decisions about homology by reference to characters themselves. Note that by referring to the "locus" of homology I do not mean a specific anatomical site within the organism, but rather a specific class of data that is believed to contain information best reflective of "true similarity" and thus homology. In other words, if only we could understand more details of character structure, interconnection, development, function, or neural control (choose one), we would be able to decide if two characters are homologous.

I have previously classified such criteria for homology as *a priori* criteria (Lauder, 1986), because the goal of using a particular class of information (such as ontogenetic, structural, or neural data) is to shed light on character homology solely by a detailed analysis of the substrate for the characters themselves (that is, prior to and independent of, a phylogenetic analysis). In contrast, the phylogenetic method discussed in Section IV recognizes homologies *a posteriori*, as a result of (after) a phylogenetic analysis of many characters, not just those of initial interest.

At various times, virtually every possible class of data has been advocated as the locus in which homologous similarity among characters is revealed. Five types of data have been especially prominent in the homology literature as proposed sources of information on character homology, and these will be considered *seriatim* below.

A. Structure as the Locus of Homology

Advocates of morphology or structure as a guide to homology have suggested that the locus of homology for nonstructural characters lies in the link between these types of characters and structure. The two most commonly used nonstructural characters are functions and behaviors, and while these two classes of characters have been viewed as labile and difficult to homologize, structural features themselves have not been seen as posing problems for applying the concept of homology.

Behaviors and functions, then, might be considered homologous to the extent that we can assign an identifiable structural correlate to these behaviors and functions. The homology of nonstructural traits depends wholly on our ability to locate homologous structures. Structures themselves are often held to be homologous on the basis of an analysis of details of those structures (such as histology, the relative position of joints, tubercles on bones, or muscle attachments).

The quotations summarized in Table I encapsulate the sentiment that dominates the literature on the importance of structural data as a basis of homology. Not only is the concept of homology often restricted to morphology (or to the morphological basis of nonstructural characters), but the methods for recognizing homologies are frequently limited to the investigation of morphological data. To be sure, a number of authors have been explicit in their recognition that structures are not the only organismal traits that might be considered homologous (Baerends, 1958; Gans, 1985; Ghiselin, 1976; Greene and Burghardt, 1978; Hinde and Tinbergen, 1958; Mayr, 1969) but even in recent treatments structure is given the dominant role in analyzing homology (e.g., Wagner, 1989).

The strongest advocates of the primacy of structure in studying the homology of traits have been Atz (1970) and Hodos (1976), and much of the debate on the application of the concept of homology to behavior has centered on the role of morphology (Lauder, 1986; Wenzel, 1992). Atz (1970) argued that the lability of behavior and function make identifying such traits difficult, and that homology is thus best applied to morphology (Table I): structure alone should be the guide to comparing behavior across species. Hodos managed to divorce the concept of behavioral homology from phylogeny (and tie it to structure):

> The degree of phyletic relatedness is not the issue in behavioral homology ... the issue is the degree to which behavior can be related to specific structural entities. (Hodos, 1976, p. 161.)

The use of a morphological foundation for the study of behavioral evolution can also be found in Simpson (1958), who used a morphological criterion to maintain that one would not expect to find homologous behaviors between taxa such as arthropods and vertebrates. As these two groups possess nonhomologous skeletal systems, how could behavior be homologous:

> Divergence and lack of homological behavior between insects and vertebrates are again illustrated, for the external skeleton-internal muscle apparatus of an insect obviously had a different origin from the internal skeleton-external muscle apparatus of a vertebrate. (Simpson, 1958, p. 509.)

Today we might note as an aside that many of the genes involved in coding for muscle proteins would be considered homologous between insects and vertebrates, and that an homologous morphological substrate does in part exist. But arguments for the primacy of structural homology in the evaluation of other types of characters (behaviors and functions) are still common.

For Atz (1970) and Hodos (1976) the relationship of behavior to a morphological substrate is of such primacy that phylogenetic evidence for or against homology is subservient to the message from structural analysis. Wagner (1989) has continued this theme, separating homology from phylogeny by defining homology in terms of structure and development:

> Structures from two individuals or from the same individual are homologous if they share a set of developmental constraints, caused by locally acting self-regulatory mechanisms of organ differentiation, (Wagner, 1989, p. 62.)

This definition of homology recalls Owen's pre-Darwinian formulation, and detaches the concept of homology from phylogeny by placing the locus of homology firmly in a structural and developmental setting.

B. The Nervous System as the Locus of Homology

The nervous system has played such a dominant role in discussions of the homology of nonstructural characters such as function and behavior that I treat it as a separate class of the structural criterion for homology described above. Behavioral biologists, in particular, have commonly referred to the nervous system as a locus of homology (Lauder, 1986). For example, in discussing how one might recognize homologous behaviors, Pribaum (1958, p. 142) says that "...uncovering a behavioral process which ... is shown to depend on homologous neural structures provides a valid criterion useful in a taxonomy of behavior...," a sentiment repeated by Hodos (1976, p. 163), who stated that "Behaviors associated with brain structures that have a common genealogical history are homologous, whether or not the behaviors are of the same type or serve the same function to the animal." Baerends (1958, p. 409) argues that the pattern of motor output from the central nervous system is critical to judgment of behavioral homology: "Our considerations lead to the conclusion that in comparative ethology it is most essential for homology that the patterns of muscle contraction should be largely identical."

There are a number of rather severe difficulties with using nervous system structure or function as the deciding criterion for behavioral homology (Lauder, 1986).

First, homologous neural structures may produce a wide variety of patterned motor output, all of which involve the same structural circuits regulated by neuromodulators (see Harris-Warrick et al., 1992; Meyrand and Moulins, 1988a,b). If protraction and retraction movements of an appendage are produced by the same (homologous) structural neural circuit in two species as a result of modulation by the presence or absence of a neuropeptide, should we insist that these two behaviors are homologous?

Second, identifying structural similarities in the nervous system is difficult; for the vast majority of behaviors we have no idea of the underlying neural pathways. Even identifying the individual neurons in a similarly designed neural circuit in two species may be a challenging task, requiring information on synaptic connections, physiological activity of the neuron, and response of the neuron to modulators (Katz and Tazaki, 1992). Indeed, the whole notion of what a neural circuit is has become more vague as the flexibility of design in circuitry underlying behavior has become apparent: neural circuits are much more than just the pattern of structural interconnections among nerve cells (Getting, 1988; Harris-Warrick, 1988). Multiple circuit designs may generate similar behaviors, and components of circuits may come and go (phylogenetically, ontogenetically, and as a result of changes in external conditions such as light and temperature) while the behavior produced remains similar.

Third, the concept of a neural criterion for behavioral homology depends on a tight link between historical changes in the structure of the nervous system and in behavior. There is now good evidence that such a linkage does not necessarily exist (Kavanau, 1990; Lauder, 1986, 1991; Paul, 1991; Striedter and Northcutt, 1991) and that there may be considerable divergence in circuit morphology, motor output, and behavior within a clade. If there is no necessary link between changes in the nervous system, physiology, and behavior, then what basis is there for assigning the locus of behavioral homology to the nervous system?

C. Developmental Patterns as the Locus of Homology

Ontogeny has long held a fascination for biologists because it is one of the few ways we have to observe structural transformation directly. Perhaps the accessibility of structural change has contributed in part to the view that a proposed homology between two structures is best examined by studying development: in order for two structures to be homologous, they must share developmental patterns or be derived from a common embryological tissue source. By a study of ontogeny we can compare the transformation of two structures; if the transformational patterns are similar we may conclude that the structures are homologous.

Roth (1984), for example, stated unequivocally her view that developmental pathways are the means by which homologous characters should be recognized:

> ... the basis of homology in the broad sense is the sharing of pathways of development,.....A *necessary* component of homology is *the sharing of a common developmental pathway.* (Roth, 1984, pp 13 and 17, emphasis Roth's.

Katz and Tazaki (1992, p. 227) state that "There are a number of ways of demonstrating homology. The most rigorous is to show that two structures have the same embryological origin ...." A similar view on the importance of development has been argued by, among others, Riedl (1978) and Wagner (1989, p. 62), who suggested that homologues are "developmentally individuated parts of the phenotype."

In 1988, Roth reversed herself and concluded that development may not be such a good guide to homology, using examples from de Beer (1971) to argue that "homologues....do not always develop in a similar fashion" (Roth, 1988, p. 5), and that both the embryological processes and precursors of characters that should be considered homologous may differ. A number of authors have discussed the problems associated with a developmental definition of homology (e.g., Rieppel, 1988), and examples of such difficulties are provided in Hall (1992) and Raff *et al.* (1990).

If adult phenotypic features in a group of related species (that to all appearances possess structure and patterns of connection similar to surrounding structural elements) differ in their pattern of development, what utility is there in assigning priority to developmental data as the locus of homological information? Similar phenotypes may result from a diversity of ontogenetic processes. Perhaps the specifics of developmental patterns used to generate an adult structure are not necessarily closely linked to the form, function, or history of the structure itself. In such cases, we would expect the history of developmental pathways in a clade to be different from that of the resulting phenotypic trait.

D. Genetic Data as the Locus of Homology

The notion that homology is a relationship written in the genes is a tempting one, given that the genetic material does contain in coded form at least part of the information needed to generate phenotypic traits. As a result, genetic data (usually gene sequences) have been advocated as a locus of phenotypic homology. Hickman *et al.* (1988, p. 114) state that "The best criterion of homology would be the identification of homologous genes." Roth (1984, p. 18) states that "...I would reject a suggestion of homology if structures are created by unrelated sets of genes." Thus, if we wished to determine if two structures are homologous we could locate the genes that code for proteins and regulatory factors in the development of those structures and examine the gene sequences.

Aside from the obvious practical difficulties in applying this research program for analyzing proposed homologies, we face the difficulty of determining if the genes themselves are homologous, using any of the above *a priori* criteria, even assuming that we can identify the correct genes. At this stage, most of the previously mentioned criteria break down; we cannot study the ontogeny of a base pair, or analyze its structure.

There are additional difficulties with ascribing the locus of homological relationship to genes: similarity in genetic sequence is no guarantee of similarity in phenotypic traits, due to the intervening complexities of epigenetic interactions and regulatory genes and pathways. In addition, quantitative genetic analyses in combination with selection experiments demonstrate that similar phenotypic end–products of selection for a specific trait (such as tail length) may be produced by different evolutionary pathways, resulting in different patterns of genetic correlation among traits in replicate lines (Rutledge *et al.*, 1974). Lenski (1988) showed that resistance to T4 virus infection in replicate lines of *Escherichia coli* may be achieved via different genetic changes: the similar phenotype of T4 resistance did not reflect similarity in underlying genetic alterations producing resistance. There is thus no tight link between patterns of genetic change and phenotypic diversity, rendering the identification of a genetic locus of homology problematical.

If we do not wish to subscribe to the idea that direct genetic coding for structures or developmental pathways is the locus of homology, perhaps a more general conception of the information content of the genome might do. Van Valen (1982) suggested that homology was "continuity of information" and this definition has been advocated by Roth (1988, p. 2) who applauded its flexibility: "the definition can be used by adherents to any school of thought by simply specifying the relevant kind of information." But this very flexibility means that it is virtually impossible to apply the definition. How are we to judge the information content of two bones in different species? Might this not be especially difficult if we accept that these two bones could have been produced by different developmental processes or even by genes (some of which might differ between the species) with differing patterns of genetic covariance and pleiotropic effect?

E. Connections among Traits as a Locus of Homology

The "principle of connections" has a strong pre-Darwinian pedigree that dates as a formal postulate from the writings of Geoffroy Saint-Hilaire (Appel, 1987), although as a guide for studying the relationships among traits the principle of connections is much older and was advocated by Goethe as a method of comparison (Rieppel, 1988). The idea that it is the pattern of connections among traits that is useful in determining homologies has been advocated by several recent authors, notably Remane (1952) whose criteria for homology have been discussed extensively (e.g. Riedl, 1978). According to Remane

(1952), two structures are homologous if they share similar patterns of connection to other structural elements. Connectivity among structures (by nerves, muscles, bone sutures, ligaments, etc.) is what reveals homology. Even though structures may move relative to one another during ontogeny, they often retain a pattern of connection by dragging with them their innervations and blood supply. The principle of connections has been viewed by several recent authors as an important *a priori* criterion for establishing homology (e.g., Beer, 1980; Golani, 1992; Jardine, 1969; Shubin and Alberch, 1986; Tyler, 1988).

One difficulty with the principle of connections has been establishing a frame of reference within which to analyze interconnections among traits. As Rieppel (1988) shows, if the frame of reference is changed, the pattern of connections among elements changes, leading to a different hypothesis of homology. In order to specify that muscles in two species are homologous because they attach to similarly positioned bones, we need to assume that the frame of reference provided by the bones is both absolute and the reference that serves as a locus of homology. Since no criteria exist for choosing a particular reference framework, there is no nonarbitrary way to decide on homology by connections. Furthermore, all the difficulties with homologizing neural circuits discussed above apply in force to the principle of connections. The implications of particular patterns of connection among neurons in circuits remains unclear, and defining an appropriate framework even to identify a "similar" connection is a daunting task (Harris-Warrick *et al.*, 1992). Indeed, evolutionary patterns in the nervous system may provide the clearest examples of both the difficulty in applying the principle of connections, and uncertainty over the significance of similar patterns of connection (Arbas *et al.*, 1991; Dumont and Robertson, 1986; Striedter and Northcutt, 1991).

F. Synthesis

As is apparent from the discussion above, recent literature on homology is replete with individual authors searching for some key characteristic or combination of characteristics that will allow them to establish homology among characters and test hypotheses of homology. Much of this literature can be summarized in the form of a multiple choice statement: homologous characters may be (recognized/defined — pick one) by reference to (structure, the nervous system, development, genetics, connections to other characters — pick one or two). The typical approach has been to choose one level of analysis, usually at a level of organization different from the traits under consideration, as the locus of information on the homology of characters in question.

Adoption of one class of data as the key to identifying homologous characters may stem from a desire to have a means of assessing homology independent from phylogeny (e.g., Beer, 1980; Golani, 1992; Goodwin, 1984; Wagner, 1989; also discussed in Donoghue, 1992). If, however, through a detailed examination of two characters by studying their genetics, development, and patterns of connection to other structural features, we could make a determination of probability of homology, then a phylogenetic analysis would be unnecessary: sufficient information would be available from a detailed study of the two traits alone.

If a nonphylogenetic view of homology seems odd nearly 140 years after the publication of *The Origin of Species*, it is increasingly common. Golani (1992), for example, in his analysis of comparative patterns of movement in animals, suggests explicitly that a definition of homology is needed for the comparative study of behavior that avoids reference to phylogeny, and argues that homology is best demonstrated through an analysis of the pattern of connection among behavioral traits in a movement sequence.

On the other hand, while many authors who advocate one particular locus of homology have recognized the importance of common ancestry to the recognition of homology among characters, in practice, phylogenetic considerations often play a small part in analyses of homology: it is *a priori* criteria that are frequently the focus of discussion when a decision is needed on the possible homology of two characters.

It is my view that there is no locus of homology, no class of data or method of examining the details of construction of individual characters that will reveal whether the relationship between two traits is homologous or analogous. Furthermore, I argue below that in order to determine if two characters are homologous, one must analyze the phylogenetic distributions, not only of those characters, but also of many other characters to provide the phylogenetic basis and historical hypothesis specifying the relationships among taxa necessary to interpreting character evolution.

IV. HOMOLOGY AND PHYLOGENY

A. *Phylogeny, Taxa, and Characters*

In this section I present a phylogenetic definition of homology and show how phylogenetic methods can be used to recognize homologous and analogous characters. One consequence of this phylogenetic approach is that *a priori* criteria for examining characters discussed above are relegated to the role of refining observations of similarity among characters and thus our proposals of homology: they do not allow tests of homology or the recognition of homology.

A second key feature of a phylogenetic analysis of homology is that, while the word *homology* may describe a relationship among two (or more) specific characters, the demonstration of that homology requires analysis of many other characters, unrelated to those of immediate interest, because possession of a trait is the property of a taxon. It is the phylogenetic relationships among *taxa* that allow us to assess the homology or analogy of individual *characters*. The criteria discussed above thus differ fundamentally from the phylogenetic approach considered here in focusing on the characters themselves rather than on the taxa possessing them.

Hypotheses of homology may be generated in a large number of ways and the supposition of homology between two characters is logically independent of a phylogenetic test of hypotheses of homology. Although hypotheses of homology are usually based on some prior knowledge of structure or development (and are often founded on "similarity" at some level, recognized by the criteria discussed above), this need not be so. A biologically uninformed observer might easily propose that the wing of an insect is homologous to the wing of a bat, or that the wing of an insect is homologous to the tail of a fish. Either of these hypotheses may be tested phylogenetically to determine the homology (or not) of the characters.

The phylogenetic definition of homology advocated here is based on the approach of Patterson (1982), who suggested that homologous similarities are those that define natural or monophyletic groups of organisms. If all five species in a clade hypothesized to be monophyletic (on the basis of a phylogenetic analysis of many characters) possess a particular developmental pathway not present in outgroup taxa, then it is most parsimonious to conclude that these species share common development by descent, and that this developmental pathway is homologous in the five species. The similar pattern of development thus is just one of many characters that suggest that these five species are a natural monophyletic taxon (see Brooks and McLennan, 1991; Eldredge and Cracraft, 1980; Fink, 1988; Wiley, 1981).

A character such as a "wing" is not homologous in insects and birds because the character "wing" is not corroborated by other evidence for the monophyly of a taxon that includes birds and insects. That is, there are very few other characters that these two taxa share uniquely that could be used to support the clade insects + birds as a natural taxon. Instead, there are many other characters that suggest a better corroborated phylogenetic hypothesis indicating that birds and insects are each more closely related to other taxa than they are to each other. On this basis, then, we conclude that wings are not homologous in insects and birds.

Figure 1 provides a schematic summary of a phylogenetic approach to homology and analogy. Taxa E, F, G, and H are considered to be a monophyletic clade on the basis of the evidence used to construct the cladogram (for example, a molecular phylogenetic analysis of proteins). Under the pattern of relationships shown in the upper panel, organismal traits 1 and 2 would be considered homologous in taxa E, F, G, and H since it is most parsimonious to assume that the common ancestor of these taxa possessed traits 1 and 2, which then provide further evidence corroborating the natural taxon E, F, G, and H. Trait 3, present in taxa A and G, would not be considered homologous because this trait is incongruent with all the other evidence that suggests that taxa A and G are not each other's closest relatives.

However, if a further study of the relationships of these taxa, perhaps by sequencing more proteins and by combining a previously gathered morphological data set with the molecular data, resulted in a different hypothesis of relationships among the 10 taxa shown (Fig. 1B), then we would have to reevaluate our conclusions about the homology of traits 1, 2,

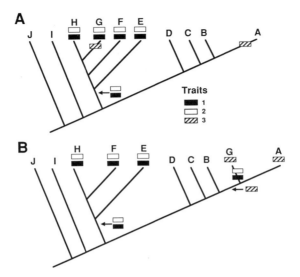

Fig. 1. The phylogenetic definition of homology. (A) Traits 1 and 2 are present in taxa E, F, G, and H and provide evidence that these taxa form a monophyletic clade (arrow). Character 3 is found only in clades A and G. While character 3 does provide evidence that taxa A and G are closely related, this evidence is overwhelmed by the other characters used to make the cladogram in the first place. Each character provides evidence of monophyly for some clades, but it is the distribution of all characters considered together that determines the most parsimonious branching topology. Note that cladograms shown in (A) and (B) are based on a phylogenetic analysis of many additional characters not depicted here. Thus, evidence supporting the branching topology is independent of characters 1, 2, and 3 under discussion here, although these characters could also be included in a phylogenetic analysis of all available data. (B) A suite of new characters has been discovered and a new phylogenetic analysis of all characters now shows that taxa A and G are now considered to be each other's closest relatives. Trait 3 is thus reinterpreted as a homology in taxa A and G and provides evidence that these taxa are a monophyletic clade (arrow). Traits 1 and 2 are still homologous in taxa E, F, and G, but are convergent with traits 1 and 2 in clade G.

and 3. Now, we have increased the amount of evidence we have to support a phylogenetic analysis of the relationships among the taxa and find there is evidence to corroborate taxa A and G as a monophyletic clade. Trait 3 thus contributes to this evidence and would be most parsimoniously interpreted now as homologous between taxa A and G. On the other hand, traits 1 and 2 in taxon G no longer corroborate taxa G, E, F, and H as a monophyletic clade: other evidence outweighs these two characters. Traits 1 and 2 are now considered to be nonhomologous between taxon G and these other taxa, but still would be homologous in taxa E, F, and H. The identification of both analogous and homologous characters thus depends on our estimate of the relationships among taxa. If we were unable to estimate relationships among the 10 taxa shown in Fig. 1, then under a phylogenetic approach to homology and analogy we would be unable to make any statement about the homology of characters present in a subset of those taxa.

A key point in the phylogenetic approach to homology is that the homology or nonhomology of two or more traits does not depend on their similarity to each other (although the initial choice of a character may often depend on a perception of its similarity to another character in a different taxon). A character in two taxa may be very similar structurally, similar in development, have a similar genetic basis, and be similar in function, but might still be nonhomologous (e.g., character 2 in Fig. 1B). It is the relationships among taxa as estimated by an analysis of all the evidence available that allows us to ascertain the homology of characters in those taxa.

B. Phylogeny, Homology, and Hierarchy

The hierarchical organization of biological systems has been discussed by many workers (Allen and Starr, 1982; Brooks and McLennan, 1991; Brooks and Wiley, 1986; Gould, 1982; Lauder, 1981; Salthe, 1985; Vrba and Eldredge, 1984) and the implications of a hierarchical view for analyses of homology have received considerable attention. Some authors have asked if the property of homology resides at any one hierarchical level (Roth, 1991). Others, embracing a hierarchical approach, have questioned whether we can analyze traits across levels by using a single methodology:

> ... methods — such as phylogenetic trees, cladograms and homologies — used for the study of phenomena at one level (say morphology) are generally not applicable to higher-level phenomena (say behavior), (Aronson, 1981, p. 37.)

It is my view that one great benefit of a phylogenetic approach to homology is its direct application to organismal traits of many kinds. Not only are phylogenetic methods applicable to behaviors, developmental sequences, and functions, but some of the most interesting questions in comparative biology arise when we explicitly consider patterns of character homology across hierarchical levels (Lauder, 1990, 1991).

For example, consider the hierarchical arrangement outlined in Table II. If we wish to analyze behavioral traits in several species, we might choose to study a sequence of display behaviors. In order to quantify the display behavior, we could analyze the precise pattern of movements of the head and body during display by measuring kinematic patterns (from films) such as bone excursions, velocities of movement, etc. In addition, we could examine by dissection of preserved individuals the topographic arrangement of muscles and bones of the structures used in the display behaviors. We could also record electrical activity from relevant muscles and thus quantify the pattern of motor output used to generate the behavior. Finally, we might undertake a study of the neuronal circuitry involved in producing the display behavior. While an investigation of all these levels in several species is probably beyond the capabilities of any reasonably finite study, analysis of a few is not (Arbas *et al.*, 1991; Harris-Warrick and Marder, 1991; Katz and Tazaki, 1992; Lauder, 1986, 1990; Paul, 1981a, b, 1991; Reilly and Lauder, 1992; Striedter and Northcutt, 1991).

Given comparative data on characters from several levels (Table II) and a phylogenetic hypothesis of the relationships among species, we can determine the mapping of homologous characters across levels of the hierarchy. As illustrated in Fig. 2, we might find that a display behavior (which appears to be kinematically similar in all species that possess it) has evolved convergently in two groups of taxa. In this case, the display illustrated by taxon H would not be homologous to that in taxa

Table II. One Possible Hierarchy of Levels (Classes) of Characters That Might Be Analyzed Phylogenetically. [a]

Hierarchical level	Example of an organismal trait that might be studied interspecifically
Behavioral	Display behavior during mating
Functional/physiological (at the level of peripheral tissues)	Kinematics of bone movement; physiological properties of muscles; biomechanical tissue properties
Structural (at the level of peripheral tissues)	Topographic arrangement of muscles and bones; tissue histology
Functional/physiological (at the level of the nervous system)	Neuronal spiking patterns; motor patterns; membrane properties; modulation by neurotransmitters
Structural (at the level of the nervous system)	Neuronal morphology; topology of neuronal interconnection; wiring of sensory and motor pathways

[a] Under a phylogenetic approach to the problem of homology it is possible that any definable pair of traits at each level might be homologous no matter what the nature of the traits is. Thus, there is no particular class of characters that serves as a locus of homology.

A, B, and C, despite the fact that behavior in these taxa is not different when we test statistically kinematic variables measured from films of the behavior. In addition, our study of morphology and motor output allows us to ask how homologous components of the mechanistic basis of display behavior relate to patterns of homology (or nonhomology) in the behavior itself. If we find that taxa B and C share a particular morphology and motor pattern (Fig. 2: morphological and motor pattern trait 2) while taxon A retains the primitive morphology and motor pattern (Fig. 2: trait 1) then we can conclude that taxa B and C possess an underlying mechanistic basis for the display behavior that is not homologous to that in taxon A. That is, the behaviors are homologous within clade A, B, and C but the morphological and motor substrates for the behaviors are not. In fact, according to the pattern shown in Fig. 2, taxon A has retained a primitive morphology and is convergent to taxon H not just in possessing the behavior, but also in the physiological basis for the behavior. Taxa D, E, F, and G possess an homologous morphological framework, but lack characters at the behavioral level: these species never evolved a display although they possess the requisite musculoskeletal structure. Phylogenetic patterns similar to those depicted in Fig. 2 are increasingly being demonstrated as comparative studies of the physiological and neural basis of behavior become more common (Katz and Tazaki, 1992; Lauder, 1990, 1991, 1993; Paul, 1991; Reilly and Lauder, 1992; Shultz, 1992; Striedter and Northcutt, 1991).

Study of the phylogenetic patterns of congruence among classes of characters at different hierarchical levels raises several general issues: 1), do some levels tend to be more conservative than others and show relatively little interspecific variation? 2), are traits at some levels more interspecifically labile and if, so, is this variation correlated with variation at another level?

Many interesting problems in the evolution of organismal design may appear where there is discordance among levels. For example, taxa in a monophyletic group might show homologous phenotypes produced by divergent developmental pathways. Or, taxa may show homologous patterns of muscle activity, but divergent behaviors because of alterations in musculoskeletal topology (Lauder, 1991). Combination of the phylogenetic definition of homology with an analysis of organismal traits at several hierarchical levels allows these issues to be addressed.

Many other types of hierarchical arrangements might be considered. Analysis of an ecological hierarchy (individuals, populations, communities, and biogeographic regions), for example, might provide a basis for examining how homologous ecological characteristics of two taxa relate to traits at other levels such as morphology or life history patterns (Brooks and McLennan, 1991). Or, we might wish to analyze a genetic/developmental hierarchy (Atchley and Hall, 1991; gene sequences, patterns of genetic covariation, epigenetic pathways, phenotypes) to discover how traits at these different levels covary phylogenetically and how homologies at one level map onto homologies at other levels.

While such approaches are still in their infancy (in part because of the view that many classes of characters are not able to be analyzed using phylogenetic methods), I believe that the historical analysis of different types of traits and the con-

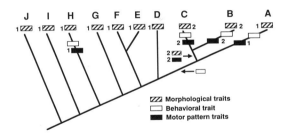

Fig. 2. Schematic illustration of a phylogenetic analysis of hierarchical characters showing how kinematically similar behaviors might be produced by different "underlying" motor patterns and structures. A behavioral trait (a display behavior, for example) is common to taxa A, B, and C and is homologous within these taxa (arrow), while the behavior in these three taxa is convergent with the similar behavior in taxon H. Motor pattern 1 is also convergent between taxa H and A. However, taxa B and C have acquired novelties in motor pattern and morphology (arrow at the internode leading to taxa B and C), so that the behavior homologous to that in taxon A is now generated by nonhomologous morphological and functional characters. Taxon A retains the primitive morphology.

comitant recognition of homologous traits at different levels will lead to the discovery of new and interesting patterns of association among characters that will guide future mechanistic investigations.

C. Phylogeny and Iterative Homology

Although *serial homology* [one form of *iterative homology* (Ghiselin, 1976)] has been an active topic of discussion in the homology literature since the time of Owen (Appel, 1987; Minelli and Peruffo, 1991; Patterson, 1982; Rieppel, 1988), recently Roth (1991) and Wagner (1989) have argued that a "biological homology concept" is needed in part because a phylogenetic approach to homology does not allow iterative traits in organisms to be homologous (Patterson, 1982).

I suggest that a phylogenetic approach to homology can easily deal with the reality of repeated traits within organisms, and thus argue that there is no necessity for a separate biological homology concept. In my view, iterative homology simply refers to homology of one or more developmental processes (or patterns of genetic covariation) at a greater level of phylogenetic generality than the individual organism. To say that cervical vertebra 4 is serially homologous to cervical vertebra 5 in an individual mammal is simply to say that species in the Mammalia share an homologous developmental pathway (or set of pathways) that produces serially arranged phenotypic structures similar in size and shape.

Figure 3 illustrates schematically how one might interpret "serial homologues" among several Recent taxa in a phylogenetic context. Taxon A might be argued on a biological homology concept to possess three serially homologous body segments. But such a statement represents a confusion of phenotypic pattern within an individual with interspecific (phylogenetic) differentiation in developmental/genetic processes. Seen in the context of its phylogenetic relatives, taxon A (Fig. 3) possesses a phenotypic condition of three repeated body segments that results from sharing an homologous novelty in developmental/genetic pattern with taxon B. Similarly, taxa A, B, and C share another (homologous) developmental/genetic novelty that gave rise to repeated segments in comparison to outgroup taxa D and E. At an even more general level, taxa A to E possess an homologous appendage on the first body segment due to novelties in development that arose in the lineage leading to the monophyletic clade A to E. The phenotypic condition of individuals in taxon A ("serial homology") thus results from a nested set of derived homologous developmental and/or genetic characteristics. It is not particularly enlightening, then, to refer to an individual in taxon A as possessing serial homology, as the components of the phenotypically repeating pattern have different phylogenetic histories that characterize increasingly inclusive monophyletic clades of taxa.

A less schematic example illustrating the above point concerns vertebrate pectoral and pelvic appendages. On the basis of a phenotypic analysis of extant tetrapod taxa one might wish to conclude that forelimbs and hindlimbs are

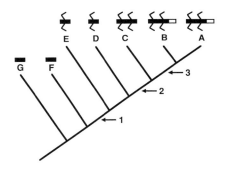

Fig. 3. Schematic illustration of how the phylogenetic definition of homology may be applied to serially homologous structures within a taxon. "Serial homology" of structures within individuals of taxon A actually reflects nested phylogenetic sets of homologous developmental and genetic novelties that define monophyletic clades. Three are shown here. Outgroup taxa (G and F) possess a single body segment with no appendages. Novelty 1: alterations in development generate a bilaterally paired appendage on the body. Novelty 2: acquisition of developmental/genetic programs (perhaps as a result of gene duplication) generate a repeated element of structural design — a second body segment. Novelty 3: further alterations in developmental/genetic processes generate a third body segment with no appendages.

serially homologous: both appendages typically possess a single proximal segment, two middle bony elements, a collection of smaller carpal or tarsal bones, and distal phalanges. Indeed, the phenotypic resemblance between fore- and hindlimbs in tetrapods is striking, and the limbs are similar in developmental characteristics also [suggesting to Roth (1984) that the forelimb and hindlimb are homologous in derived tetrapods].

But a phylogenetic analysis reveals that the pectoral and pelvic appendages in early vertebrates do not possess such detailed similarities in structure. Sharks and ray-finned fishes possess pelvic fin supports that are quite different in morphology from the endoskeletal elements supporting the pectoral fin (Jarvik, 1980). In Amia, for example, pelvic fin supports consist of a single dominant pelvic bone on each side supporting a collection of radial bones and cartilages, while the pectoral fin is supported by an elongate set of proximal radial bones, each of which articulates with a small distal radial and a large medial basal cartilage. In more derived ray-finned fishes, where the pectoral and pelvic girdles may attach to each other, there is even less morphological resemblance (Stiassny and Moore, 1992). In sharks, numerous radials articulate with pterygial cartilages in both girdles, but the number and nature of these articulations vary between the girdles.

In sharks and primitive ray-finned fishes the existence of repeated radial elements in each appendage may indicate some general level of common developmental/genetic programs for development, but these commonalities do not extend to the details of appendage construction itself. Ray-finned fishes, on the basis of the phenotypic evidence, would appear to have even less in common between pectoral and pelvic appendages.

With the origin of tetrapods and the common functional role for fore- and hindlimbs in terrestrial support and locomotion, however, developmental/genetic links between fore- and hindlimbs appear to have arisen that constitute a homology for tetrapods. The precise nature of the common developmental and genetic mechanisms that give rise to similar phenotypes in tetrapod limbs is not known, but there is ample quantitative genetic evidence (reviewed in Lande, 1978) to suggest that limb design is polygenic and that "limb genes" have many pleiotropic effects on other morphological features; see Chapter 6. It is not structural similarities in the fore- and hindlimbs of a single individual that constitutes an homology, nor is the similarity in developmental mechanisms between fore- and hindlimbs evidence of "serial homology." Rather, the homologous traits within tetrapods are the shared developmental programs and patterns of genetic covariance that characterize a monophyletic clade of taxa, the Tetrapoda.

On the basis of the comparative anatomy of the limbs in lungfishes, "rhipidistians," and early amphibians it is likely that the developmental homologies and patterns of genetic covariation that characterize tetrapod limb development arose in a series of stages and not suddenly with the origin of terrestrial life. The pelvic and pectoral appendages of *Eusthenopteron*, while similar in general pattern, possess several distinct structural features (Jarvik, 1980), while those of early amphibians show greater structural similarities in possessing a single proximal element and two larger elements that articulate distally (Coates and Clack, 1990).

Lungfishes, primitive ray-finned fishes such as sturgeons, and sharks, all possess pectoral and pelvic fins with a general structural plan in common, but only lungfishes, "rhipidistians," and tetrapods appear to share a common pattern between the two appendages of having a single proximal element articulating with the supporting girdle, and a greatly reduced number of distal elements articulating with that proximal element. A model of increasing common genetic control of both appendages prior to the origin of tetrapods at least is consistent with the phenotypic evidence, suggesting that developmental and genetic covariation between pectoral and pelvic appendages existed in taxa that were primarily aquatic, prior to the origin of tetrapods.

This scenario suggests an experiment: test for common patterns of development and genetic control (via quantitative genetic and selection experiments) in taxa such as fishes, where crosses can be easily made, and compare the covariance between fore- and hindlimb elements with the results of similar experiments in tetrapods. The prediction is that a much higher degree of correlated response to selection will be found in tetrapod limbs than in the appendages of fishes, reflecting the homologous developmental/genetic control mechanisms that may characterize many tetrapod clades.

V. HOMOLOGY AND FUNCTION

The study of function has not received much attention in systematic and comparative biology, and when it has the focus has tended to be negative (Lauder, 1990). Physiological functions have been viewed as labile, of little utility in comparative studies, and difficult to define and thus analyze. Table III provides a sampling of literature sentiment on the possibility of studying functional homology, and the consensus is not optimistic. One key issue that underlies most views of function is the proposed intimate relationship with structure: "Functions are clearly inextricably tied with the structure of the features that perform them" (Tyler, 1988; Table III).

One difficulty with this view is that functions may be tied to structure at a different hierarchical level than one might at first think (Lauder, 1990, 1993). For example, if we consider as a possible physiological function the pattern of electrical activity in a muscle, the structural substrate for this function might lie in the connections among neurons in the spinal cord, or it may be more abstract than that, lying instead in the membrane properties and sodium channels of nerve cells in the circuit (Table II). Given the increasingly abstract conceptualizations of what constitutes a structural circuit in neurobiology (Getting, 1988; Harris-Warrick *et al.*, 1992), it is difficult to understand how we can tie many functions to a structural substrate in a simple fashion. I have criticized elsewhere the view that functions are necessarily more labile than structures (Lauder, 1990, 1991). Statements on the lability of function are most commonly based on our preconceived notion that structure is solid, repeatedly observable, and definable, rather than on quantitative analyses of interspecific patterns of both structure and function.

The purpose of this section is to develop a brief case study of the evolution of structure and function to illustrate the issues discussed above. In particular, this case study of muscle morphology and function in ray-finned fishes (Actinopterygii) will be used to depict (1) the initial proposal of homology, (2) the determination of the homology of muscles, (3) an analysis of muscle function and of functional homology, and (4) phylogenetic congruence between structural and functional characters and the implications of congruence (or the lack thereof) for the recognition of homology. I will use the phylogenetic definition of homology outlined above as a tool to

Table III. Selected Statements from the Literature Commenting on the Possible Homology of Organismal Functions

Reference	Comment on organismal function and homology
1. Haas and Simpson (1946, p. 323)	"Functions, considered as abstractions and without consideration for the structures that perform these functions, should not be spoken of as homologous."
2. Atz (1970, p. 60)	it is "...impossible for nonhomologous structures to have homologous functions."
3. Riedl (1978, p. 248)	"The contraction of a biceps naturally stands beyond the limit of homology."
4. Ross (1981, p. 2157)	"Certainly, functions have phylogenetic histories but it seems that the rules are different from those which governed the patterns of evolution depicted in the familiar phylogenetic trees."
5. Tyler (1988, p. 344)	"Homology applies most appropriately to the structural features, not their functions."
6. Burggren and Bemis (1990, p. 197)	"One reason why physiologists have problems with homology is that there are no easy means to assess the homology of the quantitative features of greatest familiarity and interest to physiologists except by reference to the morphological substrates of these functions."
7. Wake (1991, p. 323)	"... many workers have difficulty accepting functional characters, both because they believe that the underlying morphology must be sought, and that morphology then provides the characters appropriate to phylogenetic analysis and because functions are viewed as associations of several potential characters ...."

examine patterns of evolution at different hierarchical levels, and to suggest new research questions in the evolution of muscle structure and function.

A. Case Study: The Evolution of Muscle Function in Ray-Finned Fishes

In 1973, D. E. Rosen, elaborating on the previous work of Holstvoogd (1965) and Nelson (1969), adduced evidence that the Neoteleostei constituted a monophyletic clade of teleost fishes. Rosen's proposal of monophyly for this clade was significant because the neoteleost fishes comprise more than half of the 24,000 species of teleost fishes, and evidence of monophyly for such a large clade represented an important step in understanding the phylogeny of fishes (Lauder and Liem, 1983). One of the characters used by Rosen (1973) was the presence of the retractor dorsalis muscle (RD) in the pharyngeal region of neotelosts. This muscle is proposed to be homologous within neotelosts (Fig. 4A) because of its consistent origin from the vertebral column and insertion on one or more of the upper pharyngeal bones (Fig. 4B): i.e., because of similarity in position and connection to surrounding elements. Because a number of other characters also corroborate the Neoteleostei as a monophyletic clade (see Johnson, 1992; Lauder, 1983b; Lauder and Liem, 1983; Stiassny, 1986), I consider the presence of a retractor dorsalis muscle to be homologous within neoteleosts (Fig. 4A) under a phylogenetic definition of homology.

However, as has been noted by a number of workers (e.g., Allis, 1897; Nelson, 1969), several other clades of ray-finned fishes also possess a muscle that appears to be very similar to the neoteleost retractor dorsalis. For example, gar (*Lepisosteus,* in the Ginglymodi) and bowfin (*Amia,* in the Amiidae) also possess a muscle that takes its origin from the vertebral column and inserts on the upper pharyngeal jaw bones (Fig. 4C). On the basis of a criterion of similarity of structure and connection to surrounding elements, one might propose that the retractor dorsalis muscle in *Amia* and *Lepisosteus* is homologous to that of neoteleost fishes. But this hypothesis is refuted by the host of other characters supporting a phylogeny in which the clades containing *Amia* and

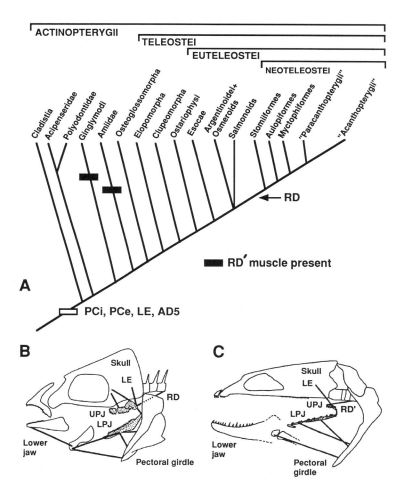

Fig. 4. (A) A greatly simplified diagram of phylogenetic relationships of the major clades of ray-finned fishes to illustrate the independent evolution of a dorsal retractor muscle (RD, arrow; and RD', black bars) in the Neoteleostei, Amiidae, and Ginglymodi. Note that many Recent and fossil clades have been omitted for clarity. A number of pharyngeal muscles are primitive for ray-finned fishes, including the pharyngocleithralis internus and externus (PCi and PCe), the external levators (LE), and the fifth branchial adductor (AD5) (Lauder

and Wainwright, 1992). A dorsal retractor has also been found in several other ray-finned fish species, although the phylogenetic position of these taxa does not affect this discussion of the convergent retractor muscle. (B) Schematic illustration of the skull in a derived ray-finned fish such as a member of the sunfish family Centrarchidae (in the "Acanthopterygii"). The RD muscle originates on the vertebral column and inserts on the upper pharyngeal jaw (UPJ). (C) Similar diagram of Amia to show the convergent retractor muscle (RD') extending from the vertebral column to the upper pharyngeal jaw. LPJ, Lower pharyngeal jaw.

Lepisosteus are not closely related to neoteleost fishes (Lauder and Liem, 1983; Nelson, 1969; Wiley, 1976).

This phylogenetic interpretation is illustrated in Fig. 4A, in which the neoteleost RD is considered as a homology for that clade, while Amia and Lepisosteus are labelled as possessing a nonhomologous dorsal retractor muscle, labelled RD'. It is highly unparsimonious to conclude that the "dorsal retractor muscle" in Amia, Lepisosteus, and neoteleosts is homologous. Under that scenario, all the characters that neoteleosts share would have to have been lost in Amia and Lepisosteus, and all the characters shared by the nested sets of clades within the Teleostei that branch off prior to the Neoteleostei (Fig. 4A) would also have to have been lost in Amia and Lepisosteus.

There are approximately seven extant species of gars in the Ginglymodi (Wiley and Schultze, 1984), and I interpret the RD' as homologous among these species. However, it is not possible to decide if the RD' is homologous in Amia and Lepisosteus (Lauder and Wainwright, 1992): the RD' muscle might have arisen once below the Ginglymodi and been lost in teleost fishes (two evolutionary "steps," in which case the RD' would be homologous in Amia and Lepisosteus) or it might have originated independently in these two clades (also two steps, indicating that the RD' is not homologous in Amia and Lepisosteus).

Given this structural pattern, what might we find if the function of the RD, RD', and surrounding muscles is investigated? Does structural homology imply functional homology, and thus that the RD and other muscles considered to be homologous within neoteleosts will possess homologous functions? Are some levels of the hierarchy illustrated in Table II more phylogenetically conservative than others? In order to propose functional homologies, we need to investigate the function of pharyngeal muscles in several ray-finned clades. By recording muscle activity patterns, using the technique of electromyography, a precise description of when each pharyngeal jaw muscle is active relative to others can be obtained (Lauder, 1983a,b; Lauder and Wainwright, 1992).

Figure 5 illustrates representative patterns of pharyngeal muscle activity in derived neoteleosts from the Family Centrarchidae (Fig. 5A and B: *Lepomis, Ambloplites*) and from *Lepisosteus* and *Amia* (Fig. 5C and D). Note that the RD' muscle possesses a grossly similar activity pattern in *Lepisosteus* and *Amia* in that the bulk of RD' activity occurs following activity of the AD5 muscle. In *Amia*, however, the levator externi muscles (LE3/4) are active during swallowing while homologous muscles in *Lepisosteus* are not (the levatores externi muscles are primitively present in ray-finned fishes). If the pattern of muscle activity in a neoteleost such as *Ambloplites* is compared to that of *Amia*, the levatores externi and retractor muscles appear to be active at similar relative times (Lauder and Wainwright, 1992).

Recordings such as those shown in Fig. 5 (also see Lauder, 1983a,b, 1993; Lauder and Wainwright, 1992) for a variety of ray-finned fish clades suggest the following three representative functional characters for consideration. (1) Activity of the posterior levator muscles originates prior to and significantly overlaps activity in the retractor muscle. (2) Onset of activity in the fifth branchial adductor (AD5) occurs prior to onset of activity in the retractor muscle during a swallowing cycle. (3) Activity in the pharyngocleithralis externus (PCe) muscle is present during prey swallowing and significantly overlaps activity in the retractor muscle.

Examining the phylogenetic distribution of these functional characters on the simplified phylogeny of ray-finned fishes shown in Fig. 4 indicates that three conclusions regarding the homology of pharyngeal motor patterns can be drawn.

First, *Lepisosteus* and *Amia* differ in the timing of activity of the external levator muscles with respect to the RD' muscle. *Amia* has a motor pattern similar to that in sunfishes, while *Lepisosteus* shows little activity in the posterior levators during swallowing. Under a phylogenetic definition of homology, this aspect of the motor pattern in *Amia* is convergent (anal-

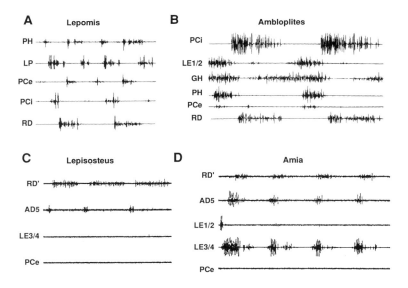

Fig. 5. Patterns of muscle activity in four taxa of ray-finned fishes. (A) and (B) show motor patterns in two genera of sunfishes (Centrarchidae: Acanthopterygii). (C) and (D) show muscle activity in two nonteleost ray-finned fishes: *Lepisosteus* (Ginglymodi) and *Amia* (Amiidae). The RD' muscle in *Amia* and *Lepisosteus* is convergent with the RD of *Lepomis* and *Ambloplites*. Abbreviations: AD5, fifth branchial adductor; GH, geniohyoideus; LE, levator externus muscles (either 1 and 2, or 3 and 4); LP, levator posterior muscle, one of the external levator series; PCe, pharyngocleithralis externus; PCi, pharyngocleithralis internus; PH, pharyngohyoideus; RD, retractor dorsalis muscle of neoteleostean fishes; RD', the dorsal retractor muscle of *Amia* and *Lepisosteus*.

ogous) to that of neoteleosts, while *Lepisosteus* is uniquely derived in this character. Thus, despite structural homology of the external levator muscles in ray-finned fishes there has been considerable evolutionary differentiation at a functional level.

Second, both *Lepisosteus* and *Amia* possess relative timing of the AD5 and RD' muscles similar to that between the AD5 and retractor dorsalis (RD proper) in neoteleostean fishes.

Phylogenetically, then, these taxa are convergent in sharing a similar (but not homologous) motor pattern.

Third, *Amia* and *Lepisosteus* share a lack of activity in the PCe muscle during swallowing, and are divergent from neoteleostean fishes which possess such activity. However, within the neoteleostean fish family Centrarchidae, the genus *Ambloplites* differs from the other closely related sunfish taxa studied to date in the timing of PCe activity relative to the RD (Fig. 5A and B; Lauder, 1983a,b). Thus, despite structural homology of the PCe and RD within the Centrarchidae, there has been divergence at the functional level.

Comparison of activity patterns in structurally homologous branchial muscles among the three major clades of percomorph teleosts (a large derived neoteleostean clade within the "Acanthopterygii"; Fig. 4A) that have been studied to date (the Centrarchidae, Cichlidae, and Haemulidae) shows that numerous functional specializations have occurred in the activity patterns of homologous muscles. For example, in haemulid fishes, the sternohyoideus muscle is strongly active during prey transport (Wainwright, 1989a,b) while activity is never seen in the sternohyoideus muscle during prey transport in centrarchid fishes (Lauder, 1983a,b). Phylogenetic differentiation in muscle function (motor output) among percomorph fish clades has thus occurred within a structurally homologous muscular framework.

Results of this case study are typical in illustrating the complexity of the relationship between structure and function obtained when one measures functional characters and compares their distribution to associated structural features (Lauder, 1990, 1993). At the same time, in the course of past work on the evolution of muscle function in ray-finned fishes, several extremely conservative functional characteristics have been identified as retained despite extensive reorganization of musculoskeletal topology (Lauder, 1990, 1991; Wainwright and Lauder, 1992). Both structures and functions may be conserved phylogenetically, and both types of characters are capable of extensive transformation.

B. Synthesis: Homology and Function

While Riedl (1978) has suggested explicitly that muscle contraction may not be analyzed using the concept of homology (Table III), I would argue that a definable pattern of muscle activity is an organismal trait just like any structural feature. Just like a structural character, functions may be considered homologous if they characterize a natural, monophyletic clade of taxa. So, the common pattern of activity in the AD5 muscle in *Amia* and neoteleosts is one character providing evidence, along with the presence of a dorsal retractor muscle, that neoteleost fishes and *Amia* are a monophyletic clade. But I reject this hypothesis because there are many other characters that support a different grouping of monophyletic clades — that shown in Fig. 4A. If further characters came to light indicating that the most parsimonious interpretation of all characters bearing on ray-finned fish phylogeny shows *Amia* to actually be a neoteleost, then I would reverse my interpretation of the RD' and muscle activity characters, and conclude them to be homologous between *Amia* and sunfishes. The claim of homology (in this case of a functional character) stands or falls on the basis of the phylogeny as a whole.

A corollary of the phylogenetic approach to analyzing organismal traits is that even if further investigation showed that the RD' of *Amia* and neoteleosts has a very similar pattern of development, and even if it was shown that the genes producing these muscles were homologous and present in all ray-finned fishes, I would still regard the phenotypic structure of a retractor muscle (and related motor patterns) as convergent between *Amia* and neoteleosts.

The approach taken here — examining the distribution of muscle structure and function on a phylogeny of ray-finned fishes — suggests a number of avenues for further study. In particular, if muscle activity patterns are convergent among *Amia*, *Lepisosteus*, and neoteleosts, are the neural circuits that produce the motor pattern similar in structure? In other words, does convergence at one level of the hierarchy of Table II (muscle structure) entail convergence at other levels? In addition, in cases within neoteleosts where structurally homologous muscles are shown to possess analogous (convergent) functional traits, one might ask if the neural circuitry that generates the motor output has been conserved phylogenetically, or which aspects of the circuit have changed — the wiring pattern, or perhaps neuromodulators or membrane properties. Finally, if changes have occurred in muscle structure and function, what does this imply for the behavioral (kinematic) level? Are the observed behaviors generated by analogous muscles and their functions similar too? As I have discussed elsewhere, interspecific changes in musculo-skeletal design and function may still produce identical behaviors and, conversely, a change in behavior may result from changes at any of several hierarchical levels (Lauder, 1991). Understanding the diversity of transformational patterns among hierarchical levels and the mechanistic bases for these patterns is a major challange for the future.

VI. CONCLUSIONS

My goals in this chapter are twofold. First, I wish to demonstrate the prominent role that *a priori* criteria for the recognition of homology have played in the literature. Despite wide recognition that the concept of homology has strong phylogenetic underpinnings, much of the recent homology literature is devoted to considering the extent to which one nonphylogenetic criterion or another provides information on the homology of characters. The promise of such an approach is that the homology or analogy of organismal traits might be determined by a detailed examination of the characters themselves, without reference to the phylogenetic distribution of the characters in ingroup and outgroup clades. There is a clear trend in the literature to divorce the definition and recognition of homology from a phylogenetic basis (Beer, 1980; Golani, 1992; Goodwin, 1984; Roth, 1991).

Second, in the context of presenting a phylogenetic definition of homology and of providing an example of the analysis of structural and functional homology in the musculoskeletal system of ray-finned fishes, I suggest that the search for a "locus" of information on character homology is misguided. Such searches actually provide data on the similarity of characters we might use in proposing an hypothesis of homology. This proposal is then tested via an explicit phylogenetic analysis which examines the distribution of characters on a phylogeny and estimates the relationships of taxa. The result of the phylogenetic analysis alone determines if two characters are homologous. In the absence of a phylogenetic analysis, one can only propose homologies based on character similarity; one cannot test hypotheses of homology. It is interesting to note that, in the literature advocating a particular class of data as a locus of homology, that locus is most often claimed to reside at a different hierarchical level than the traits under consideration. The class of data that is proposed to contain homological information on a phenotypic trait (e.g., development, genetic data, the structure of the nervous system) is often relatively inaccessible and limits the utility of such *a priori* criteria for contributing to an understanding of historical patterns to different types of traits.

The adoption of an explicitly phylogenetic definition of homology leads naturally to the analysis of congruence among hierarchical classes of characters: if no one level or class of traits serves as a locus of homological information, then all may be equally subject to historical analysis. For example, how congruent are behavioral changes in a clade with morphological and physiological features involved in the production of those behaviors? How congruent are developmental pathways and epigenetic associations with the phenotypic end products of those processes? How concordant are patterns of genetic correlation in life history traits among taxa with homologous ecological or trophic patterns? I regard the documentation of patterns of historical concordance and divergence among classes of data, levels of organization, and developmental, genetic, and physiological processes as one of the most exciting areas of investigation in organismal biology.

Since we lack even a modest number of case studies of this type, it is still unclear what patterns will emerge and what kinds of causal hypotheses will be suggested by the results of such phylogenetic analyses of behavior, function, development, and genetics. The analysis of the mechanisms driving patterns of character association is a virtually untapped field that may well be a future focus of comparative biology.

ACKNOWLEDGMENTS

Preparation of this chapter was supported by NSF IBN 9119502 and NSF DCB 8710210. Gary Gillis provided many insightful comments on the manuscript, and Don Miles pointed me toward a lost reference.

REFERENCES

Allen, G. (1975). "Life Science in the Twentieth Century." Wiley, New York.

Allen, T. F. H., and Starr, T. B. (1982). "Hierarchy: perspectives for ecological complexity." Univ. of Chicago Press, Chicago.

Allis, E. P. (1897). The cranial muscles and cranial and first spinal nerves in *Amia calva. J. Morphol.* **12**, 487-772.

Appel, T. A. (1987). "The Cuvier-Geoffroy Debate: French biology in the decades before Darwin." Oxford Univ. Press, New York and Oxford.

Arbas, E. A., Meinertzhagen, I. A., and Shaw, S. R. (1991). Evolution in nervous systems. *Annu. Rev. Neurosci.* **14**, 9-38.

Aronson, L. R. (1981). Evolution of telencephalic function in lower vertebrates. *In* "Brain Mechanisms of Behaviour in Lower Vertebrates" (P. R. Laming, ed.), pp. 33-58. Cambridge Univ. Press, Cambridge.

Atchley, W. R., and Hall, B. K. (1991). A model for development and evolution of complex morphological structures. *Biol. Rev. Cambridge Philos. Soc.* **66**, 101-157.

Atz, J. (1970). The application of the idea of homology to animal behavior. *In* "Development and Evolution of Behavior: essays in honor of T. C. Schneirla" (L. Aronson, E. Tobach, D. S. Lehrman, and J. S. Rosenblatt, eds.), pp. 53-74. Freeman, San Francisco.

Baerends, G. P. (1958). Comparative methods and the concept of homology in the study of behavior. *Arch. Neerl. Zool. Suppl.* **13**, 401-417.

Beer, C. G. (1980). Perspectives on animal behavior comparisons. *In* "Comparative Methods in Psychology" (M. H. Bornstein, ed.), pp. 17-64. Erlbaum, Hillsdale, N.J.

Brooks, D. R., and McLennan, D. A. (1991). "Phylogeny, Ecology and Behavior." Univ. of Chicago Press, Chicago.

Brooks, D. R., and Wiley, E. O. (1986). "Evolution as Entropy: Toward a Unified Theory of Biology." Univ. of Chicago Press, Chicago.

Burggren, W. W., and Bemis, W. E. (1990). Studying physiological evolution: paradigms and pitfalls. In "Evolutionary Innovations" (M. H. Nitecki, ed.), pp. 191-228. Univ. of Chicago Press, Chicago.

Carroll, R. L. (1988). "Vertebrate Paleontology and Evolution." Freeman, San Francisco.

Coates, M. I., and Clack, J. A. (1990). Polydactyly in the earliest known tetrapod limbs. *Nature (London)* **347**, 66-69.

Coleman, W. (1977). "Biology in the Nineteenth Century: Problems of Form, Function, and Transformation." Cambridge Univ. Press, Cambridge.

de Beer, G. R. (1971). "Homology, an Unsolved Problem." Oxford Biol. Readers, No.11. Oxford Univ. Press, London.

Desmond, A. (1982). "Archetypes and Ancestors: Paleontology in Victorian London 1850-1875." Univ. of Chicago Press, Chicago.

Donoghue, M. J. (1992). Homology. In "Keywords in Evolutionary Biology" (E. F. Keller and E. A. Lloyd, eds.), pp. 170-179. Harvard Univ. Press, Cambridge, MA.

Dumont, J., and Robertson, R. M. (1986). Neuronal circuits: An evolutionary perspective. *Science* **233**, 849-853.

Eldredge, N., and Cracraft, J. (1980). "Phylogenetic Patterns and the Evolutionary Process." Columbia Univ. Press, New York.

Fink, W. L. (1988). Phylogenetic analysis and the detection of ontogenetic patterns. *In* "Heterochrony in Evolution" (M. L. McKinney, ed.), pp. 71-91. Plenum, Press, New York.

Gans, C. (1985). Differences and similarities: Comparative methods in mastication. *Am. Zool.* **25**, 291-301.

Getting, P. A. (1988). Comparative analysis of invertebrate central pattern generators. *In* "Neural Control of Rhythmic Movements in Vertebrates" (A. H. Cohen, S. Rossignol, and S. Grillner, eds.), pp. 101-127. Wiley, New York.

Ghiselin, M. T. (1976). The nomenclature of correspondence: a new look at "Homology" and "Analogy." *In* "Evolution, Brain and Behavior: Persistent Problems" (R. B. Masterton, W. Hodos, and H. Jerison, eds.), pp. 129-142. Erlbaum, Hillsdale, NJ.

Golani, I. (1992). A mobility gradient in the organization of vertebrate movement: the perception of movement through symbolic language. *Behav. Brain Sci.* **15**, 249-309.

Goodwin, B. C. (1984). Changing from an evolutionary to a generative paradigm in biology. *In* "Evolutionary Theory: Paths into the Future" (J. W. Pollard, ed.), pp. 99-120. Wiley, New York.

Gould, S. J. (1982). The meaning of punctuated equilibria and its role in validating a hierarchical approach to macroevolution. *In* "Perspectives on Evolution" (R. Milkman, ed.), pp. 83-104. Sinauer Associates, Sunderland, MA.

Greene, H., and Burghardt, G. M. (1978). Behavior and phylogeny: Constriction in ancient and modern snakes. *Science* **200**, 74-77.

Haas, O., and Simpson, G. G. (1946). Analysis of some phylogenetic terms, with attempts at redefinition. *Proc. Am. Philos. Soc.* **90**, 319-349.

Hall, B. K. (1992). "Evolutionary Developmental Biology." Chapman & Hall, London.

Harris-Warrick, R. M. (1988). Chemical modulation of central pattern generators. *In* "Neural Control of Rhythmic Movements in Vertebrates" (A. H. Cohen, S. Rossignol, and S. Grillner, eds.), pp. 285-331. Wiley, New York.

Harris-Warrick, R. M., and Marder, E. (1991). Modulation of neural networks for behavior. *Annu. Rev. Neurosc.* **14**, 39-57.

Harris-Warrick, R. M., Marder, E., Selverston, A. I., and Moulins, M. (eds.). (1992). "Dynamic Biological Networks: The Stomatogastric Nervous System" MIT Press, Cambridge, MA.

Hickman, C. P., Roberts, L. S., and Hickman, F. M. (1988). "Integrated Principles of Zoology." Mosby, St. Louis, MO.

Hinde, R. A., and Tinbergen, N. (1958). The comparative study of species-specific behavior. *In* "Behavior and Evolution" (A. Roe and G. G. Simpson, eds.), pp. 251-268. Yale Univ. Press, New Haven, CT.

Hodos, W. (1976). The concept of homology and the evolution of behavior. *In* "Evolution, Brain, and Behavior: Persistent Problems" (R. B. Masterton, W. Hodos, and H. Jerison, eds.), pp. 153-167. Erlbaum, Hillsdale, NJ.

Holstvoogd, C. (1965). The pharyngeal bones and muscles in Teleostei, a taxonomic study. *Proc. K. Ned. Akad. Wet. Ser. C* **68**, 209-218.

Jardine, N. (1969). The observational and theoretical components of homology: A study based on the morphology of the dermal skull roof of rhipidistean fishes. *Biol. J. Linn. Soc.* **1**, 327-361.

Jarvik, E. (1980). "Basic Structure and Evolution of Vertebrates," Vol.1. Academic Press, London.

Johnson, G. D. (1992). Monophyly of the Euteleostean clades — Neoteleostei, Eurypterygii, and Ctenosquamata. *Copeia*, **1992**, 8-25.

Katz, P. S., and Tazaki, K. (1992). Comparative and evolutionary aspects of the crustacean stomatogastric system. *In* "Dynamic Biological Networks: The Stomatogastric Nervous System" (R. M. Harris-Warrick, E. Marder, A. I. Selverston, and M. Moulins, eds.), pp. 221-261. MIT Press, Cambridge, MA.

Kavanau, J. L. (1990). Conservative behavioural evolution, the neural substrate. *Anim. Behav.* **39**, 758-767.

Lande, R. (1^78). Evolutionary mechanisms of limb loss in tetrapods. *Evolution (Lawrence Kans.)* **32**, 73-92.

Lauder, G. V. (1981). Form and function: Structural analysis in evolutionary morphology. *Paleobiology* **7**, 430-442.

Lauder, G. V. (1983a). Functional and morphological bases of trophic specialization in sunfishes. *J. Morphol.* **178**, 1-21.

Lauder, G. V. (1983b). Functional design and evolution of the pharyngeal jaw apparatus in euteleostean fishes. *Zool. J. Linn. Soc.* **77**, 1-38.

Lauder, G. V. (1986). Homology, analogy, and the evolution of behavior. *In* "The Evolution of Behavior" (M. Nitecki and J. Kitchell, eds.), pp. 9-40. Oxford Uniersity Press, Oxford.

Lauder, G. V. (1990). Functional morphology and systematics: studying functional patterns in an historical context. *Annu. Rev. Ecol. Syst.* **21**, 317-340.

Lauder, G. V. (1991). Biomechanics and evolution: integrating physical and historical biology in the study of complex systems. *In* "Biomechanics in Evolution" (J. M. V. Rayner and R. J. Wootton, eds.), pp. 1-19. Cambridge Univ. Press, Cambridge.

Lauder, G. V. (1993). On the inference of function from structure. *In* "Functional Morphology in Vertebrate Paleontology" (J. Thomason, ed.). Cambridge Univ. Press, Cambridge (in press).

Lauder, G. V., and Liem, K. F. (1983). The evolution and interrelationships of the actinopterygian fishes. *Bull. Mus. Comp. Zool.* **150**, 95-197.

Lauder, G. V., and Wainwright, P. C. (1992). Function and history: the pharyngeal jaw apparatus in primitive ray-finned fishes. *In* "Systematics, Historical Ecology, and North American Freshwater Fishes" (R. Mayden, ed), pp. 455-471. Stanford Univ. Press, Stanford, CA.

Lenski, R. (1988). Experimental studies of pleiotropy and epistasis in *Escherichia coli.* I. Variation in competitive fitness among mutants resistant to virus T4. *Evolution (Lawrence, Kans.)* **42**, 425-432.

Mayr, E. (1969). "Principles of Systematic Zoology." McGraw-Hill, New York.

Meyrand, P., and Moulins, M. (1988a). Phylogenetic plasticity of crustacean stomatogastric circuits. I. Pyloric patterns and pyloric circuit of the shrimp *Palaemon serratus. J Exp. Biol.* **138**, 107-132.

Meyrand, P., and Moulins, M. (1988b). Phylogenetic plasticity of crustacean stomatogastric circuits. II. Extrinsic inputs to the pyloric circuit of the shrimp *Palaemon serratus. J Exp. Biol.* **138**, 133-153.

Minelli, A., and Peruffo, B. (1991). Developmental pathways, homology and homonomy in metameric animals. *J. Evol. Biol.* **3**, 429-445.

Nelson, G. J. (1969). Gill arches and the phylogeny of fishes, with notes on the classification of vertebrates. *Bull. Am. Mus. Nat. Hist.* **141**, 475-552.

Patterson, C. (1982). Morphological characters and homology. *In* "Problems of Phylogenetic Reconstruction" (K. A. Joysey and A. E. Friday, eds.), pp. 21-74. Academic Press, London.

Paul, D. H. (1981a). Homologies between body movements and muscular contractions in the locomotion of two decapods of different families. *J. Exp. Biol.* **94**, 159-168.

Paul, D. H. (1981b). Homologies between neuromuscular systems serving different functions in two decapods of different families. *J. Exp. Biol.* **94**, 169-187.

Paul, D. H. (1991). Pedigrees of neurobehavioral circuits: Tracing the evolution of novel behaviors by comparing motor patterns, muscles and neurons in members of related taxa. *Brain, Behav. Evol.* **38**, 226-239.

Pribaum, K. (1958). Comparative neurology and the evolution of behavior. *In* "Behavior and Evolution" (A. Roe and G. G. Simpson, eds.), pp. 140-164. Yale Univ. Press, New Haven, CT.

Raff, R. A., Parr, B. A., Parks, A. L., and Wray, G. A. (1990). Heterochrony and other mechanisms of radical evolutionary change in early development. *In* "Evolutionary Innovations" (M. H. Nitecki, ed.), pp. 71-98. Univ. of Chicago Press, Chicago.

Raup, D. M., and Stanley, S. M. (1971). "Principles of Paleontology." Freeman, San Francisco.

Reilly, S. M., and Lauder, G. V. (1992). Morphology, behavior, and evolution: comparative kinematics of aquatic feeding in salamanders. *Brain, Behav., Evol.* **40**, 182-196.

Remane, A. (1952). "Die Grundlagen des naturlichen Systems, der Vergleichenden Anatomie und der Phylogenetik." Akad. Verlag, Leipzig.

Riedl, R. (1978). "Order in Living Organisms." Wiley, New York.

Rieppel, O. C. (1988). "Fundamentals of Comparative Biology." Birkhäuser, Basel and Boston.

Rosen, D. E. (1973). Interrelationships of higher euteleostean fishes. *In* "Interrelationships of Fishes" (P. H. Greenwood, R. S. Miles, and C. Patterson, eds.), pp. 397-513. Academic Press, London.

Ross, D. M. (1981). Illusion and reality in comparative physiology. *Can. J. Zool.* **59**, 2151-2158.

Roth, V. L. (1984). On homology. *Biol. J. Linn. Soc.* **22**, 13-29.

Roth, V. L. (1988). The biological basis of homology. *In* "Ontogeny and Systematics" (C. J. Humphries, ed.), pp. 1-26. Columbia Univ. Press, New York.

Roth, V. L. (1991). Homology and hierarchies: problems solved and unresolved. *J. Evol. Biol.* **4**, 167-194.

Russell, E. S. (1916). "Form and Function: A Contribution to the History of Animal Morphology." John Murray, London. (Reprinted in 1982 with a new Introduction by G. V. Lauder. Univ. Chicago Press, Chicago.)

Rutledge, J. J., Eisen, E. J., and Legates, J. E. (1974). Correlated response in skeletal traits and replicate variation in selected lines of mice. *Theor. Appl. Genet.* **45**, 26-31.

Salthe, S. N. (1985). "Evolving Hierarchical Systems: their structure and representation." Columbia Univ. Press, New York.

Shubin, N., and Alberch, P. (1986). A morphogenetic approach to the origin and basic organization of the tetrapod limb. *Evol. Biol.*, **30**, 319-387.

Shultz, J. W. (1992). Muscle firing patterns in two arachnids using different methods of propulsive leg extension. *J. Exp. Biol.* **162**, 313-329.

Simpson, G. G. (1958). Behavior and evolution. *In* "Behavior and Evolution" (A. Roe and G. G. Simpson, eds), pp. 507-535. Yale Univ. Press, New Haven, CT.

Stiassny, M. L. J. (1986). The limits and relationships of the acanthomorph teleosts. *J. Zool. (Lond) B* **214**, 411-460.

Stiassny, M. L. J., and Moore, J. A. (1992). A review of the pelvic girdle of acanthomorph fishes, with comments on hypotheses of acanthomorph intrarelationships. *Zool. J. Linn. Soc.* **104**, 209-242.

Striedter, G. F., and Northcutt, R. G. (1991). Biological hierarchies and the concept of homology. *Brain, Behav. Evol.* **38**, 177-189.

Tyler, S. (1988). The role of function in determination of homology and convergence — examples from invertebrate adhesive organs. *Fortsch. Zool.* **36**, 331-347.

Van Valen, L. (1982). Homology and causes. *J. Morphol.* **173**, 305-312.

Vrba, E. S., and Eldredge, N. (1984). Individuals, hierarchies and processes: towards a more complete evolutionary theory. *Paleobiology* **10**, 146-171.

Wagner, G. P. (1989). The biological homology concept. *Annu. Rev. Ecol. Syst.* **20**, 51-69.

Wainwright, P. C. (1989a). Functional morphology of the pharyngeal jaw apparatus in perciform fishes: an experimental analysis of the Haemulidae. *J. Morphol.* **200**, 231-245.

Wainwright, P. C. (1989b). Prey processing in haemulid fishes: patterns of variation in pharyngeal muscle activity. *J. Exp. Biol.* **141**, 359-375.

Wainwright, P. C., and Lauder, G. V. (1992). The evolution of feeding biology in sunfishes (Centrarchidae). *In* "Systematics, Historical Ecology, and North American Freshwater Fishes" (R. Mayden, ed.), pp. 472-491. Stanford Univ. Press, Stanford, CA.

Wake, M. H. (1991). Morphology, the study of form and function, in modern evolutionary biology. *Oxford Surv. Evol. Biol.* **8**, 289-346.

Wenzel, J. W. (1992). Behavioral homology and phylogeny. *Annu. Rev. Ecol. Syst.* **23**, 361-381.

Wiley, E. O. (1976). Phylogeny and biogeography of fossil and Recent gars (Actinopterygii: Lepisosteidae). *Univ. Kans. Mus. Nat. Hist. Misc. Publ.* **64**, 1-111.

Wiley, E. O. (1981). "Phylogenetics: The theory and practice of phylogenetic systematics." Wiley (Interscience), New York.

Wiley, E. O., and Schultze, H.-P. (1984). Family Lepisosteida (gars) as living fossils. *In* "Living Fossils" (N. Eldredge and S. M. Stanley, eds.), pp. 160-165. Springer-Verlag, New York.

Wilson, E. O., Eisner, T., Briggs, W. R., Dickerson, R. E., Metzenberg, R. L., O'Brien, R. D., Susman, M., and Boggs, W. E. (1973). "Life on Earth." Sinauer Associates, Sunderland MA.

5

CAN BIOMETRICAL SHAPE BE A HOMOLOGOUS CHARACTER?

Fred L. Bookstein

Center for Human Growth and Development
University of Michigan
Ann Arbor, MI 48109-0406

I. HISTORICAL INTRODUCTION

Correctly interpreted, morphometric data analysis refers only to the purposes of prediction and summary for which Karl Pearson originally created biometrics a century ago (Pearson, 1894). This narrowness of competence is far more fundamental and ineluctable than is generally realized. In this essay I review earlier, partial exegeses of the fundamental difficulty and argue the unacceptability of several earlier attempts, including some of my own, at evading the natural limits of biometrical statistical inference. In my view, the answer to the question asked in my title must be negative. Morphometrics cannot supply homologous shape characters, but must be informed about them in advance. The languages of systematics and biometrics must remain incommensurate.

> Science, no less than theology or philosophy, is the field for personal influence, for the creation of enthusiasm, and for the establishment of ideals of self-discipline and self-development. No man becomes great in science from the mere force of intellect, unguided and unaccompanied by what really amounts to moral force. Behind the intellectual capacity there is the devotion to truth, the deep sympathy with nature, and the determination to sacrifice all minor matters to one great end. (Karl Pearson, 1906, p. 1)

Thus, Karl Pearson (1857-1936), the central figure in the historical development of my own professional field, summarized the Victorian scientific spirit in a eulogy for his great biometric colleague W. F. R. Weldon (1860-1906). While Pearson did not quite discover the correlation coefficient (stumbled on by his great mentor Francis Galton), it was Pearson who fixed upon the canonical "product-moment" formula, discovered the partial correlation coefficient, which corrects the zero-order coefficient for common antecedent causes, and first turned the language of regression analysis to the understanding of causes and effects in the natural sciences. Throughout the first 15 years of work in this area, Weldon was Pearson's constant source of data and criticism, his coeditor in the great labor of love that was their new journal *Biometrika*, and the man who, more than any other of his generation, typified to Pearson "the devotion to truth, the deep sympathy with nature."

In his eulogy, Pearson claims that Weldon virtually invented biometry in one go — the great paper of 1894/95. The paper, Pearson declared, "was absolutely novel at the time, and embraces...the best manner still of testing the truth of the Darwinian theory. ... It formulates the whole range of problems which must be dealt with biometrically before the principle of selection can be raised from hypothesis to law." Here is the crux of Weldon's argument from this "Report of the Committee for Conducting Statistical Inquiries into the Measurable Characteristics of Plants and Animals":

> The value of a merely empirical expression for the relation between abnormality[1] of one organ and that of another is very great. It cannot be too strongly urged that the problem of animal evolution is essentially a statistical problem: that before we can properly estimate the changes at present going on in a race or species we must know accurately (a) the percentage of animals which exhibit a given amount of abnormality with regard to a particular character; (b) the degree of abnormality of other organs which accompanies a given abnormality of one; (c) the difference between the death rate per cent. in animals of different degrees of abnormality with respect to any organ; (d) the abnormality of offspring in terms of the abnormality of parents and *vice versa*. These are all questions of arithmetic; and when we know the numerical answers to these questions for a number of species we shall know the deviation and the rate of change in these species at the present day — a knowledge which is the only legitimate basis for speculations as to their past history, and future fate. (Weldon, 1895, as quoted in Pearson, 1906, p. 19)

For a paragraph such as this to have been published a century ago, before the rediscovery of Mendel's paper, before path analysis, before factor analysis, indeed before the mathematics of evolution had emerged out of the stage of animal husbandry — to find so prescient a prophecy in the exuberance of high Victorian science is quite uncanny. In the century since Weldon's invention of the method of "organic correlation," correlation among disparate measures of the single form (as distinct from Galton's original innovation, correlation of the same trait between parent and child). all applications of Pearson's multivariate statistical methods within biology have been along the lines of this

---

[1] By "abnormality" here is meant "different from type," i.e., variation about a mean.

original biometric design: studies of cause and effect, selection and response. as codified fundamentally in *covariances*, the "arithmetic" to which Weldon was referring.

It is crucial for my argument here that the study of evolutionary history *per se* is *not* among those for which biometry was originally intended. It is the "principle of selection," not the actual historical record, that is Weldon's concern, and Pearson's (and Wright's, and mine). Pearson was quite consistent about this: evolutionary biology was a matter of covariances among "particular dimensions." It is the calculus of correlations, not the branching implicit in the record, that epitomizes this thrust.

After Weldon's sudden and untimely death Pearson never again found the collaborator he needed to focus on scientifically realistic aspects of evolutionary biology. Instead, infatuated by the moral tone of the eugenic movement, he narrowed the scope of "evolutionary theory" to one single topic with which he had been concerned since youth: the heritability of "human worth." Consequently the empirical papers of his last decades are on the whole quite awful. The clarion call in the epigraph to this section served instead as a trap: The papers that sacrificed "all minor matters," including objectivity, to the "one great end" of eugenics are acutely embarrassing to the modern statistician, however fascinating they are for sociologists of science.[2]

Classification and evolutionary history were not relevant to this redirected pursuit, and Pearson did not explore clustering or any other graphical tactic that might bridge the biometrical and taxonomical styles of *Naturwissenschaft*. He seems to have no essays on this specific point of methodology — for instance, the use of scatterplots in his great *Grammar of Science* (Pearson, 1892), is restricted to their appearance when error about regressions is rather small. This implicit stance is clearest, if most paradoxical, in his last great work, which is biographical rather than biological.

Pearson's 1400-page tribute to Galton (Pearson, 1914-1930) is one of the great works of high Victorian scholarship. Here, more than in any other writings I have seen, Pearson leaves evidence of his clear determination, representing that of the entire founding generation, that the

---

[2] For a construction of Pearson's work in eugenics as a folie à deux with his surrogate father Galton, see Bookstein (1993a).

proper domain of biometry be restricted to the study of covariances alone. In Volumes II and III of *The Life, Labours,
and Letters of Francis Galton*, the works of Galton's last
decades are reviewed not in chronological order but under
several thematic rubrics: Psychological Investigations. Photographic Researches and Portraiture, Statistical Investigations, Correlation, Personal Identification, and Eugenics.
The discussion of "Measurements of Resemblance" falls only
under Portraiture. Writing in 1930, Pearson laments:

> What is needed is that some one should take up the subject where
> Galton was forced to leave it. ... What are the average degrees of
> resemblance of parent and child, of brothers and sisters, of first
> cousins, etc.? And would the results obtained from Galton's Index of
> Mistakability correspond with those found by the principle of corre-
> lation from a single character in kinsmen of various degrees?
> (Pearson, *Life of Galton*, vol. II, p. 333)

Galton's Index is a direct measure of dissimilarity, not a
geometric summary of any sort; Pearson was regretting the
difficulty of converting it into a regression coefficient for
sib-sib similarity. As a mere measure of resemblance between units, what is now called a "distance," it would be of
no biometric use at all.

This clear separation between the purposes of correlation analysis and identification is patent throughout the
biography: for instance, fingerprints and profiles are each
treated as sources of individual data without regard for any
biometric analysis to come. In this way. the extraction of
individual identifiers — characters, if you will — is kept
wholly separate from the issue of correlational analysis of
causes and effects, both in Galton's own thinking and in
Pearson's *Life*. As Pearson was working on the third volume
of the biography, for instance, he was also taking up the
cudgels for eugenics one last time as the founding editor of
Annals of Eugenics. His foreword to the first volume
(Pearson, 1924) embraces one of Galton's slogans, "Probability the Foundation of Eugenics," and goes on to insist that
only correlation-based (i.e., biometric) studies can make any
contribution to the study of "racial differences in man."
When the word "similarity" arises in Pearson's biometrics, it
is only as a useful name for the regression coefficients that
arise in the study of correlations among different degrees of
relatives. Otherwise there is no operationalization of

"similarity" in biometrics. Units of analysis are exchangeable; only covariances among traits are real.

[I could not leave this subject without sketching a splendid irony that relates to my own career with an unmistakable piquancy. Immediately preceding this discussion of "mistakability," the *Life* spends a few pages recounting Galton's method for "numeralisation of portraits." This method is a combination of two techniques of truly astonishing insight. In an article in *Nature* of 1907 (see Fig. 1), decades before statistical methods existed for coping with such data, the 84-year-old Galton presents *precisely* my method of "shape coordinates" (Bookstein, 1984, 1986), thus scooping my original publication of these remarkably convenient pairs of shape measures by more than three-quarters of a century. Three years later, again in *Nature*, Galton sets out a method for classifying curvatures of profile segments between landmarks wholly consistent with the computer-graphic resolution of this problem in the 1960s. But, as in the parallel case of fingerprint analysis, the "numeralisation" was for the sole purpose of simplification without loss of identity. As Pearson notes (*Life*, vol. II, p. 324), Galton "apparently threw over any idea of measuring resemblance by likeness of formulae." The purpose was, instead, to enable the compression of profiles for transmission by telephone, telegraph, or wireless [radio]. Galton thus failed (perhaps intentionally) to anticipate numerical taxonomy even as he leaped ahead over three full generations of intermediate morphometric innovations. While this whole episode is wonderfully suggestive of roads not taken in the history of statistics, it distracts too far from my principal theme, and so I must leave it here].

The argument of the present paper is simple: Pearson and Galton were correct in reserving detailed shape data for "identification" rather than explanation. In our hands, as in theirs, shape data are indefinitely rich and can be elaborated far beyond the power of biometric methods, whether theirs or ours, to summarize genealogically in any convincing way. As I said in the first sentence of my 1991 treatise, "Morphometrics is the study of the covariances of biological form."

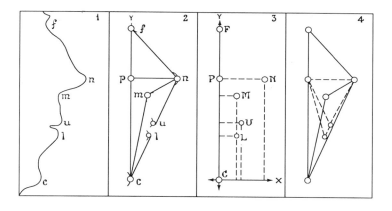

Fig. 1. The original "shape coordinates." From Galton (1907), as reproduced in Pearson, *Life*, Vol.II, p. 325. The point *c* is at (0,0) and *f* at (0,1).

Even today, no biometric method developed for the analysis of measurable characteristics of single organisms, their correlations and regressions — and this category includes everything in modern morphometric practice, whether multivariate or "geometric" — can be turned validly to the analysis of "resemblances" or, more specifically, to the reconstruction of evolutionary history. I have hinted at this point of view in several earlier essays that have been, in my view, inappropriately ignored. The invitation to contribute to this sesquicentennial celebration of Owen's great paper on the evidentiary basis of homology seems an ideal opportunity to show how morphometric data do *not* contribute to that evidentiary base.

My subject here, then, is the profound incompatibility between biometric and systematic styles of reasoning from data. In the next section I will briefly review my earlier attempts at this argument in their original contexts. the unexpectedly stubborn unmeasurability of D'Arcy Thompson's Cartesian grids, the impossibility of ordination in shape spaces, and the apparent requirement that to be statistically cogent a biometrics of shape must be invariant against specific selections of shape variables. In Section III, I attempt to bring all of these failures together under one single explanation, that of the irreducible curvature of any parametric space in which we choose to represent biometric shape, whether from lengths or from landmarks. I sketch, albeit too briefly, the facts of differential geometry that frustrate our attempts to treat the organism and the statistics of its measurable features in a single rhetoric. In particular, I show precisely how and where the biotheoretical notion of homology is blocked from passing to the biometric context. A closing discussion emphasizes how deep the difficulty runs: Owenian homology is fundamentally incompatible with any metric approach, bio- or otherwise. My argument is designed not as a criticism of either morphometrics or systematics, only as a declaration that the attempt to link the one to the other is futile.

II. EARLIER VERSIONS OF THE ARGUMENT

A. On Not Being Able to Get Past D'Arcy Thompson

The impossibility of bridging biometrics and systematics is a homiletic peculiarly mine. Others who contemplated the difficulty of finding quantitative equivalent for the classic version of homology treat the problem as a challenge to be met by one or another ingenious contrivance. D'Arcy Thompson's celebrated method of Cartesian transformations, for instance, implicitly replaces the biological notion of homology between *parts* by the geometric notion of a point-to-point mapping. For a general discussion of Thompson's construct, see Bookstein (1978, Chapter 5); Bookstein (1993b) incorporates an update.

Thompson noted, correctly, that this subterfuge (which of course he did not label as a subterfuge) allows the extraction of (some) features of such differences, though not without a considerable amount of additional analysis, and supports algorithms for extrapolation and interpolation of series of forms. Modern extensions of this method, such as the method of thin-plate splines that I currently prefer, support the same thrusts — feature extraction and extrapolation — along with some other biometric novelties, such as the drawing-out of forms "predicted" according to covariances with exogenous causes. Neither Thompson nor any modern epigone, however, has been able to extend the Cartesian technique further to incorporate convincing criteria of "resemblance." The method of grids, no matter in whose hands, is suited neither for the computation of "magnitude" of shape change nor for the clustering of individual specimens according to degree of similarity of shape.

We easily forgive Thompson his failure to anticipate late twentieth century biometrics, in view of his apparently total disdain for early twentieth century biometrics. Thompson was a Platonist, not an Aristotelian: his concern was the type, not the variant. More troublesomely, however, every subsequent attempt to extract biometrically usable quantities from his maps has ended in frustration. Huxley's method of "growth-gradients," for example, extracted parameters corresponding to specific ratios of instantaneous growth rates along presumably homologous "segments" (transects of the organism). But as Huxley himself noted, even if these parameters exactly described the relative rates of growth of, say, a length and a width, there was no parameter that described the relative growth of a diagonal line with respect to either over any finite interval; no parameter captured the growth of the full segment of which two sub-segments grew at different relative rates, a composite of two areas growing at different rates, and so on. The model of allometric growth, then, was ambiguous; one's findings were a function of the choice of "homologous regions" across which to place one's ruler or around which to run the planimeter. Nothing in the ensuing 60 years vitiates this problem; for instance, my biorthogonal grids of 1977 likewise would not extrapolate or interpolate to the plane without an unpleasantly *ad hoc* relaxation. Whether the parameter was Huxley's scalars or Bookstein's tensors, failure to behave like a good covariance — to explicitly calibrate an expected consequence of a divisible change — exemplifies the general impossibility of doing biometrics correctly in the presence of the intrinsic curvature of all biometric shape representations.

B. *The Shape Nonmonotonicity Theorem*

In the years after publication of this early suggestion of mine for regularizing the computation and display of Cartesian grids, I was repeatedly frustrated in the attempt to erect a truly biometric method on the principles and parameters of deformation. Regardless of the specific aims of my grants, instead of making progress I would uncover more and more good reasons to think the task impossible. The first of these, originally published as a "Shape Non-monotonicity Theorem" (Bookstein, 1980), was mislabelled: essentially, it is the elementary observation that the shape space of a triangle of landmarks is curved.

The "theorem" asserts that for any three triangles of landmarks no two of which have exactly the same shape, and for any ordering of the three triangles (e.g., 123, 231, 321), there exist indefinitely many shape measures consistent with that ordering. Notice the subversive quality of this observation. It implies, for instance, that for data in more than a single dimension (e.g., planar landmarks, or more than one ratio) no multistate shape characters can be assigned a "natural" order *a priori.* There is no possibility of "gap-coding" the even more stringent model of a scatter in two or more geometric "dimensions."

An alternative phrasing of the problem (exemplified in Fig. 2) is that there are simply too many possible "shape variables." Shapes which appear to lie on a straight line according to one pair of parameters lie along a curve in the spaces of other pairs of parameters. In the triangle that three shapes form there is always another shape variable "along the altitude," so that the form judged to be intermediate according to one shape measure can always be made to seem an end-member according to another measure *of the same three shapes.* Hence, even when there is a very convenient canonical solution (a shear) for the mapping problem. ordination of the resulting forms is irreducibly ambiguous as soon as one is free to use both ruler and protractor. Unless this liberty is restrained, there seems to be no consistent biometrics of "Cartesian transformations," even for triangles.

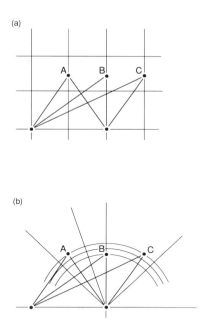

Fig. 2. The shape non-monotonicity theorem exemplified by three triangular forms. (a) In shape coordinates (Bookstein, 1991), the three form a straight line, along which form *B* is obviously intermediate. (b) To the grid given by another pair of shape parameters, form *B* is obviously an extremum. Lines: Isopleths (loci of equal values) of the angle at the lower right landmark. Circles: Isopleths of ratio of sides through the lower right landmark. For any set of three different shapes such paradoxical relations among shape variables can be constructed *ad libitum*.

C. Shape Space for a Set of Landmarks

Perhaps we are being too generous in our embrace — there may be a biometry of the raw data even if there is none for this riotous profusion of derived "shape variables." Ambiguity of shape orderings would dissolve if we could toss a sample of forms into a classic biometric space without having to worry about further nonlinearities of "variables" — ordinate first, and only later construct the variables underlying the ordination. Escape from the paradox, in other words, is via a representation of shapes that restricts the sorts of "shape variables" one is authorized to report. Multivariate statistics typically achieves restrictions of this type by modeling variation (here, landmark-based shape) in a *formally linear* measurement space in which the only acceptable findings are those that are independent of change of basis.

This combination of the descriptive and the statistical aspects of landmark-based shape is the celebrated (at least by me) *morphometric synthesis* that emerged more or less fully formed between 1983 and 1986 in papers of David Kendall, Colin Goodall, and myself. Its development has been described at length elsewhere (Bookstein, 1991, Section 2.1) and will not be elaborated at length here, although an overview may prove helpful.

Any biometric approach requires restriction of the numerical data to a finite-dimensional parametric manifold, a "feature space." For morphometrics it proves convenient to begin with data in the form of *landmark configurations* archived as sets of Cartesian coordinate pairs or triples. The feature space is then constructed by algebraic operations that explicitly remove size, position, and orientation: that is, by explicitly extracting "shape" out of the raw Cartesian coordinates. The curvature of the space enters irreversibly in the course of these operations, as when, for instance, 36 repetitions of the same 10° rotation bring you back to where you started. To arrive at a specifically biometric space, a linear vector space, the construction must proceed through one final technical step, arriving at the (unique) space tangent to the elliptic geometry of shapes in the vicinity of the sample mean form.[3]

There results an approach to all the classic biometric thrusts — studies of group difference, growth, and general relationships between shape and its causes or effects — for which the ambiguity of choice of variables is circumvented and the paradox of interjacency is made impotent by restricting one's attention to a single highly restricted family of mutually linear shape measures. The assortment of measures allowed is sufficient to encompass any familiar choice, such as a distance-ratio or an angle, "in the small" but not "in the large" (see Bookstein, 1991, Section 5.2 and Appendix 2). By ensuring as canonical precisely those findings Pearson and Weldon would have recognized as appropriately "biometric" — findings based on covariances or coordinations of small changes — the approach via shape

---

[3] The mean can be constructed by operations prior to the tangent space itself; it can be extracted, for instance, via the first principal component of the interspecimen distance matrix, where distance is taken by a Euclidean construction, such as the Procrustes fit.

space and its associated diagrams makes it unnecessary to assign any basis of "shape variables" for the description of those findings. The variables arise instead quite a bit later in the scientific cycle, just before the generation of actual reports of findings (see Bookstein, 1991, Chapter 7).

All these assertions extend without change from landmark configurations to data sets that link the landmarks by outline information (Bookstein and Green, 1993). To simplify the diction in this chapter, I shall nevertheless go on referring to "landmark configurations."

D. Partial Warps: A Tentative Basis of Shape Descriptors

While the purposes of biometrics *sensu stricto* are achieved by the morphometric synthesis, the first audiences to whom this resolution was revealed were not noticeably pleased. Morphometrics had once been considered the private preserve of systematics. For instance, the first appearance of the word that I can find (Blackith, 1965) refers to morphometrics as a subfield of systematics, as if nobody else would have any reason to care about the statistics of shape. At workshops at Michigan (1988) and Stony Brook (1990) it became obvious that reduction to biometrics — the bland morphometrics of mere covariances — was unacceptable to this distinctive community. For the problems of evolutionary history out of which the demand for these methods first arose, it seemed essential that there be a mechanism for expressing biometric findings in terms of *characters.* The argument of this paper, namely that that purpose is mathematically inaccessible, had not yet been faced squarely; instead, some of us attempted to approach characterization more closely within the biometric framework.

The morphometric synthesis of 1983-1986 assigns no specific basis to shape space. It circumvents the problem of shape *variables* by reducing all possibilities to a unique linear space of their *differentials* ("small" changes and their covariances) leaving open the specification of a basis to be "as helpful as possible" (consistent with the rules about mutual linearity of variables). By 1989 we had stumbled upon a choice, *partial warps,* that came as close to a character set

as we imagined we would ever come. Again, this essay is not
the place for a technical exposition of the solution; see
Bookstein (1991, Section VII.5).

To appreciate the characteristics of this particular
canonical basis for shape space in the vicinity of a mean
landmark configuration, it is necessary to reverse the
abstraction of the landmark configuration into a "point of
shape space." Instead, every form of the sample must be
considered just as Thompson originally had done: a Cartesian deformation of the typical or average form, taking
landmarks to landmarks and visually evocative in-between.
The particular mapping that carries out this transform is
unknowable between the landmarks. We can, nevertheless,
imagine the mapping that effects that match of landmark
set to landmark set with the least amount of regional distortion, by which is meant the double integral of summed
squared second derivatives (rates of change of first derivatives, gradient magnitude of the local "shearing") across the
picture. In what is really a mathematical miracle, the quantity minimized here turns out to be an ordinary quadratic
form in the original Cartesian coordinates (the raw data),
and the mapping we seek can be written as a simple
(though neither symmetrical nor antisymmetrical) matrix
function of the two landmark configurations involved. In an
inexplicable gift from the muse of Geometry, the principal
axes of that quadratic form serve in the vicinity of the mean
form as an orthonormal basis of dimensions of shape, complete with diagrams of the associated "homology maps," in
strictly decreasing order of spatial scale. These dimensions
are the *partial warps*; their formulae depend on the mean
form, and each configuration in the sample is characterized
by its full set of "scores."

The system of partial warps attempts a compromise
between the classic notion of homology — the correspondence of parts by reason of common descent — and the
geometry of successively smaller regions of the organism as
they emerge from a study of the covariation of all possible
regionalizations scaled inversely by something like their
area. Analysis by warps (for group differences, partial warps;
for intragroup ordination, relative warps) allows for the
empirical uncovering of partitions of the form by biometric
criteria, irrespective of homology *sensu stricto*.

Of the spectrum of possibly homologous parts of an extended shape, the method of warps supplies some, but not all, for which the "partition" is consistent with the empirical basis of biometry. There result suggestive ordinations of specimens or taxa, "character" by "character," which can be very helpful in interpreting the nature of evolutionary radiations given a true cladogram (Zelditch *et al.*, 1992, 1993). For instance, the warps, or their ontogenetic changes, are most like characters when they assort themselves distinctively one from another along separate branches of a cladogram. In such cases the patterns of their radiation may be examined in light of hypotheses of ecophenotypy or selection in the usual biometric fashion.

III. GETTING TO THE ROOT OF THE DILEMMA

At this point we do well to return to the epigraph, from Pearson's eulogy to Weldon. It would appear that we may have fallen into the Pearsonian trap of "sacrificing all minor matters to one great end" — extension of the biometric style, the study of variation, to include shape, to the great detriment of the purposes for which shape was originally considered. But the semantics of a mathematical representation is no "minor matter." In this section I wish to show how in turning shape into a multivariate descriptor vector we have made impossible any match to the classic language of homology.

Classic multivariate statistics works as well as it does because it incorporates a geometry all its own, that of the linear *N*-dimensional spaces you read about in textbooks. Notions of multiple group centroids, discriminant functions, prediction, and the like really pertain to those linear spaces, not to the real world. In most cases, that arrogation is acceptable, inasmuch as the "real world" proffers no geometry by which to contradict the biometrician's. This liberty is, however, incompatible with sound morphometric practice.

There is already a geometry of the phenomenon — the geometry of real two- or three-dimensional space, with its groups of transformations, including the transformation to shape that is explicit in the construction of "shape space." These geometries supply an *a priori* substratum for meaningful description reaching farther than any semantics that biometrics can impose. Even omitting the evolutionary-historical aspects of the Owenian definition of homology, aspects that clearly cannot be encoded in the biometric formalism of "variables," there is a richness to the strictly operational notion of homology — a freedom of choice of quantifications, that is incompatible with every finite-dimensional biometric analysis of a shape phenomenon, whether by lengths. ratios, shape coordinates, outlines, or any other strictly geometric approach. That measured lengths, for instance, should not be considered homologous as characters is argued in Bookstein (1991, Section III.1).

The fundamental problem making impossible any biometrics of homology is the same one circumvented in the linearization of the preceding discussion. Because evolutionary history requires consideration of forms at an arbitrary level of detail, the biometric approach is not capable of stating with confidence that two transformations are the same. If this is not possible, then we cannot decide the difference between synapomorphy and convergence, and hence there is no logical possibility of generating cladistically usable characters no matter how persuasive the evidence of an ordination that was, of necessity, arbitrarily selected.

I noted above that the geometric construct which is the "shape space" of a set of landmarks is curved, so that to arrive at a canonical biometrics we had to linearize (turn to the tangent space of the manifold) in the vicinity of a sample average. I have not yet said what it means for a space to be "curved": it turns out that it means that there are no rectangles.[4] But for the concept of homology to apply in morphometrics, it is quite necessary that there be rectangles in shape space. Without rectangles, one will never find "the same" shape change twice. Our conceptual impasse is as simple as that, and it is absolute.

---

[4] By "rectangle" is meant the usual construction of a two-dimensional "brick," with opposite sides equal and all angles 90°. It is not meant, of course, that rectangles are not shapes, but that no four shapes can be coherently assigned the required sets of pairwise equal distances and 90° angles between those distances.

A. *The Notion of Curvature in Differential Geometry and Cladistics*

In the limited space of a celebration of Owen it is not possible to explain as effectively as I'd like just why it is that non-zero curvature entails the nonexistence of rectangles.[5] But I can hint at the reasoning that leads to such an argument by reminding you of the curving surface you know best — the surface of the earth — and of why there are no rectangles down here. (In fact, the shape space of triangles is very nicely identified with the shape of a sphere — see the discussion of Kendall in my 1991 monograph — and so some of this argument goes over to triangles without any change.)

Stand on the equator, and travel west 1000 miles. Then travel north 1000 miles, then east 1000 miles, then south 1000 miles. As Fig. 3 reminds you, you do not get back to where you started. It can be shown that this nonclosure property is equivalent to the facts that the angles of a spherical triangle do not total 180°, that area of a circle on a sphere does not increase as fast as the square of the diameter, that all straight lines through a point on a sphere meet again (at the antipodal point), and a great many other propositions you already know; so I ask your indulgence in the generalization of the square equatorial walk to any other parallelogram. Taken to an extreme, if you travel 6200 miles north from either of two points on the equator, you get to the North Pole. If one of the other two sides of the "rectangle" is a walk along the equator, then the "fourth side" is of length 0. We have a spherical triangle, which seems to embody a 100% failure to be a rectangle. At the North Pole, the character state of longitude is not "absent" — it is undefinable. This blank where data should be is forced by the geometry of spheres, not by any aspect of the evolutionary path taken from equator to pole.

[5] In the literature of geometry for the layman there is one great work, David Hilbert and S. Cohn-Vosen's *Geometry and the Imagination* of 1932. The best way for the novice to begin to understand curvature is to browse through that source. It is still in print.

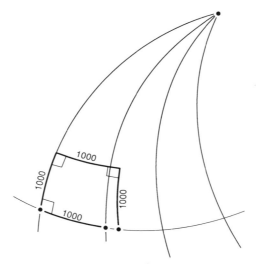

Fig. 3. The nonexistence of rectangles in curving shape space. On a sphere such as the earth, the "square" drawn does not close. See text for details.

Suppose there were homologous shape change characters, singly parameterized deformations X (taking A to B) and Y (taking A to C). Let D be the result of applying transformation Y to B. Then, except in unusual cases, D is not the same shape as the result of applying transformation X to C. A typical failure of this system of two "characters" to be commutative is sketched in Figure 4. To imagine applying "the same" transformation to two forms is biometric language, vector language; and outside of the "biometric" version of morphometrics, the study of covariation "in the vicinity of a mean form," a shape change can be specified *only* as the vector linking two specific forms. In geometric language, it is a "bound" vector, not a "free" vector. It cannot serve as a cladistic "character" because it will never be found in the same state twice. If forms A and B are different, then the fully specified geometric meanings of applying X to A and to B are different, as in any curving space used for their representation there is *no* change from form B that is identical to the change from A to a third form C. To adequately represent the geometric realities of shape in our two- or three-dimensional world, the representation of shape has

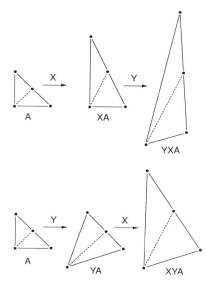

Fig. 4. Exemplifying dependence of the meaning of change of one shape variable on the values of all others: shape changes expressed using arbitrary parameters do not commute. Operation X increases the height of a triangle by 50% to the baseline indicated. Operation Y doubles lengths in the direction of the median to the long side shown. The result of applying X and then Y to triangle A (top) is not the same as the result of applying Y and then X (bottom).

no choice but to curve. Use your intuition about the earth again: the actual (solid, three-dimensional) segment connecting any two cities — say, Detroit and Vancouver — cannot be put down in any other position than what it now occupies if both its end points are to lie on the surface of the globe. The vector from Detroit to Vancouver irreducibly characterizes *both* cities, not just one.

The incompatibility of biometrics with cladistics is thereby a matter as much of logic as of algebra or geometry. The crux is the cladistic axiom that characters can be assigned states in any order. A form is described by an unordered list of characters. x1, y1, z0 and so on, but not x1 and then y1 and then z0, and so on. Here observation of the states of x are independent of the states of y, z, etc., and *vice versa*. The geometry of a biometric shape space never cooperates with that requirement; no two changes of the "state" of shape are ever quite the same. In cladistic language: *morphometric shape variables cannot possibly form a hierarchy.*

Let us consider the very simple, practical problem of Fig. 5. We have two transformation series of triangles, A→B and

C→D. Are these the same transformation? Because of the curvature of the shape space of triangles, they cannot possibly be considered the same, no matter what they are. The transformation as shown doubles the height of triangle A to the horizontal baseline, and likewise that of C. And it changes the angle of A at landmark 1 from 45° to 90°, and that of C likewise. Since there are only two "dimensions" of this shape space, this paired agreement ought to be sufficient for identity. But I chose those two shape descriptors quite arbitrarily: there are many other variables on which these changes *fail* to agree. For example, the angle at the lower right is invariant between A and B at 45°, whereas it changes from 90° in C to 65.9° in D. We will always be able to find variables whose values are the same in such paired transformations and variables whose values are unpaired: it's just an application of the shape nonmonotonicity theorem. Then there is just no gain in using the change A→B as a "character" — we cannot ever hope to unearth it again.

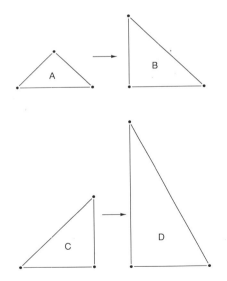

Fig. 5. Exemplifying the proposition that no two shape changes may be considered the same. The changes of triangular form from *A* to *B* and from *C* to *D* both double the aspect ratio to the horizontal baseline, and both double the angle at the lower left landmark; but the angle at the lower right landmark behaves differently in the two changes.

There is another way to phrase the same paradox: a deformation from A to B, when specified in sufficient detail, cannot be applied to any other starting form C. Suppose, for instance, we have a set of measured lengths $1...m$ between landmarks, and forms A and B differ by ratios 1:2, 1:3, 2:3, ...,.m-1:m. If we apply these ratios to any other form than A, the resulting network usually does not lie flat (if A was planar), or does not exist in three-dimensional space (if A was three-dimensional). Therefore, however elegantly we can carry out statistical analysis of these ratios, we cannot use the set of all ratios as a character for the description of relative growth. The set does not represent a generalizable shape difference, but, rather, it explicitly (and with considerable algebraic complexity) encodes two fully specified shapes.[6] Of course, it is not satisfactory to settle for a subset of these ratios, even a complete subset numbering $2m$-3, since our cladistics, *though not our biometrics*, would vary as a function of that arbitrary choice.

This curvature of shape space, and the consequent impossibility of observing the same shape change more than once, apply to all sufficiently complex representations of shape that begin from biometric data: to landmark shape

---

[6] This argument extends in an instructive way past single vectors of ratios to the principal components that embody the "single-factor allometric model" (Bookstein *et al.*, 1985; Bookstein, 1991). That model computes the first principal component from a covariance matrix of logged distances, and then considers the residuals from the loglinear allometric regressions specified by that component (for study of group differences or for extraction of further factors). But, especially when the single-factor model is most powerful, some irreducible minimum of deviations from loglinearity is forced by the geometry of Euclidean space, *even in the absence of all processes of biological heterochrony or heterotopy.* Synchronized deviations from loglinear allometry of lengths, then, are not evidence for secondary regulation of growth changes, and differences across taxa in such secondary factors automatically confound shape regulation with details of the mean form. It is thus not tenable to compare within-group loadings of factors of log-distances across taxa differing in mean form, which is to say, across any interesting pair of taxa. *Factor loadings of distances or log-distances are not characters*; neither are the covariance matrices, genotypic or phenotypic, from which they are derived. One must substitute a *geometric* approach to allometry, explicitly constructing the distributed regression of a shape basis on size, as suggested by Mosimann (1970). For the application to landmark data, see Bookstein (1991, Sec. 5.4). Even then, all the rest of the problems reviewed here remain.

and outline shape and ratio shape, in one, two, or three dimensions, and regardless of the underlying biometrics. We can never call two biometric shape changes of different starting forms the same. Nor do we have a biometric language for quantifying their differences.[7] Instead, the biometric approach consists of formally effacing the possibility of such differences, by computing covariances with causes or effects only of the part of shape description that is common to all sets of sufficiently many variables "in the vicinity of the mean form."

IV. THE INCOMMENSURABILITY OF BIOMETRICS AND SYSTEMATICS

We might hope to escape from these problems by accepting a certain degree of looseness in the conception of "equality" for shape transformations. Hikes in the woods can be steered by flat maps on paper — over sufficiently small patches of terrain, the curvature of the globe does not entail confusion. Perhaps the match between morphometrics and cladistics is like that in some sense, restricted to small patches of morphospace.

A landmark-based study of closely related taxa (for instance, shape comparisons among ape crania diagrammed in Bookstein, 1978) may show shape ratios in the range of 5:1 or so. This range is close to that noted in Appendix 2 of

----

[7] In the classic differential geometry of Riemannian manifolds, there is a scheme, called *parallel transport*, that moves differentials (here, shape variables) from the vicinity of one form to the vicinity of another. In effect, we hold constant the angles made by the shape measure with *geodesics* (shortest paths) in that space. This "solution" is not acceptable in morphometrics because homology of characters is defined operationally on the organism prior to the metric geometry that underlies this ``transport." The metric, hence the "transport," is a function of the (psychological) choice of landmarks, not the selective landscape of the organism over its evolutionary history. Nevertheless, for the special case of triangles (see Bookstein, 1991, Appendix 2), it is possible to homologize ratios of perpendicular transects in this way, if the metric be taken as log anisotropy rather than the more convenient Procrustes formula. I know of no extension to more complex data for which parallel transport preserves what we wish to have preserved of ruler-and-protractor shape measures. Current research by Colin Goodall may result in a greater role for geodesics in morphometrics.

Bookstein (1991), beyond which linearizations underlying any biometric approach begin to lose their credibility even for biometric purposes. Means become unstable, heteroscedasticity becomes more than a nuisance, and the possibility of making inferences from covariance structures, which is all that biometrics is competent to do, becomes clouded.

In the typical context of successful intraspecific applications, such as studies of vertebrate skulls, hearts, or major features of brains, shape ratios range over a span of 2 : 1 or 3 : 1. In applications to botany, where growth can be over the range of factors of many hundreds, landmark-based biometrics appears remarkably unhelpful; likewise in applications to embryogenesis over substantial numbers of successive doublings. Any growth allometry, sufficiently iterated, will eventually render unreliable the linearizations required for any sort of biometric analysis at all.

Morphometrics, then, applies to modest changes, changes for which no length-ratios are altered by factors outside the range of, say, 0.1 to 10. For cladistic applications, this limits its use, in practice, to lineages as closely observed as the descent of *Homo*, for which evidentiary weight of landmark configurations is contested (or in any case swamped) by the more recondite and subtle data of biomechanics and paleoarchaeology and by the additional "geometry" that is the geography of the fossil record. In such contexts, morphometrics is hardly worth pursuing, at least, not for the mere understanding of descent. Rather, the problems of descent to which morphometrics would conceivably contribute new insights of potentially very great weight would be those related to triples of groups much more widely separated in shape; unfortunately, that very separation of shapes renders the evidence of morphometrics much less persuasive owing to the problem of linearization just reviewed. The hard problems of evolutionary history are hard, in part, precisely because the biometric approach cannot work. When shapes vary over greater ranges than those for which covariances can be imagined practically independent of the suite of shape characters used, we cannot make much headway using covariances.

I would submit this as a sort of "Bookstein's Law of Morphometrics": if the biometric approximation is adequate — if you can treat shape variation by Pearson's sort of methods, by variances and covariances — then you are studying either ontogeny, or ecophenotypy, or evolution at the sub-species level: small modifications of a shared ontogenetic program, most of the variation of which may be captured in a small number of geometrizable factors. No important evolutionary change can be captured persuasively in the language of biometrics. Ontogenetic processes are smooth in the small, even if diverted by bifurcations of dynamic systems; but evolutionary change is rough, the summation of a great many bifurcations superimposed toward chaos. The methods of biometrics are restricted in practice to smooth changes. They are powerful, but specialized.

Figure 6 is a sketch of how you might imagine shape space as it applies in most evolutionary applications. (cf. Fig. 7.5.1 of Bookstein, 1991). The geometry of the biometric shape space for the general set of landmarks is very highly curved. It is not like a sphere, but rather more like a very high-dimensional ellipsoidal cylinder, the axes of which are as different in "length" as the spacing of landmarks dictates. There is no way to guarantee that the evolutionary changes we see in practice — differences accumulated over thousands or millions of generations of mutation of ontogenetic programs — stay "close to the tangent planes" of this model. Remember, it is not sufficient to remain "nearly flat" in some or even most of the dimensions of shape space. For a proper cladistic analysis, one must understand most carefully those characters which are varying fastest; and those are almost certainly the ones for which the biometric approximation fails.

The flaw in any attempt to quantify Owenian homology, then, is the incoherence of any legible geometric metaphor for "morphospace" above the species level. The fundamental analogy between a morphospace on printed paper (or computer screen) and a vector space of shape descriptors quickly becomes untenable as the relevant range of shapes broadens. Biometrics is the coloring of that vector space by a covariance, or a list of covariances, along each axis; it stands as much on the realism of those axes as on the significance of those covariances. The only acceptable geometric analogy for morphospace above the species level is a "surface" like broccoli or the more corrugated varieties of

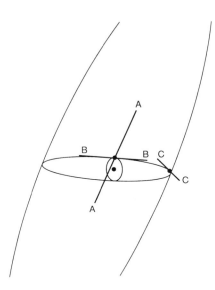

Fig. 6. A schematic of shape space may be expected to curve as a function of mean form and direction of change. Direction A represents the differentials (covariances) of rearrangements of landmarks with respect to others at a great distance; direction B, rearrangements with respect to those at intermediate distance; direction C, the "same" rearrangements of the same landmarks about a nearby mean for which the relevant neighbors are configured differently.

lettuce: a surface that, by its obvious fractal character, clearly renders absurd any notion of changing "in the same direction" at different points.

Owen, who lived a hundred years before multivariate analysis claimed even to handle races, may be forgiven for not anticipating the fundamental problem underlying the extension of statistics into evolutionary history. But living proponents of views deriving from his ought to know better; this view of homology is far too stripped of information from the specimen for any possible biometric analysis. At its core, the notion of homology allows no approximations. Structures either are or are not "the same" — no notion of "near miss" can apply. Owen's homology, at root, is thus *not a metric concept*, biometric or otherwise.[8] And, as is implied by the very construction of the word, there is no way to carry out any sort of biometric analysis without a metric. All multivariate statistics rests on a distance function, either the χ^2 or one of its multivariate normal approximations. We could not decide when two changes of a "shape character" are or are not "close enough" without a standard for that closeness; but such a notion of distance can come only from a biometric approach, which would thus need to specify a finite-dimensional space of shape variables (such as is derived from the Cartesian coordinates of a configuration of landmarks) before homology could hope to be established. It is clear that any such foundation for a metrization of "homology" is an unacceptable begging of the evolutionary question.

Homology is not a biometrizable construct because the language of possible shape changes consistent with homology is far too rich. The transformation group against which homology is judged is at least as large as the group of all *diffeomorphisms* (transformations that are smooth and have smooth inverses).[9] Diffeomorphisms cannot be "parameterized" in any useful way. There is no natural metric by which they can be summarized in a "magnitude" and a "direction"; rather, they bear information at every scale of spatial detail and at every order of differentiation. *Pace* Sneath (1967), there was never any possibility of a useful biometrics of these mappings in the general case. Biometrics is the study

[8] The difficulty here is not peculiar to notions of homology, but extends to most appearances of truly categorical variables in biology. Pearson lost his battle with Bateson over the proper statistical methods for evolutionary studies mainly because biometric methods of covariance and correlation are unable to incorporate abstract Mendelian genes, that is, the fact that a single trait can sometimes be reliably declared "the same" or not between a pair of individuals. Pearson's beloved "Law of Ancestral Heredity," essentially a multiple regression or factor prediction, was expeditiously refuted by actual data on a great variety of obviously segregating traits of great eugenic relevance. This topic is treated by most of the standard histories of eugenics; or see MacKenzie (1981).

[9] Jardine (1969) argues that it can be extended to certain groups of graph-theoretic transformations as well. I know of no attempts since this one to reduce these graphs to computational practice

of variation; variation requires a metric beforehand; but Owen's notion of homology claims to judge sameness well in advance of any assignment of metric, and so is incommensurate with any later biometrical work-out of a data set of forms, however closely related.

Then Owen's notion of homology, the hundred-fiftieth anniversary of which we are celebrating in this volume, must remain the incomprehensible link between the geometry of biological form and evolutionary history. There is no way to extend the methods of biometrics across this divide: no way to test assertions of homology by biometric methods, and no way, also, to incorporate more than the brute assertion of homology into biometric analyses of causes and effects of form. Modern morphometrics is the biometric technology for the study of causes and effects of form, and "common ancestry" is not the sort of "cause" to which these methods can be applied. The classic notion of homology is not a parameter — a homologous part is or is not there, 1 or 0, with no quantification of intermediacy. (You can't have "some" of a part, and probabilistic approaches do not lead to a version useful for landmarks). In morphometrics, causal factors act with subtly calibrated weights given, ultimately, by regression coefficients. The causes of form then must be in units of their own: typically, units of biomechanics, physiology, energetics, or some other metric of function. A shape variable is justifiable only if it is a complete summary of the effects on shape of some separately identifiable causal agent. (When the partial warps "work," it is because their segregation along branches of a cladogram suggests that they serve as the record of the branching event in that sense). Absent this knowledge, there is (literally) no way to tell which of the uncomfortably broad variety of shape variables could be consistent with actual history.

The causes to which biometric techniques are applicable are *factors*, influences that can be calibrated (at least, in principle) by a consideration of their conjoint consequences using experience with other data independent of the evolutionary history in question. For instance (Crespi and Bookstein, 1989), the outcome of a selection experiment, in which organisms are found to be either "alive" or "dead," is to be treated as an estimate (the crudest possible) of the continuous underlying factor of "fitness" as corrected by an adjustment for phenotypic "general size." Taxon can never be a cause in that sense — an estimate of a continuous underlying factor — and hence species names cannot serve as causes in the formulas of covariance.

This is no criticism of morphometrics *per se*, which serves very well in a great variety of applications from computer vision to surgery. Nor, for that matter, is it a criticism of systematics or cladistics. I am asserting only that reconstruction of evolutionary history, by a cladogram or any other formalism, is not a biometrical problem and is not susceptible to a biometrical solution. The languages of homology and of morphometrics are mutually incomprehensible. Instead of attempting to bridge them, by such *ad hoc* methods as parsimony or minimum spanning trees of gapcoded "characters," we should instead revel in the mystery of their separateness. Evolutionary history is not the proper subject of morphometric investigation. It is, rather, the substrate that makes possible all subsequent morphometric investigations into the causes and consequences of form, investigations that are empowered to use all the power of Pearson's multivariate biometric machine.

V. CONCLUSIONS

The power of modern multivariate statistics for summary and prediction of complex quantitative morphological data masks a devastating incompatibility between its language of biometrics, which arose from the Victorian preoccupation with eugenics, and the terminology of modern systematics. The rhetoric by which the modern biologist determines whether aspects of two organisms, two ontogenies, or two taxa are "the same character state" is not compatible with the algebra or geometry of the curving manifolds that house our morphometric measurements. Once a set of biological forms (or their images) is parameterized, whether that parameterization be by lengths, by shape, or by any other geometric approach, one can proceed with biometric analysis of variation and covariation of "variables" (quantitative traits). But the statistics of these parameters, univariate or multivariate, exist in vector spaces that are geometrically distinct from the space of shapes in crucial ways. Such measurement vectors can be compared unambiguously only in the infinitesimal vicinity of a single form. Absent this approximation, a "character" representing change in a geometric shape trait is irrevocably a function of both forms, not just the descendant, and hence cannot contribute to resolution of uncertainties about ancestry.

Correctly interpreted, morphometric data analysis refers only to the purposes of prediction and summary for which Karl Pearson originally created biometrics a century ago. This narrowness of competence is far more fundamental and ineluctable than is generally realized. In this essay I review earlier, partial exegeses of the fundamental difficulty and argue the unacceptability of several earlier attempts, including some of my own, at evading the natural limits of biometrical statistical inference. In my view, the answer to the question asked in my title must be negative. Morphometrics cannot supply homologous shape characters, but must be informed about them in advance, and the languages of systematics and biometrics must remain incommensurate.

ACKNOWLEDGMENTS

Preparation of this manuscript has been supported in part by USPHS Grant GM-37251 to the University of Michigan (F. L. Bookstein, principal investigator) and by a faculty fellowship from the Institute for the Humanities of the University of Michigan.

REFERENCES

Blackith, R. (1965). Morphometrics. In "Theoretical and Mathematical Biology" (T. H. Waterman and H. J. Morowitz, eds.), pp. 225-249. Blaisdell, New York.

Bookstein, F. L. (1978). "The Measurement of Biological Shape and Shape Change," Lect. Notes Biomath., vol. 24., Springer-Verlag, Berlin.

Bookstein, F. L. (1980). When one form is between two others: An application of biorthogonal analysis. Am. Zool. **20**, 627-641.

Bookstein, F. L. (1984). Tensor biometrics for changes in cranial shape. Ann Hum. Biol. **11**, 413-437.

Bookstein, F. L. (1986). Size and shape spaces for landmark data in two dimensions (with discussion and rejoinder). *Stat. Sci.* **1**, 181-242.

Bookstein, F. L. (1991). "Morphometric Tools for Landmark Data." Cambridge Univ. Press, New York.

Bookstein, F. L. (1993a). Utopian skeletons in the biometric closet. *In* "Utopian Visions" (T. Siebers, ed.). Univ. of Michigan Press, Ann Arbor (in press).

Bookstein, F. L. (1993b). "The Morphometric Synthesis: A Brief History. Lect. Notes Biomath., vol. 100., Springer-Verlag, Berlin (in press).

Bookstein, F. L., Chernoff, B., Elder, R., Humphries, J., Smith, G., and Strauss, R. (1985). "Morphometrics in Evolutionary Biology: The Geometry of Size and Shape Change, with Examples from Fishes." Academy of Natural Sciences of Philadelphia, Philadelphia.

Bookstein, F. L., and Green, W. D. K. (1993). A feature space for edgels in images with landmarks. *J. Math. Imag. Vision* (in press).

Crespi, B. J., and Bookstein, F. L. (1989). A path-analytic model for the measurement of selection on morphology. *Evolution (Lawrence, Kans.)* **43**, 18-28.

Galton, F. (1907). Classification of portraits. *Nature (London)* **76**, 617-618.

Hilbert, D., and Cohn-Vossen, S. (1932). "Geometry and the Imagination." Chelsea, New York.

Jardine, N. (1969). The observational and theoretical components of homology: A study based on the morphology of the dermal skull-roofs of rhipistidian fishes. *Biol. J. Linn. Soc.* **1**, 327-361.

MacKenzie, D. (1981). "Statistics in Britain 1865-1930." Edinburgh Univ. Press, Edinburgh.

Mosimann, J. (1970). Size allometry: size and shape variables with characterizations of the log-normal and generalized gamma distributions. *J. Am. Stat. Assoc.* **65**, 930-945.

Pearson, K. (1892). "The Grammar of Science." W. Scott, London (Reprinted in 1937 by J. M. Dent, London, and in 1969 by P. Smith, Gloucester, MA.).

Pearson, K. (1894). Contributions to the Mathematical Theory of Evolution. *Philos. Trans. R. Soc. A* **185**, 70-110.

Pearson, K. (1906). Walter Frank Raphael Weldon, 1860-1906. *Biometrika* **5**, 1-52.

Pearson, K. (1914-1930). "The Life, Labours, and Letters of Francis Galton," Three volumes bound as four. Cambridge Univ. Press, Cambridge.

Pearson, K. (1924). Foreword. *Ann. Eugen.* **1**, 1-4.

Sneath, P. H. A. (1967). Trend-surface analysis of transformation grids. *J. Zool.* **151**, 65-122.

Weldon, W. F. R. (1895). Attempt to measure the death-rate due to the selective destruction of *Carcinus moenas* with respect to a particular dimension. *Proc. R. Soc. London* **57**, 360-379.

Zelditch, M. L., Bookstein, F. L., and Lundrigan, B. (1992). Ontogeny of integrated skull growth in the cotton rat *Sigmodon fulviventer. Evolution (Lawrence, Kans.)* 46, 1164-1180.

Zelditch, M. L., Bookstein, F. L., and Lundrigan, B. (1993). The ontogenetic complexity of developmental constraints. *J. Evol. Biol.* (under review).

6

HOMOLOGY, DEVELOPMENT,

AND HEREDITY

Brian Goodwin

Department of Biology
The Open University
Milton Keynes, MK7 6AA
England

Homology: The Hierarchical Basis of Comparative Biology
Copyright © 1994 by Academic Press, Inc.
All rights of reproduction in any form reserved.

I. INTRODUCTION

One of the most long-lasting and fruitful dialogues in biology arises from the tension between explanatory modes based either on history or on structure, now often referred to as process and pattern (Hall, 1992). These two ways of making sense of both the diversity and the unity of species revealed through evolution — the one stressing contingency, adaptation, and inheritance, the other emphasizing intrinsic necessity, form, and generative process — has animated many biological debates. The most famous of these was undoubtedly that between Cuvier and Geoffroy Saint-Hilaire in 1830 concerning the explanatory principles of animal morphology. Cuvier's focus was on the functional unity of species within fixed types that have a contingent creation, while Geoffroy emphasized intrinsic structural order and the transformational unity underlying all biological forms. It is clear that both made very important contributions to our understanding of morphology and its connections with taxonomy. Goethe, who was passionately interested in the outcome of their debate, considered that Geoffroy's position was closer to the truth as he perceived it through his work on metamorphosis:

> If we see at the outset the regularity of nature we are apt to think that it is necessarily so, and was so ordained from the very first, and hence that it is something fixed and static. But if we meet first with the varieties, the deformities, the monstrous misshapes, we realize that although the law is constant and eternal it is also living; that organisms can transform themselves into misshapen things not in defiance of law, but in conformity with it, while at the same time, as if curbed with a bridle, they are forced to acknowledge its inevitable dominion [Goethe (1830) cited by Cassirer, 1950, p. 140.)

Goethe was actually going beyond both Cuvier and Geoffroy to a fully dynamic conception of biological form and transformation that defined a path toward their reconciliation.

Some 160 years later, the dialogue is still very much alive, or this volume would never have been written. We are still exploring conceptual schemes that make sense of the vastly greater amount of empirical information now available about evolution and development that is relevant to understanding morphology. Homology is the key concept used to examine relationships between history and structure, between evolution and development. I follow Goethe in using a dynamic approach to the problem and in emphasizing the range of *possible* forms that can be produced by morphogenetic rules. Homology, connecting development and taxonomy, will then relate to invariant properties of the patterns resulting from particular generative processes.

II. HOMOLOGY: AN UNSOLVED PROBLEM

A. *The Historical Definition*

de Beer's (1971) fascinating little treatise, whose title this section bears, exposed very clearly the difficulties faced by different definitions of homology. That which comes to us from Darwin is historical: two structures are homologous if they share descent (with modification) from a common ancestral form. Applying this to an example such as tetrapod limbs, the proposition is that the arm of a human, the wing of a chick, and the forelimb of a newt are homologous because these animals are descended from a common ancestor with the same basic forelimb bone pattern. This ancestor was assumed to have had five digits, but the most ancient tetrapods now identified had seven (*Ichthyostega*) and eight (*Acanthostega*) digits (Coates and Clack, 1990). So the significance of polydactyly lies elsewhere than in evolutionary origins. I shall return to this issue later.

de Beer pointed out that this historical definition of homology works for forelimb and hindlimb taken separately, but not for the structural relations between them "as fore limb and hind limb cannot be traced back to any ancestor with a single pair of limbs." The evidence in fact is that forelimbs and hindlimbs evolved independently from pectoral and pelvic fins in primitive gnathostomes.

William Bateson (1894) had earlier made a similar point to de Beer's about the limitations of historical definitions of homology because of the difficulty of applying them to structural similarities between different parts of the same organism, such as the different body segments of insects or the leaves and flower parts of plants, which can undergo transformation (homeosis). Insects, for example, have not arisen from a common ancestor with a single segment, so the different segments are not related by homology defined as descent from a common ancestral structure. It does seem desirable to have a definition of homology that can be applied to the comparison of structures within an organism as well as between species.

B. *Homology and Genes*

de Beer considered the possibility that homology might have a genetic basis. This would preserve an historical perspective while allowing for comparisons of similar structures within organisms in terms of similar patterns of gene activity. He observes that mutation in a single gene can cause morphological changes in characters that are not homologous. Pleiotropy is in fact the hallmark of gene action. So the same genes can be involved in the development of nonhomologous characters.

Conversely, *different* genes can result in the formation of the *same* characters. This happens whenever a mutation alters or eliminates a character (e.g., the *eyeless* gene in *Drosophila*) and in subsequent generations the character returns by the action of "modifier genes" despite the continuing presence of the mutant gene. So we are forced to the conclusion that homology cannot be defined in terms of invariant gene action.

Recent studies of gene activity during the formation of tetrapod limbs require, however, that we look more closely at this general conclusion.

Homeobox-containing genes (the *Hox*-4 cluster) are active in tetrapod limb formation. As expected from the mutant effects of such genes in *Drosophila*, altered patterns of expression in these genes result in homeotic transformations: for instance, digit I of the chick hindlimb is transformed to a digit II-like element when *Hox*-4.6 is overexpressed (Morgan *et al.*, 1992). This accords with the description of the digits of a limb as an example of *serial homology*, which Bateson used for repeated but not identical parts of an organism. As the person who introduced the term homeosis into comparative morphology, as well as the term gene into biology, Bateson would have been very interested in this relationship between gene expression and homeotic transformation, particularly as he had a different definition of gene from that which we use today. For Bateson, the gene referred to a generative unit in morphogenesis — i.e., a dynamic generator that was inherited. This links with Goethe's conception, and relates to a concept of homology that I shall soon introduce.

Hox-4.6 is one of five genes involved in formation of tetrapod limbs examined so far. *Hox* genes are expressed in a spatial pattern that provides unique combinations for each of the five digits. An interesting case has been made by Tabin (1992) that pentadactyly is to be identified with the constraint imposed by this combinatorial genetic code: there cannot be more than five distinct digits because there are only five distinct combinations of the genes. If limbs have more than this number, as they do in conditions of polydactyly and in fossil tetrapods, then digit identities must be duplicated, as Tabin suggests for the case of *Acanthostega*; see also the discussion in Chapter 2. This looks like an important case of a genetic constraint, an historical invariant. The question arises: does this lead to a general definition of homology in terms of invariant gene action? I suggest that it does not. Rather, this constraint finds its place naturally within another conception of homology which is closely related to Bateson's way of describing it.

III. HOMOLOGY AS EQUIVALENCE

It is time for a definition. I start with the following. Homology is defined by an equivalence relation over the members of a set, defined by a transformation that takes any member into any other member within the set (see also Goodwin, 1982).

This concept of equivalence under transformation is commonly used in mathematics. For example, we can consider the set of shapes produced by transformations that preserve connectivity and so do not introduce or eliminate holes in some initially given shape. For example, a sphere can be distorted into any shape that preserves the connectedness of all points in the sphere; or a doughnut (a torus) can be transformed into any shape that conserves the hole.

This class of transformations define topological equivalence. The invariant is simply connectedness. We can consider mappings (transformations) that preserve the differentiable properties of functions (essentially, their smoothness). This defines equivalence under the action of diffeomorphisms. The mappings can obey the constraint that the angles between intersecting lines in one space are preserved in the image space. These are equivalences under conformal mappings. Or distances can be preserved, giving equivalence under metric transformations. Each equivalence set has an invariant property that is preserved by the transformation. A hierarchy of such equivalence classes can be defined, each successive invariant including but imposing more constraints than the previous one.

A. Hierarchies and Sets

Developmental processes are hierarchical. So are biological classification schemes. I shall follow a well-trodden route in attempting to relate them. This is the route toward a theory of biological forms and their logical relationships (a rational taxonomy).

Considering first tetrapod limbs, let us start with the highest level of the hierarchy, the organization of the whole limb, and proceed downward to smaller units, which will bring us back to digits. Classical embryology established the limb field as the developmental unit of spatiotemporal organization at this level (Harrison, 1921), although of course the limb field is itself initiated within a larger field, the whole embryo. Its relative autonomy comes from the observations that, at a particular stage of development, a limb field can be transplanted to an ectopic site in the embryo and produce a supernumerary limb.

The property of tetrapod limbs first identified as the invariant preserved despite transformations was spatial arrangement of the limb elements. These were described by Geoffroy Saint-Hilaire in terms of his principle of connections. This is the familiar order of humerus, tibia and fibula, carpals, metacarpals, and phalanges of the forelimb, or the equivalent elements of the hindlimb.

Richard Owen defined limb homology as the set of forms derived by transformations of these elements that preserves their basic connections. He introduced the concept of an archetypal pentadactyl limb , an ideal tetrapod limb that underlies the various transformations seen in actual tetrapods. This was the forerunner of the common ancestral pentadactyl limb that Darwin proposed as a real historical instantiation of the archetype.

However, neither the notion of an archetype *sensu* Owen, nor of a common ancestor is necessary in the concept of homology as equivalence. All members of a set are equivalent under transformation; none has any special status. Furthermore, what the transformation defines is a set of *possible* forms, not simply those that have been realized in known species. Issues of historical contingency (inheritance) and functional stability (natural selection) have to be recognized as particular aspects of evolution as it actually happened. But these should not be confused with attempts to define the set of possible forms that could be generated. We find out about these by the study not only of genetic mutants but also by the study of teratologies — abnormal forms resulting from disturbances to the limb field, as in embryological studies or the examination of spontaneous congenital abnormalities — Goethe's "monstrous misshapes" that nevertheless obey constraints imposed by rule-governed transformations. Bateson's *Materials for the Study of Variation* (1894; recently reprinted by Johns Hopkins University Press), is dedicated to precisely the task of exploring the set of possible forms and deducing the constraints that limit them, as suggested by the range of genetic and congenital abnormalities.

B. Rules of Transformation

This leads us to the attempt to identify the rules or the organizing principles that underlie the production of tetrapod limbs and their transformations. These are embodied in morphogenetic processes arising within the limb field. What can we say about the constraints on these processes?

It used to be thought that all limb buds initially generated the archetypal or ancestral pentadactyl limb pattern, subsequently undergoing transformation to the different tetrapod forms by secondary modifications such as fusion and differential growth. This is now known to be incorrect. In the chick, for example, the number and relations of the chondrogenic condensation sites in limb-bud mesenchyme are the same as the elements of the adult limb (Hinchliffe and Hecht, 1984; Hinchliffe, 1990). So we need to examine this condensation process to discover both constraints responsible for the regularity of tetrapod limbs and variations within these constraints that result in the range of possible forms constituting the transformation set.

Now it becomes necessary to use a model of the condensation process that suggests what the origin and nature of the constraints might be. This is because the enterprise that we are engaged in is theoretical in the sense that the set of possible tetrapod limb forms is to be understood in terms of a generator whose rules of operation can be described and used to actually generate the patterns representing the members of the transformation set. This is a modelling exercise and proceeds tentatively, in dialogue with experimental study.

IV. GENERATIVE INVARIANTS

A. *The Model*

The model that I shall utilize to illustrate the argument is that proposed by Shubin and Alberch (1988) and by Oster *et al.* (1983, 1988) to describe the mechanics of the mesenchyme aggregation process giving rise to the elements of the developing limb. This is based on experimental studies of the behavior of mesenchyme cells during their condensation to produce the cartilaginous elements that later become the bones of the limb, and on a mathematical model that demonstrates how these processes could occur by plausible mechanisms of cell adhesion and the behavior of aggregates as they grow (Oster *et al.*, 1985).

In the model, condensation is initiated by a local change in osmotic properties of the extracellular matrix that results in its collapse, bringing cells closer together. Adhesive forces between cells then result in a focus of cell aggregation. This grows by recruitment of cells which become polarized along tension lines in the extracellular matrix and move toward the aggregate. Their movement generates traction forces that reinforce polarity and directed movement.

During growth of the condensate, whose direction is partly dependent on the geometry of the limb bud, and partly by growth along the proximodistal axis of the limb, the condensation process can become unstable and bifurcate in one of only two ways. It can branch to produce a Y-shaped structure; or it can segment, producing another discrete condensation that is on the same axis as the first.

The arrangement of elements in a limb is a result of these processes: initial or focal *condensation*, branching *bifurcation*, and *segmentation*. These are suggested as the primary components of the generative process at the level of limb element production, according to the model. They explain well, not only normal limb patterns but also structures such as those observed in limbs whose mesenchyme cells have been removed from the bud, spatially scrambled (disaggregated and pelleted), then reinserted into the limb under an ectodermal jacket and allowed to develop (Pautou, 1973). Such treatment destroys spatial organizing influences such as those from the zone of polarising activity (Balcuns *et al.*, 1970; Fallon and Crosby, 1977) and the *Hox*-4 complex gene products that result in ordered patterns of distinct digits.

We can thus recognize different levels in the hierarchy of limb-forming processes, including: (1) individuation of the limb field as an autonomous domain; (2) establishment of axial order within the limb field; (3) formation of the limb elements by condensation, branching, and segmentation; and (4) modification of the elements along the antero-posterior axis by influences such as those deriving from the zone of polarizing activity and the *Hox*-4 genes. These overlap in time, but can be distinguished within the unitary process that is the unfolding of spatial order in the limb field.

B. Fish Fins and Tetrapod Limbs

Fish fins are produced by the action of a subset of the generators involved in producing tetrapod limbs.

In the condensation process there are no branching bifurcations, and there is an absence of strong antero-posterior (AP) polarity. At the same time, the AP axis is spatially extended over more somites. The result is an appendage that is made up of a meristic series of very similar repeating elements along the AP axis, with segmentation along the proximodistal (PD) axis.

Fish fins and tetrapod limbs can thus be defined as transformations of one another within these generative rules; i.e., they are homologous structures at a particular level of the generative hierarchy. By adding further generative constraints, namely restriction of the number of somites involved, adding branching bifurcations and antero-posterior polarity, tetrapod limbs are produced. This is presumably the route followed in the evolution of tetrapod limbs from fish fins, via primitive gnathostomes (Jarvik, 1980).

Fish fins and tetrapod limbs thus fall within particular parameter domains of the same generative space. Exploring the size of the domains in parameter space that gives fins as compared to limbs is part of the research program investigating the sizes of the dynamic attractors for different types of biological form. If the domains are small, then the forms are intrinsically improbable and the continued persistence of the forms in fishes and tetrapods must be due to strong stabilizing factors (e.g., genes, natural selection). If the domains are large, the structures are highly probable, easy to find,and have an intrinsic robustness (i.e., they are generic — see Goodwin, 1989; Newman and Comper, 1990).

C. Toward a Theory of Biological Form

Obviously these issues are very important for our understanding of evolution. It tends to be assumed that most biological structures (eyes, brains, limbs, flowers) are improbable, that they have been discovered by effective genetic search procedures in a vast space of possible genetic programmes, and are stabilized by natural selection. However, there is evidence from the study of complex nonlinear systems describing morphogenetic processes that these have an intrinsic robustness that gives resultant structures the status of generic forms, arising within large attractors in morphospace (Goodwin, 1993; Goodwin *et al.*, 1993).

Developmental constraints, widely recognised as imposing limitations on evolutionary transformations of biological structures, may then find an explanation in intrinsic dynamic properties of morphogenetic fields producing quantized attractors (Oster and Alberch, 1982) rather than historical inertia of some kind. This would make possible the study of a rational taxonomy of organismic morphologies, an obvious goal of any theory of biological form. Relationships between hierarchical morphogenesis and a hierarchical taxonomy thus begin to emerge.

The emphasis of such a study is rather different from taxonomy as genealogy, the actual history of life on earth, as Darwin conceived the goal of systematics. History is not explanatory of form because it does not describe the generative processes that make different forms possible. Natural selection is equally deficient because it addresses the question of persistence, of stability (including, of course, instability) of characters in relation to environments; it neither explains generative origins of characters nor why they are possible; see Webster (1993) for a more comprehensive critique of Darwinian genealogy and the conceptual foundations of a theory of biological form.

V. GENE ACTION

We can now return to the role of genes in morphogenesis, and how this relates to homology.

A theory of biological forms and their transformations that I am exploring in this essay is clearly concerned with the definition of *sufficient* conditions for generating structures, as required for a model that actually simulates the process of limb formation. Such a model is necessarily a field theory that describes the changing patterns of forces acting within and between cells and the extracellular matrix in the production of, say, limb elements within the limb mesenchyme, as does that proposed by Oster *et al.* (1985). I use this model because it illustrates clearly how various factors can enter into the determination of structural patterns.

A morphogenetic field theory must involve a description of the physical forces that generate the shape being modeled and whatever biochemical influences act on these, such as enzymes like hyaluronidase that affect the osmotic state of the extracellular matrix, strength of the adhesive forces between cells, materials of the extracellular matrix that transmit strain and orientate cells, effect of strains in cells which, acting via the cytoskeleton, influence rates of translation and of transcription (Ben-Ze'ev *et al.*, 1980, 1988), and many other factors. For every one of these, many genes are involved.

But a knowledge of how genes influence morphogenesis, while extremely useful, is not sufficient to construct a morphogenetic field model, which is what we need to understand the process. For this we need to understand such principles as the physical laws describing viscoelastic media, stresses and strains, osmotic principles and how they act in the extracellular matrix, the behavior of cells in response to strain fields, diffusion, and many other aspects of the macroscopic field of the tetrapod limb bud that together provide a working model.

Genes affect primarily the parameter values of such fields, and in so doing can alter details of the form generated. For instance, retinoids mimicking some of the effects produced by the zone of polarizing activity (Maden, 1982; Tickle *et al.*, 1982) appear to exert their influence by acting on gap junctions, hence changing the effective diffusion constants of materials involved in the global organization of the limb field (Mehta *et al.*, 1989). Other influences such as those that result in congenital abnormalities also affect the particular form produced. So what can we say about pentadactyly, which appears to be intimately connected with inherited patterns of *Hox*-4 gene transcriptions?

By affecting parameter values such as effective diffusion constants in the extracellular matrix, rates of osmotic change, and cell adhesiveness, genes can modify condensation patterns and the characters of the limb elements. A stable spatial transcription pattern of *Hox*-4 and other genes across the limb bud can thus produce a stable sequence of digit characters in any species. Digits are equivalent under such transformations of character within a limb, between forelimb and hindlimb, or between species, so they can be homologized. But, according to the definition of homology as equivalence, it is not clear what it would mean to homologize particular digits of the limb of one species with those of another. What are the generative invariants that identify what is called digit III as distinct from digit II, say?

In the cat hindlimb there are four fully-formed toes, usually designated II, III, IV, and V, each with three phalanges. Digit I is rudimentary, consisting of a small cylindrical bone. In the chick hindlimb, digit I has two phalanges, digit II has three, and digits III and IV each have four phalanges. Similar variations can be found across other species. What is it that identifies digits III as a homologous set that excludes the other digits? It is not at all clear that there is an identifiable constraint that allows us to designate these digits as a distinct homology class (see also Goodwin and Trainor, 1983). Action of the *Hox*-4 genes does not provide this evidence. What this shows is that digits constitute an equivalence set under homeotic transformations, and so are homologous structures. This was known to Bateson on the basis of other evidence, and it is satisfying to have some of the genetic factors identified that are responsible for determining the different characters that constitute the various members of the equivalence set. But to homologize digits III of all tetrapods requires evidence that these digits can be transformed one into the other without affecting the other digits. So far as I am aware, such transformations have not been demonstrated. They are *assumed* by the concept of historical descent by an argument that uses circular logic: historical constraint on individual digits is included in the definition of their homology.

What the *Hox*-4 data supports in addition to the general homology of digits within and between species is that the overall transformation set is constrained by gene action to a maximum of five distinct digit characters in any one species. This is again a very interesting and important result, though conjectural as far as fossil tetrapods with more than five digits are concerned. But the observation of genetic constraint is not itself novel, since gene action normally constrains morphological expression in any species by stabilizing particular morphogenetic trajectories and hence defining a particular norm of reaction for the variants that arise due to congenital and environmental influences. What is additionally interesting about the *Hox*-4 genes is the ordered spatial pattern in which they are expressed in the limb, with which we have become familiar in the well-defined spatial and temporal patterns of segmentation gene transcription in *Drosophila*. These results, combined with models of limb mesenchyme condensation or insect segmentation, bring closer our understanding of how genes exert their action on morphogenetic fields in terms of the actual parameters affected and how these act via physical properties of the extracellular matrix and cell behavior, relevant to morphogenesis.

VI. UNITARY MORPHOGENETIC FIELDS

Hox-4 genes are sometimes described as giving identity to digits (Tabin, 1992). But is it only *Hox* genes that do this?

"Digit III" of the chick and "digit III" of the cat are different — i.e., they have different identities, and the same is true of "digit III" in any other species. Clearly other genes are involved in specifying digit character than just the *Hox*-4 cluster. Genes are involved in all aspects of morphogenesis, influencing all levels in the hierarchy of stages that starts with the initiation of a field and ends with detailed morphological structure. Differences between digit III of the chick and digit III of the cat are almost certainly due to the action of different genes influencing different levels of the limb field, from initiation and axial specification through condensation, branching, and segmentation events of chondrogenic aggregate formation, to the lengths and detailed shapes of the various elements. At no stage can we talk of a separate and distinct process conferring identity on elements.

Some confusion about this seems to arise from a curious convention about gene action during development that has become very prominent in the literature but which seems to me to obscure rather than clarify what is going on. This is the separation of "positional information" from "interpretation" in embryonic development (Wolpert, 1969, 1989). Using the *Drosophila* segmentation genes as probably the clearest example to illustrate the point, we can ask if the gradient in *bicoid* protein is an interpretation of a previous positional information pattern or part of the positional information for gap gene expression. Are the gap gene patterns an interpretation of *bicoid* (together with *oscar* and other anteroposterior genes), or are they part of the positional information system for the pair-rule genes? Are the pair-rule patterns an interpretation of gap gene activity or are they part of the positional information system for segment polarity genes? Or are they always both, so that these aspects of gene action are not separable? I can see no sensible case for their separation.

The notion that the identity of a structure is conferred by gene action at a particular level seems to confound the confusion by separating out that level as in some sense decisive and by suggesting that "interpretation" has a meaning other than influencing the spatiotemporal dynamics of the morphogenetic process. We have seen, for example, that the "morphogen" that arises from the zone of polarizing activity appears to act as a modifier of a parameter, the effective diffusion constant of substances involved in limb field organization. It thus has a direct influence on morphogenesis and does not need to be "interpreted" by genes.

VII. CONCLUSIONS

A hierarchical morphogenetic field model focuses on the dynamic properties of morphogenesis and the possible patterns that can be generated by change of any component of the process, while preserving as invariant certain properties of relational order over the field. At one level of order, fish fins and tetrapod limbs can be defined as homologous: they are equivalent under transformations that preserve proximodistal axial order, condensation, and segmentation of elements that form a meristic series in anteroposterior order. Tetrapod limbs constitute a distinct equivalence class within this larger set of vertebrate appendages by having additional properties that arise from the operation of further invariants — branching bifurcations in the condensation process occur — and there is a distinct anteroposterior asymmetry so that digits are distinguishable. Thus, tetrapod limbs are generated by at least two additional symmetry-breaking processes in the morphogenetic field to those occurring in fish fins, and hence have a more complex structure.

By identifying such dynamic properties of the generative process, it is possible to define logical principles of hierarchical inclusion of morphological structures based on their generative principles. In this way, homological relations defined as equivalence classes can be identified by invariant generative rules, so that a hierarchical taxonomy of biological forms can be constructed on the basis of morphogenetic processes. These principles can be used for the classification of teratologies as well as for normal morphology, as shown in a very instructive example by Ho (1990). Developmental dynamics thus become the basis of the structural relations between morphological types, and a rational taxonomy of biological forms becomes possible.

REFERENCES

Balcuns, A., Gasseling, M. T., and Saunders, J. W. (1970). Spatio-temporal distribution of a zone that controls anteroposterior polarity in the limb bud of the chick and other bird embryos. *Am. Zool.* **10**, 323.

Bateson, W. (1894) "Materials for the Study of Variation". Cambridge Univ. Press, Cambridge (Reprinted in1992 by Johns Hopkins Univ. Press, Baltimore, MD).

Ben-Ze'ev, A., Farmer, S. R., and Penman, S. (1980). Protein synthesis requires cell-surface contact while nuclear events respond to cell shape in anchorage-dependent fibroblasts. *Cell* **21**, 365-372.

Ben-Ze'ev, A., Robinson, G. S., Bucher, N. L. R., and Farmer, S. R. (1988). Cell-cell and cell-matrix interactions differentially regulate the expression of hepatic and cyto-skeletal genes in primary cultures of rat hepatocytes. *Proc. Nat. Acad. Sci. U.S.A.* **85**, 2161-2165.

Cassirer, E. (1950) "The Problem of Knowledge." Yale Univ. Press, New Haven, CT.

Coates, M. I., and Clack, J. A. (1990). Polydactyly in the earliest known tetrapod limbs. *Nature (London)* **347**, 66-69.

de Beer, G. R. (1971) "Homology: an Unsolved Problem," Oxford Biol. Readers, No. 11. Oxford Univ. Press, London.

Fallon, J. F., and Crosby, G. M. (1977). Polarizing zone activity in limb bud amniotes. *In* "Vertebrate Limb and Somite Morphogenesis" (D. A.Ede, J. R. Hinchliffe, and M. Balls, M, eds.), pp. 55-69. Cambridge Univ. Press, Cambridge.

Goethe, J. W. (1830) Principes de Philosophie zoologiques discutés en mars, 1830, au sein de l'Académie Royale des Sciences. *Naturwiss.Schr., Werke* **7**,189.

Goodwin, B. C. (1982). Development and evolution. *J. Theor. Biol.* **97**, 43-55.

Goodwin, B. C. (1989). Evolution and the generative order. *In* "Theoretical Biology: Epigenetic and Evolutionary Order from Complex Systems" (B. C. Goodwin, and P. T.Saunders, eds.) pp. 89-100 , Edinburgh Univ. Press, Edinburgh.

Goodwin, B. C. (1993). Development as a robust natural process. *In* "Thinking About Biology" (W. D. Stein, and F. Varela, eds.). pp. 123-148. Santa Fe Institute Studies in the Sciences of Complexity, Addison-Wesley.

Goodwin, B. C., and Trainor, L. E. H. (1983). The ontogeny and phylogeny of the pentadactyl limb. *In* "Development and Evolution" (B. C.Goodwin, N. J. Holder, and C. C.Wylie, eds.). pp. 75-98. Cambridge Univ. Press, Cambridge.

Goodwin, B. C., Kauffman, S. A., and Murray, J. D. (1993). Is morphogenesis an intrinsically robust process? *J. Theor Biol.* (in press).

Hall, B. K. (1992) "Evolutionary Developmental Biology." Chapman & Hall, London and New York.

Harrison, R. G. (1921). On relations of symmetry in transplanted limbs. *J. Exp. Zool.* **32**, 1-136.

Hinchliffe, J. R. (1990). Towards a homology of process: evolutionary implications of experimental studues on the generation of skeletal pattern in avian limb development. *In* "Organizational Constraints on the Dynamics of Evolution" (J. Maynard Smith, and G. Vida, eds.), pp. 119-131. Manchester Univ. Press, Manchester.

Hinchliffe, J. R., and Hecht, M. (1984). Homology of the bird wing skeleton: embryological versus palaeontological evidence. *Evol. Biol.* **30**, 21-39.

Ho, M.-W. (1990). An exercise in rational taxonomy. *J. Theor Biol.* **169**, 43-57.

Jarvik, E. (1980). "Basic Structure and Evolution of Vertebrates," Volume 2. Academic Press, London.

Maden, M. (1982). Vitamin A and pattern formation in the regenerating limb. *Nature (London)* 295, 672-675.

Mehta, P., Bertram, J. S., and Loewenstein, W. R. (1989). The actions of retinoids on cellular growth correlate with their actions on gap junctional communication. *J. Cell Biol.* **108**, 1053-1065.

Morgan, B. A., Izpisua-Belmonte, J. C., Duboule, D., and Tabin, C. J. (1992). Targeted misexpression of *Hox-4.6* in the avian limb bud causes apparent homeotic transformation. *Nature (London)* **358**, 236-239.

Newman, S. A., and Comper, W. D. (1990). 'Generic' physical mechanisms of morphogenesis and pattern formation. *Development (Cambridge, UK)* **110**, 1-18.

Oster, G. F., and Alberch, P. (1982). Evolution and bifurcation of developmental programs. *Evolution (Lawrence, Kans.)* **36**, 444-459.

Oster, G. F., Murray, J. D., and Harris, A (1983). Mechanical aspects of mesenchymal morphogenesis. *J. Embryol. Exp. Morphol.* **78**, 83-125.

Oster, G. F., Murray, J. D., and Maini, P. (1985). A model for chondrogenic condensations in the developing limb: the role of extracellular matrix and cell tractions. *J. Embryol. Exp. Morphol.* **89**, 93-112.

Oster, G. F., Shubin, N., Murray, J. D., and Alberch, P. (1988). Evolution and morphogenetic rules: the shape of the vertebrate limb in ontogeny and phylogeny. *Evolution (Lawrence, Kans.)* **42**, 862-884.

Pautou, M.-P. (1973). Analyse de la morphogénèse du pied des oiseaux à l'aide de mélanges cellulaires interspécifiques. *J. Embryol. Éxp. Morphol.* **69**, 1-6.

Shubin, N. H., and Alberch, P. (1986). A morphogenetic approach to the origin and basic organization of the tetrapod limb. *Evol. Biol.* **20**, 319-387.

Tabin, C. J. (1992). Why we have (only) five fingers per hand: *Hox* genes and the evolution of paired limbs. *Development (Cambridge, UK)* **116**, 289-296.

Tickle, C., Alberts, L., Wolpert, L., and Lee, J. (1982). Local application of retinoic acid to the limb bud mimics the action of the polarizing zone. *Nature (London)* **298**, 564-566.

Webster, G. C. (1993). Causes, fields, and forms. *Acta Biotheor.* (in press)

Wolpert, L. (1969). Positional information and the spatial pattern of cellular differentiation. *J. Theor Biol.* **25**, 1-47.

Wolpert, L. (1989). Positional information revisited. *Development (Cambridge, UK)* **107**, Supplement, 3-12.

7

HISTORY, ONTOGENY, AND

EVOLUTION OF THE ARCHETYPE

Neil H. Shubin

Department of Biology
University of Pennsylvania
Philadelphia, PA
19104-6017

Homology: The Hierarchical Basis of Comparative Biology
Copyright © 1994 by Academic Press, Inc.
All rights of reproduction in any form reserved.

I. INTRODUCTION

In many ways, the concepts of homology and "the archetype" are as vital today as when Owen first discussed them; see e.g., Hall (1992), Slack et al., (1993) and Chapter 1. Owen, Geoffroy Saint-Hilaire, and others recognized a commonalty of organismic design reflecting structural order within animal diversity. While Darwin's recognition of common descent did much to provide a material explanation for homology and "archetypes," both of these concepts remain controversial, with their recognition, mechanistic basis, and even their very existence matters of current debate. One factor that underlies these controversies is the relationship between common descent and the developmental/genetic mechanisms that produce organic structures in each generation.

Analysis of homology both informs and is informed by the study of biological processes. One process, ontogeny, has always held a central role. It is oftentimes assumed that the generative mechanisms and patterns that underlie development of complex structures are more conserved than the structures themselves. That is, features that share a common phylogenetic history are frequently supposed to share similar sequences and/or mechanisms of development. There are, however, numerous cases where ontogenetic and historical data appear to conflict. Many features that, by all accounts, appear to be homologous undergo different patterns of development (de Beer, 1971; Streider and Northcutt, 1991). Fundamental components of the vertebrate body plan such as neural tubes, pronephros, and alimentary canals, can develop in several different ways. One can conclude that either these structures are not homologous or that these structures are homologous but that the developmental trajectories have evolved. To evaluate the evolutionary significance of this phenomenon we need to review the manner in which ontogenetic data can be integrated with historical approaches to homology.

A. Ontogenetic Data

Ontogenetic data have been used to address several key problems of homology.

The first is the distinction between homologous and homoplastic similarities. This issue reflects the close ties between so called taxic and *transformational* approaches to homology (Eldredge and Cracraft, 1980; Patterson, 1982; Rieppel, 1988; de Pinna, 1991). That is, analysis of the evolutionary transformations that underlie the origin of individual structures is intimately tied with understanding the hierarchical relationships between taxa. To say that "limbs are homologous as the paired appendages of tetrapods" is as much a statement about the genesis of limbs (their derivation from the paired appendages of nontetrapod vertebrates) as it is a statement about the composition and definition of the group, Tetrapoda. The problem investigated here is the role these transformational and taxic approaches play in analysis of the evolution of developmental sequences.

The second major problem in the analysis of homology rests with the causal explanation of conservatism and morphological transformations. The explanation of homology rests with analysis of those mechanisms that serve to both constrain morphological features and to individuate them during ontogeny and phylogeny (Roth, 1988; Wagner 1989ab).

B. Phylogenetic and Biological Homology

Many definitions of homology exist, differing chiefly in the types of comparisons they involve and the issues they attempt to explain. Recent discussions focus on two closely related concepts, *phylogenetic homology* and *biological homology* (Roth, 1984, 1988, 1991; Wagner, 1989a,b; Minelli and Peruffo, 1991; Haszprunar, 1992).

There are many definitions of phylogenetic homology (see references in Rieppel, 1980, 1988; Patterson, 1982, 1988; Bock, 1989; de Pinna, 1991; and earlier chapters). Two features are phylogenetically homologous if they are derived from the same feature in a common ancestor. In the framework of cladistics, homology has been equated with synapomorphy. This approach to homology has also been termed a taxic one because phylogenetic homologies imply a hierarchy of groups (Eldredge and Cracraft, 1980; Patterson, 1982; Rieppel, 1988; de Pinna, 1991).

Recent attention to the concept of homology has focused on the causal basis of morphological conservatism. Van Valen (1982, p. 305) proposed that homology is "correspondence caused by a continuity of information." In this context, "information" describes the genetic, epigenetic, and other mechanisms that specify phenotypes. The manner in which "information" is transmitted defines the types of comparisons that are needed. The central issue in causal analysis of homology is recognition of the classes of biological processes that cause the invariance, lability, and individuality of morphological structures.

Biological homologies are defined as features that share a set of specific developmental constraints. Biological homology has been defined as follows:

> Structures from two organisms or from the same individual are homologous if they share a set of developmental constraints, caused by locally acting self-regulating mechanisms of organ differentiation. These structures are thus developmentally individualized parts of the phenotype. Wagner (1989a, p. 62).

Biological and phylogenetic approaches to homology exemplify different roles for ontogenetic data in the analysis of homology. At first glance, these homology concepts appear to conflict; yet, when we distinguish between the different issues involved in the analysis of homology, they represent two complementary approaches to understanding the evolution and stability of structure.

II. THREE ISSUES IN THE ANALYSIS OF HOMOLOGY

To discuss the role of ontogenetic data in the analysis of homology we must distinguish between the criteria used to formulate, test, and mechanistically explain these hypotheses. The criteria used in each of these levels of analysis may at times differ, but they are fundamentally interrelated as discussed below.

A. Formulation of Hypotheses of Homology

There are few guidelines for the discovery of new scientific hypotheses, and the methodology for the formulation of hypotheses of homology is not well constrained. The distinction between the acts of formulating and testing hypotheses of homology has long been recognized (e.g., Jardine, 1970; Rieppel, 1980, 1988; Kluge and Strauss, 1985; Bock, 1989; de Pinna, 1991). One line of analysis seeks specific criteria that allow for proposal of hypotheses of homology (e.g., Remane, 1952). This approach identifies different classes of similarities that may suggest homologous correspondences. In addition, classical evolutionary morphologists frequently identify "key" features that provide special information about morphological homology and phylogenetic relationships. Certain functional, developmental, and structural properties are seen as fundamental and able to change only in certain directions. Dollo's Law, Haeckel's Biogenetic Law, Morse's Law (see below), and other theories of transformation have provided inspiration for the discovery of homologies. Whereas these theories of transformation are legitimate criteria for the formulation of hypotheses of homology, they neither test nor validate hypotheses themselves. The ability to test theories of transformation must rely on noncircular criteria. Generalizations about phylogenetic transformations cannot be used to both recognize and test hypotheses of homology and phylogeny.

Ontogenetic data can greatly aid the formulation of hypotheses of homology. Comparative analyses of ontogenetic sequences and mechanisms can suggest likely and unlikely character definitions, correlations, and morphological transformations. Experimental manipulations of ontogeny, comparative studies of ontogenetic sequences, and the analysis of intraspecific patterns of variability can all provide information about potential homologues. These data can provide clues that enable the formulation of robust hypotheses of homology—ones that can be tested by historical data as discussed below.

B. *Tests and Evaluation of Hypotheses of Homology*

A major methodological constraint in the analysis of homology lies in the test or the validation of proposals of homology. The test of an hypothesis of homology relies on a knowledge of phylogenetic history. Recognition and test of phylogenetic homology is a two-step procedure: hypotheses of homology are proposed and later tested by phylogenetic analysis (de Pinna, 1991 and references therein). The proposal of an hypothesis of homology carries implicit phylogenetic information because it is as much a statement about character evolution as it is about the phylogenetic relationships of the taxa involved. That is, when two features are deemed homologous an explicit prediction is made about the phylogenetic relationships of the taxa under comparison. These phylogenetic predictions can be used to test an hypothesis of homology. If a particular hypothesis of homology is not supported by phylogenetic analysis, then the original hypothesis of homology may be in error. The phylogenetic tests of a particular hypothesis of homology rely on congruence with a robust phylogeny (Patterson, 1982, 1988; de Pinna, 1991, and see Chapters 2–4).

Phylogenetic homology is an explicit statement about common descent and is only indirectly related to *similarity*. Not all similarities are homologous and, conversely, homologous features may be very dissimilar. Similarity has been cited as being both a recognition criterion for homology and a test of hypotheses of homology. The use of similarity as a cue in the recognition of homologous features is not controversial. Similarity alone is an exceedingly weak test of an hypothesis of homology (Bock, 1989). Dissimilarity of structures or morphogenetic processes are not strong falsifiers of hypotheses of homology.

Criteria used to recognize and test homologies are one and the same for many developmental studies of homology because many of these approaches are ultimately based on overall similarity of genetic and epigenetic interactions. An hypothesis of biological homology, for example, is a statement about the manner in which a set of features is able to vary. If features share a set of conserved developmental constraints, they will maintain similar norms of reaction and will have similar patterns of biased variation. Hypotheses of biological homology will contain a phylogenetic "signal" if the developmental mechanisms that are involved in the formation of a biological homologue are so complex that they are unlikely to have evolved twice. As a consequence, these biological homologues should be congruent with robust phylogenetic hypotheses. Unfortunately, we have no *a priori* way of knowing how many times a structure, or a set of epigenetic interactions, has evolved. There is no rigorous way in which biological homology concepts can be used to distinguish between phylogenetic homology and homoplasy. This does not imply that the biological approach to homology carries no phylogenetic information. Analyses of biological homologies play a central role in the formulation and mechanistic explanation of hypotheses of homology. Phylogenetic tests of these hypotheses can reveal the extent to which these developmental processes are conserved.

Biological and phylogenetic hypotheses of homology may not be congruent; i.e. experimental and descriptive studies of development may suggest that a particular structure is homologous, but phylogenetic information may suggest multiple independent origins for these biological homologues. Incongruence between biological homologues and phylogenetic ones can result from:

1. incorrect hypotheses of biological homology
2. incorrect phylogenetic hypotheses
3. homoplasy of form-generating mechanisms

The discrimination between these three outcomes involves an ongoing interplay between the biological analysis of putative homologues and analysis of their phylogenetic implications.

C. Mechanistic Explanation of Homology

The mechanistic study of homology and homoplasy entails the biological explanation for morphological continuity and change. Wagner (1989a,b) describes three important features of homologues: (1) their evolutionary conservation, (2) their individuation relative to other aspects of the phenotype, and (3) their uniqueness (i.e., their origin in monophyletic groups). A true mechanistic understanding of homology entails a unified explanation of each of these three fundamental properties of homologies. A host of biological processes, including (in part) stabilizing selection, genetic and epigenetic constraints, and lack of additive genetic variability, can produce evolutionary conservatism. Furthermore, the developmental and genetic mechanisms that generate a particular homologue should be sufficiently individuated that this particular feature can express its own characteristic features (Wagner, 1989a,b).

If we acknowledge that developmental systems and developmental constraints can evolve, then phylogenetic analysis plays a key role in mechanistic analyses of homology. Phylogeny informs about the evolutionary conservation, uniqueness, and individuation of homologues. The mechanistic explanation of homology, however, relies on more than a knowledge of phylogenetic history — it involves an analysis of the mechanisms that underlie evolutionary conservation and individuation. The demonstration of developmental and genetic biases on the production of new phenotypes entails the analysis of intra- and interspecific variation, teratology, and experimental manipulation of developing systems. Historical analysis of these data can provide information about the nature of the constraints acting on form and reveal the extent to which these mechanisms have evolved.

III. DEVELOPMENTAL APPROACHES TO HOMOLOGY

Two different perspectives dominate current developmental approaches to the problem of homology and these are reviewed below.

A. *Recapitulation: Homology and Polarity*

In strict Haeckelian interpretations, recapitulation was used to recognize, validate and mechanistically explain homologies. Strict notions of recapitulation either implicitly or explicitly assume a conservatism of development. Two types of comparisons can be performed under the recapitulationist framework: (1) comparison of "corresponding" features during the ontogeny of a single individual and (2) comparison of both ontogenies and adult features in different taxa. Haeckelian interpretations assume that one can homologize a single structure during the ontogeny of an individual and that this *ontogenetic homology* (*sensu* Haszprunar, 1992) can be extended to comparisons between taxa. The framework of recapitulation allows both recognition of homologues and polarization of different character states.

A major example of this approach comes from studies of tetrapod limbs. Morphological comparisons, using both extant and fossil forms, suggest that the reduction or loss of different components of the limb skeleton is a dominant mode of limb evolution (Hinchliffe and Johnson, 1980). Under the guise of recapitulation, the development of these limbs was seen to parallel the phylogenetic pattern of reduction. In its early development the limb of tetrapods was seen to recapitulate the exact number of limb elements seen in primitive fossil taxa (Holmgren, 1955). These comparative studies of development provided a framework for analysis of homology of the elements of the limb skeleton. Similarity of the early stages of the ontogeny of the limb skeleton in diverse tetrapod taxa suggested homology between adult carpal and tarsal elements even though the adult elements may share few phenotypic similarities.

Recent embryological studies have contradicted a strict application of this view. Limbs do not recapitulate the exact number of elements of fossil taxa in their early development (Hinchliffe, 1977; Hinchliffe and Griffiths, 1983; Shubin and Alberch, 1986). Although reduction or loss of elements occurs during evolution synapomorphies of higher level taxa can be seen very early in development (Shubin, 1991). Tetrapod limbs share many features of limb development, but these shared features do not suggest either von Baerian or Haeckelian recapitulation.

B. Structuralist Perspectives

Mechanistic approaches to ontogeny can lead to a rejection of the notion of homology altogether (Goodwin and Trainor, 1983). Small changes in developmental parameters can produce coordinated changes across the entire organism. Goodwin and Trainor (1983) argue that the recognition of homologous features involves the atomism of phenomena that are only part of integrated dynamic systems. Individual anatomical features, to Goodwin and Trainor, may not have an independent identity. This lack of individuality is a product of global morphogenetic mechanisms that cannot necessarily be atomized into smaller discrete processes; see discussion by Goodwin in Chapter 6.

Tetrapod limb reduction serves as an example of the structural approach to comparative biology. Comparative morphologists have proposed homologies between different digits and phalangeal elements of tetrapod limbs. These homologies are typified by the comparison of phalangeal formulae of the limbs of different tetrapods. The loss of a digit, or a single phalanx, is a phylogenetic pattern seen independently in many tetrapod taxa. When comparative morphologists compare a reduced, four-digited limb with a primitive pentadactyl pattern they propose that a specific digit (usually either digit I or V) has been lost. This approach implicitly assumes that the different digits have an independent phylogenetic identity and that this individuality can be used to atomize the limb skeleton into discrete parts.

Goodwin and Trainor (1983) suggest that a developing limb arises within a developmental field that is controlled by globally acting developmental parameters. Evolutionary change, in this perspective, is the consequence of the perturbation of these global parameters. Consequently, Goodwin and Trainor propose that individual digits do not have a discrete developmental identity. The act of homologizing individual digits involves an artificial atomism with no relevance to developmental biology.

The applicability of the structuralist argument depends on the extent to which developmental systems are individuated or "compartmentalized." Goodwin and Trainor (1983) clearly accept that one can homologize whole limbs in tetrapods. If they accepted purely global mechanisms of morphogenesis, then they could not discuss and compare heads or appendages in different vertebrates. Wagner (1989ab) suggests that the most meaningful comparisons (and in fact the only possible ones) are between developmentally individuated parts of the phenotype. The degree to which developmental systems are individuated dictates the resolution at which homologies can be drawn and the extent to which Goodwin and Trainor's notion applies in particular cases (Shubin and Alberch, 1986; Wagner, 1989a.b; Minelli and Peruffo, 1991).

Phylogenetic analysis and experimental embryology reveal instances where elements may be developmentally individuated. Digital reduction in the theropod hand follows a consistent pattern; digits from the posterior side of the limb tend to be reduced while other digits remain unchanged (see below). This suggests that the posterior digits can vary independently of those of the anterior side. Similar examples of individuation come from studies of experimental embryology. Application of mitotic inhibitors to developing amphibian limbs results in nonrandom, taxon-specific, patterns of digital loss (Alberch and Gale, 1985). One digit will be lost with no phenotypic modification of other limb elements. If, in these experimental manipulations a single digit was lost and the remaining digits were also altered, the reconstruction of homology of these teratologies would not be possible. The fact that we can make highly corroborated hypotheses of homology using discrete features provides important information on the developmental individuation of morphological structures. The historical analysis of ontogeny (and phylogenetic tests of hypotheses of homology) can tell us about the conservatism, lability, and resiliency of developmental mechanisms.

IV. CONFLICTS BETWEEN HISTORY AND ONTOGENY

A. Homology of the Digits of the Avian Wing

The homology of the digits of the avian limb has been cited as an example of conflict between ontogenetic and historical approaches to homology (see review in Hinchliffe and Hecht, 1984). Several interpretations of the homology of the avian digits have been debated over the past 170 years. The three digits of birds wings have been proposed to be homologous to digits I-II-III, II-III-IV, I-II-IV or III-IV-V of other archosaurs. Hinchliffe and Hecht (1984) argued that paleontological data supports the I-II-III view, while embryological information supports the II-III-IV hypothesis. While this is in large part correct, these different interpretations reflect different approaches to the analysis of homology and phylogenetic relationships as well.

Some embryologists and many paleontologists have proposed that the digits of the avian wing are homologous to digits I, II, and III of diapsids (Meckel, 1821; Gegenbaur, 1864; Huxley, 1868 Rosenberg, 1873; Jeffries, 1881; Furbringer, 1888; Parker, 1888; Zehntner, 1890; Broom, 1906; Siegelbaur, 1911; Heilmann, 1926; Steiner, 1934; Romer, 1955; Ostrom, 1976; Gauthier, 1986).

Several lines of evidence are used in support of the I-II-III hypothesis: (1) similarity of the phalangeal formula of *Archaeopteryx* to that of other archosaurs, particularly theropod dinosaurs, (2) phylogenetic analysis of digital reduction in theropods, and (3) phylogenetic analyses that suggest that birds are derived from three-toed theropods.

The most widely cited support of the I-II-III hypothesis comes from comparison of the manus of *Archaeopteryx* with the manus of archosaurs (e.g., Gauthier, 1986). The phalangeal formula of the manus of *Archaeopteryx* is 2-3-4-x-x (designations of phalangeal formulae follow the suggested terminology of Padian, 1992). The primitive archosaur phalangeal formula is a matter of debate (Steiner, 1929; Romer, 1955). Romer (1955) hypothesizes that the phalangeal formula of "reptiles" is 2-3-4-5-3 and that this formula is also a general characteristic of archosaurs as well. The three digits of *Archaeopteryx* have similar phalangeal formulae as digits I, II, and III of these primitive pentadactyl archosaurs (Fig. 1).

· If the digits of *Archaeopteryx* are I-II-III, and have a phalangeal formula of 2-3-4-x-x, then the most parsimonious hypothesis for the homologies of the digits of bird wing is I-II-III. This hypothesis implies that several phalanges have been lost during the evolution of modern birds. Further evidence comes from analysis of the trend of digital reduction in the outgroups of birds — various theropod dinosaurs. Theropods tend to lose or reduce the postaxial digits of the manus (Fig. 1). The loss of metacarpal V is a synapomorphy of Theropoda (including Aves) and digit IV is reduced in several taxa (including *Archaeopteryx* and birds) independently (Figs 1 and 2).

The II-III-IV interpretation of the digits of the avian manus has been supported by both embryological and paleontological data (Owen, 1836; Morse, 1872; Leighton, 1894; Prein, 1914; Montagna, 1945; Holmgren, 1955; Tarsitano and Hecht, 1980; Thulborn and Hamley, 1982; Hinchliffe and Hecht, 1984; Shubin and Alberch, 1986; Müller and Alberch, 1989; Müller,

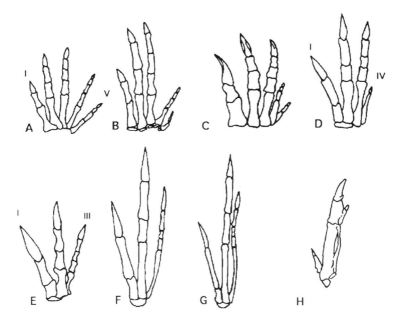

Fig. 1. The manus of selected archosaurs: (A) *Crocodylus* sp. (Crocodylomorpha), (B) *Heterodontosaurus* (Ornithischia), (C) *Massospondylus* (Sauropodomorpha), (D) *Syntarsus* (Ceratosauria), (E) *Allosaurus* (Carnosauria), (F) *Deinonychus* (Dromaeosauria), (G) *Archaeopteryx* (Aviale), (H) *Gallus* (Aviale). Drawings not to scale. (Modified from Gauthier, 1986.)

1991). Several criteria have been cited as evidence: (1) comparative analysis of the vascular supply of the avian manus (Owen, 1836), (2) recapitulation (Holmgren, 1955), (3) topological comparisons of precartilagenous blastema of chicks and other amniotes (Hinchliffe and Hecht, 1984), (4) sequence of chondrogenesis (Shubin and Alberch, 1986; Muller and Alberch, 1989), and (5) the relative size of the digits of the manus (Thulborn and Hamley, 1982).

Under the guise of recapitulation, some embryologists have identified a limb archetype that makes a transient appearance during early development of the hand and foot. Holmgren (1933, 1955) proposed that all five metacarpals appear during development of the avian hand, but that digits I and V are resorbed in later stages. Most embryologists have been able to observe only four digits, so identifications rely on other comparative evidence. Montagna (1945) identified the four digits as being II-III-IV-V and this hypothesis relies on the fact that most other amniotes have a tendency to lose digit I.

Morse's (1872) "law of digital reduction" serves as the basis for many supporters of the II-III-IV view. Morse (1872) noted that the first digits to be lost in most amniotes are digits I and V. These outer digits are reduced or lost many times independently in squamates and mammals and the expectation of this trend has been extended to analysis of the avian wing (Hinchliffe and Hecht, 1984). Hinchliffe and Hecht (1984) support the II-III-IV interpretation because of the presence of anterior and posterior necrotic zones that would tend to remove digits I and V. These authors also suggest that the extreme postaxial border is defined by the position of the pisiform and that digit IV cannot lie along this axis. As a consequence, they identify the most postaxial digit as digit V.

These hypotheses of homology can be tested by analyzing their phylogenetic consequences. Phylogenetic evidence (Fig. 1; Gauthier, 1986) strongly suggests that (1) Aviale (including *Archaeopteryx* and extant birds) is a monophyletic group, and (2) that the outgroups of Aviale are various nonavian theropod dinosaurs. This phylogenetic hypothesis is supported by numerous skeletal characters. No matter whether one accepts either the I-II-III or the II-III-IV homology scheme, the distribution of characters supporting this hypothesis must be taken into consideration. This phylogenetic scheme suggests that loss of the fifth metacarpal and reduction of the fourth metacarpal is a derived character for theropods (Fig. 2). This homology is based on the conserved phalangeal formulae (2-3-4-x-x), size and shape of the anterior three digits (Figs. 1 and 2). Derived theropods have only three digits, and these three digits mostly retain a phalangeal formula of 2-3-4-x-x. These patterns strongly support the notion that theropods possess digits I, II, and III.

If one accepts the I-II-III hypothesis of homology, a homoplasy of developmental sequence must be accepted (Muller and Alberch, 1989). Furthermore, acceptance of this scheme suggests that the hand of theropods exhibits a very unusual sequence of digital reduction. Digital reduction in most amniotes (in both the manus and pes) follows a stereotypical pattern such that the first digits to be lost are digits I and V (Hinchliffe, 1989ab; 1991). In extreme cases (such in those of the hands of lizards of the genus *Lerista*, Greer, 1990), digits II and then III will be lost (Hinchliffe, 1991). The apparent stability of digit one in the theropod manus is unusual, especially in light of the fact that the theropod foot shows the standard sequence of digital reduction. The phylogenetic interpretation of Gauthier (1986) unambiguously supports this view of nonavian theropod digital reduction. The extension of this view to birds involves homoplasy either in developmental patterns (in the I-II-III interpretation) or in numerous other skeletal characters (in the II-III-IV interpretation).

What are the phylogenetic implications of retaining the II-III-IV hypothesis? One could use the II-III-IV interpretation to specify a new phylogenetic hierarchy. The II-III-IV interpretation could be used to hypothesize that birds (including *Archaeopteryx*) are not related to theropod dinosaurs and that they are more closely related to other archosaurs. To accept this view, one would have to accept a very large amount of homoplasy of characters of the skull, axial skeleton, and limbs. With regard to limbs, similar phalangeal formulae and digital proportions of the manus of *Archaeopteryx* and theropods would be seen to evolve in parallel.

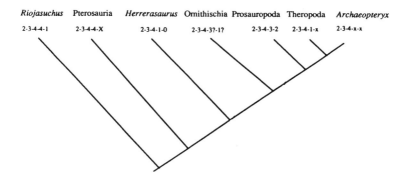

Fig. 2. Phylogenetic relationships (after Gauthier, 1986 and Sereno, 1991) of selected archosaurs are included with the generalized phalangeal formulae of the terminal taxa. The taxon, Aviale, is a monophyletic group that includes *Archaeopteryx* and birds.

The II-III-IV hypothesis of homology could also be interpreted in light of the phylogenetic hierarchy proposed by
Gauthier (1986). One could accept that the digits of theropods
are I-II-III and the digits of birds (including *Archaeopteryx*) II-
III-IV. This scheme would involve (1) loss of digit I, (2) phylogenetic reappearance of digit IV, including the metacarpal and
four phalanges, (3) loss of two phalanges on digit II, (4)
addition of one phalange on digit III. Furthermore, one would
have to accept that digits II-III-IV of *Archaeopteryx* convergently acquire the same proportions, size, and shape as
digits I-II-III of nonavian theropods. Alternatively, one could
accept that the digits of *Archaeopteryx* and nonavian
theropods are I-II-III and that extant birds share digits II-III-IV.
This case involves a similar reappearance of digit IV as well as
the loss of phalanges on several of the interior digits. If one
used developmental information to reinterpret the digits of the
primitive theropod hand as II-III-IV-V, similar difficulties
would be encountered.

Phylogenetic analysis that relies on the total evidence
reveals the implications of each hypothesis of homology and
supports the I-II-III view. It is interesting that the unique trend
in the loss of digits IV and V of theropods would be present
regardless of whether one identifies the digits of birds as being
I-II-III or II-III-IV.

This example underscores the importance of fossils in the
interpretation of developmental homologies. The interpretation
of developmental patterns of birds must, perforce, be strongly
affected by paleontological data. Without fossil theropods, and
the phylogenetic interpretations that they imply, the II-III-IV
interpretation of the homologies of the avian digits would,
perhaps, be a more likely interpretation.

B. Digits of the Urodele Hand

Comparative and experimental embryology appear to
suggest a different set of homologies for the digits of the
salamander limb than do phylogenetic criteria.

It has long been known that limb development in urodele
amphibians is different from that seen in other tetrapods
(Holmgren, 1933; Hinchliffe and Johnson, 1980; Shubin and
Alberch, 1986). The general sequence of digital formation in

larval urodeles proceeds in an anterior to posterior sequence while the opposite sequence is observed in other tetrapods. A strict use of developmental criteria suggests that the digits in urodeles are not homologous with those of other tetrapods and that tetrapods are polyphyletic.

Holmgren (1933) and Jarvik (1980) use developmental evidence to support notions of tetrapod polyphyly. In particular, Holmgren argued that salamanders were phylogenetically related to dipnoans whereas frogs and amniotes were derived from rhipidistians. Jarvik argued that anurans and urodeles arose from two different groups of rhipidistian fishes. In these schemes, the limbs arose twice. Both Holmgren and Jarvik use observed differences in the patterns of limb development between urodeles and anurans/amniotes as one of the criteria supporting their polyphyletic schemes. Holmgren (1933, 1939) cites several differences between urodele and anuran limb development used to support his polyphyletic scheme. He placed particular emphasis on the fact that the external shape of the urodele limb bud, and its pattern of centralia, recapitulates the dipnoan fin pattern.

Experimental manipulations also suggest that digital reduction in urodeles is different from that of anurans. Urodeles tend to lose digit V during treatments with the mitotic inhibitor, colchicine, whereas anurans lose digit I (Alberch and Gale, 1985). These homologies between teratological and "normal" limbs are suggested by the conserved phalangeal formulae of digits I-IV in normal and teratological limbs. These experiments suggest that the pattern of developmental constraints are different in both anuran and urodele limbs (Alberch and Gale, 1985).

Experimental and comparative embryology can suggest that either (1) urodele digits are not homologous to those of other tetrapods because tetrapods are polyphyletic, (2) there is not enough developmental individuation for us to delineate morphological homologies of the digits, or (3) all tetrapod digits are homologous, but underlying patterns of development and developmental constraints have evolved.

Monophyly of tetrapods is a well-supported phylogenetic hypothesis (Rosen *et al.*, 1981; Panchen and Smithson, 1987; Schultze, 1992). If one were to use differences in the constraints and patterns of development to propose that tetrapods are polyphyletic, one would have to accept a great deal of parallel evolution of other features. The distinction between the latter two hypotheses rests on how individuated tetrapod digits are during development and evolution.

Both phylogenetic and developmental studies attest to the difficulty in assigning a homologous number to each particular digit of urodele and anuran limbs. Anurans and urodeles were derived from a common ancestor that possessed more than four digits in the hand — the loss of digits is one of several characters that are a synapomorphy of a larger group that contains urodeles, anurans, and other tetrapods. No matter whether one can homologize *individual* digits with one another or not, the presence of digits is a taxic synapomorphy of the group Tetrapoda. Furthermore, a four-digited forelimb is seen in temnospondyls, frogs, and anurans, suggests that this characteristic is a synapomorphy of a group that contains these taxa (Lombard and Sumida, 1992). It is clear that during evolution of the group, the generative process of digital development has evolved to the point where there are different sequences of digital formation, and different developmental constraints acting on the limbs. A strict use of ontogenetic criteria to recognize and validate the homology of the four-digited hand would never have revealed this resiliency of developmental programs because tetrapods would have been seen to be polyphyletic.

V. CONCLUSIONS

Analysis of homology consists of three interrelated steps: the formulation of hypotheses of homology, the test of these hypotheses, and their mechanistic explanation. Comparative and experimental analysis of ontogeny provides important information necessary to propose homologies and to mechanistically explain the origin, individuation, and the evolutionary conservation of structures.

Evolutionary morphologists are confronted by holistic approaches (that suggest that there is not enough developmental and genetic individuation for homology to exist) and atomistic approaches (that decompose organisms into numerous traits without consideration of their biological basis) (Mayr, 1983). The historical analysis of ontogeny provides a median course, one that can reveal the extent to which features are individuated and the manner by which these integrating mechanisms evolve. Genetic and epigenetic interactions are examples of a class of mechanisms that can circumscribe limits to morphological integration and variation. An understanding of the developmental and genetic basis of morphological integration enables a biologically meaningful atomism of features into discrete characters. These developmental criteria enable the formulation of mechanistic hypotheses of homology that can be tested by phylogenetic analysis. Both ahistorical and historical analyses of development are vital components in the analysis of form. Ahistorical approaches provide hypotheses about likely and unlikely modes of morphological transformation. Historical data not only provide tests of these hypotheses but provide information about the manner in which the archetype has, itself, evolved.

ACKNOWLEDGMENTS

M. Balsai, D. Hauser, M. Seidl, and B. Slikas provided helpful comments on the manuscript. The author' work has been supported by NSF BSR 9006800.

REFERENCES

Alberch, P., and Gale, E. (1985). A developmental analysis of an evolutionary trend. Digital reduction in amphibians. *Evolution (Lawrence, Kans.)*, **39**, 8-23.

Bock, W. (1989). The homology concept: its philosophical foundation and practical methodology. *Zool. Beitr.* **32**, 327–353.

Broom, R. (1906). On the early development of the appendicular skeleton of the ostrich. *Trans. S. Afr. Philos. Soc.* **16**, 355-369.

de Beer, G. R. (1971). "Homology: An Unsolved Problem," Oxford Biol. Readers, No. 11, Oxford Univ. Press, London.

de Pinna, M. (1991). Concepts and tests of homology in the cladistic paradigm. *Cladistics* **7**, 367-394.

Eldredge, N., and Cracraft, J. (1980). "Phylogenetic Patterns and the Evolutionary Process. Method and Theory in Comparative Biology." Columbia Univ. Press, New York.

Furbringer, M. (1888). "Untersuchungen zur Morphologie und Systematik der Vogel." Amsterdam.

Gauthier, J. (1986). Saurischian monophyly and the origin of birds. *Mem. Calif. Acad. Sci.* **8**, 1-55.

Gegenbaur, H. (1864). "Untersuchungen zur vergleichenden Anatomie der Wirbeltiere. I. Carpus und Tarsus." Leipzig.

Goodwin, B. C., and Trainor, C. (1983). The ontogeny and phylogeny of the pentadactyl limb. *In* "Development and Evolution"(B.C. Goodwin, N. Holder, and C. C. Wylie, eds.), pp. 75-98. Cambridge Univ. Press, Cambridge.

Greer, A. E. (1990). Limb reduction in the Scincid lizard genus, *Lerista*. Variation in the bone complements of the front and rear limbs and the number of post-sacral vertebrae. *J. Herpetol.* **24**, 142-150.

Hall, B. K. (1992). "Evolutionary Developmental Biology." Chapman & Hall, London and New York.

Haszprunar, G. (1992). The types of homology and their significance for evolutionary biology and phylogenetics. *J. Evol. Biol.* **5**, 13-24.

Heilmann, G. (1926). "Origin of Birds." Witherby, London.

Hinchliffe, J. R. (1977). The chondrogenic pattern in chick limb morphogenesis: a problem of development and evolution. *In* "Vertebrate Limb and Somite Morphogenesis" (D. Ede, J. R. Hinchliffe, and M. Balls, eds.), pp. 293-309. Cambridge Univ. Press, Cambridge.

Hinchliffe, J. R. (1989a). An evolutionary perspective of the developmental mechanisms underlying the patterning of the limb skeleton in birds and other tetrapods. *In* "Ontogenese et Evolution"(J. Chaline, ed.), pp. 119-131. Geobios, Dijon.

Hinchliffe, J. R. (l989b) Reconstructing the archetype: Innovation and conservatism in the evolution and development of the pentadactyl limb. *In* "Complex Organismal Functions: Integration and Evolution in Vertebrates"(D. Wake, and G. Roth, eds.), pp. 171-188. Wiley, London.

Hinchiffe, J. R. (1991). Developmental approaches to the problem of transformation of limb structure in evolution. *In* "Developmental Patterning of the Vertebrate Limb "(J. Hinchliffe, J. Hurle, and D. Summerbell, eds.), pp. 313-324. Plenum, London.

Hinchliffe, J. R., and Griffiths, P. (1983). The pre-chondrogenic patterns in tetrapod limb develoment and their phylogenetic significance. *In* "Development and Evolution" (B. C. Goodwin, N. Holder,and C. C. Wylie, eds.), pp. 99-121. Cambridge Univ. Press, Cambridge.

Hinchliffe, J. R. and Hecht, M. (1984). Homology of the bird wing skeleton. *Evol. Biol.* **20**, 21-37.

Hinchliffe, J. R, and Johnson, D. R. (1980). "The Development of the Vertebrate Limb." Oxford Univ. Press, London.

Holmgren, N. (1933). On the origin of the tetrapod limb. *Acta Zool.* **14**, 185-295.

Holmgren, N. (1939). Contribution to the question of the origin of the tetrapod limb. *Acta Zool.* **20**, 89-124.

Holmgren, N. (1955). Studies on the phylogeny of birds. *Acta Zool.* **36**, 243-328.

Huxley, T. (1868). On the animals which are most nearly intermediate between birds and reptiles. *Ann. Mag. Nat. Hist.* [4] **4**, 2.

Jardine, N. (1970). The observational and theoretical components of homology: a study based on the morphology of the dermal skull roofs of rhipidistian fishes. *Biol. J. Linn. Soc.* **1**, 327-361.

Jarvik, E. (1980). "Basic Structure and Evolution of Vertebrates," Vol. 2. Academic Press, London.

Jeffries, J. A. (1881). On the fingers of birds. *Bull. Nuttall Ornithol. Club* **6**, 6-11.

Kluge, A. G., and Strauss, R. E. (1985). Ontogeny and systematics. *Annu. Rev. Ecol. Syst.* **16**, 247-268.

Leighton, A. (1894). The development of the wing of *Sterna wilsoni. Am. Nat.* **28**, 681-774.

Lombard, R. E., and Sumida, S. S. (1992). Recent progress in understanding early tetrapods. *Am. Zool.* **32**, 609-622.

Mayr, E. (1983). How to carry out the adaptationist program? *Am. Nat.* **121**, 324-334.

Meckel, J. F. (1821). "System der vergleichenden Anatomie," Vol. 2, Rengerschen Buchhandlung, Halle.

Minelli, A., and Peruffo, B. (1991). Developmental pathways, homology, and homonomy in metameric animals. *J. Evol. Biol.* **3**, 429-445.

Montagna, W. (1945). A reinvestigation of the development of the wing of the bird. *J. Morphol.* **76**, 87-118.

Morse, E. (1872). On the carpus and tarsus of birds. *Ann. Lyc. Nat. Hist., New York* **10**, 3-22.

Müller, G. (1991). Evolutionary transformations of limb pattern: heterochrony and secondary fusion. *In* "Developmental Patterning of the Vertebrate Limb" (J. R. Hinchliffe, J. Hurle, and D. Summerbell, eds.), pp. 395-405. Plenum, New York.

Müller G., and Alberch, P. (1989). Ontogeny of the limb skeleton in *Alligator mississippiensis*: developmental invariance and change in the evolution of archosaur limbs. *J. Morphol.* **203**, 151-164.

Ostrom, J. (1976). *Archaeopteryx* and the origin of birds. *Biol. J. Linn. Soc.* **8**, 91-182.

Owen, R. (1836). Aves. *In* "Todd's Cyclopedia of Anatomy and Physiology." Vol. 1, pp. 265-358.

Padian, K. (1992). A proposal to standardize tetrapod phalangeal formula designations. *J. Vert. Paleont.* **12**, 260-262.

Panchen, A. L. and Smithson, T. R. (1987). Character diagnosis, fossils and the origin of tetrapods. *Biol. Rev Cambridge Philos. Soc.* **62**, 341-438.

Parker, W. K. (1888). On the structure and development of the wing in the common fowl. *Phil. Trans. R Soc. London, Ser. B.* **179**, 385-395.

Patterson, C. (1982). Morphological characters and homology. *In* "Problems of Phylogenetic Reconstruction" (K. Joysey, and A. E. Friday eds.), pp. 21-74. Academic Press, London.

Patterson, C. (1988). Homology in classical and molecular biology. *Mol. Biol. Evol.* **5**, 603-625.

Prein, F. (1914). Die Entwicklung des vordern Extremitatenskellettes beim Haushuhun. *Anat. Hefte* **51**, 643-690.

Remane, A. (1952). "Die Grundlagen des Natürlichen Systems der Vergleichenden Anatomie und der Phylogenetik." Geest & Portig, Leipzig.

Rieppel, O. (1980). Homology, a deductive concept? *Z. Zool. Syst. Evolutionsforsch.* **18**, 315-319.

Rieppel, O. (1988). "Fundamentals of Çomparative Biology." Birkhäuser, Basel and Boston.

Romer, A. (1955). "Osteology of the Reptiles." Univ. of Chicago Press, Chicago.

Rosen, D. E., Forey, P., Gardiner, B., and Patterson, C. (1981). Lungfishes, tetrapods, paleontology and pleisiomorphy. *Bull. Am. Mus. Nat. Hist.* **167**, 159-276.

Rosenberg, A. (1873). Uber die Entwicklung des vordern Extremitatenskelettes bei einigen durch Reduction ihrer Gliedmassen charakteristischen Wirbeltiere. *Z. Wiss. Zool.* **23**, 116-166.

Roth, V. L. (1984). On homology. *Biol. J. Linn. Soc.* **22**, 13-29.

Roth, V. L. (1988). Homology revisited. *In* "Ontogeny and Systematics" (C. J. Humphries, ed.), pp. 1-26, Columbia Univ. Press, New York.

Roth, V. L. (1991). Homology and hiearchies: problems solved and unresolved. *J. Evol. Biol.* **4**, 167-194.

Schultze, H.-P. (1992). A comparison of controversial hypotheses on the origin of tetrapods. *In* "Origins of the Higher Groups of Tetrapods Controversy and Consensus" (H.-P. Schultze, and L. Trueb, eds.), pp. 27-67. Cornell Univ. Press (Comstock), Ithaca, N.Y.

Sereno, P. (1991). Basal archosaurs: phylogenetic relationships and functional implications. *J. Verteb. Paleont.* **11**, 1-53.

Shubin, N. (1991) The implications of "The Bauplan" for the development and evolution of the tetrapod limb. *In* "Developmental Patterning of the Vertebrate Limb" (J. R. Hinchliffe, J. Hurle, and D. Summerbell eds.), pp. 411-421. Plenum, New York.

Shubin, N., and Alberch, P. (1986). A morphogenetic approach to the origin and basic organization of the tetrapod limb. *Evol. Biol.* **20**, 319-387.

Siegelbaur, F. (1911). Zur Entwicklung der Vogelextremitat. *Z. Wiss. Zool.* **97**, 262-313.

Slack, J. W., Holland, P. W. H., and Graham, C. F. (1993). The zootype and the phylotopic stage. *Nature (London)* **361**, 490-492.

Steiner, H. (1922). Die ontogenetische und phylogenetische Entwicklung des Vogel-Flugelskelettes. *Acta Zool.* **3**, 1-54.

Streider, G. F., and Northcutt, R. G. (1991). Biological hiearchies and the concept of homology. *Brain Behav. Evol.* **38**, 177-189.

Tarsitano, S. and Hecht, M. (1980). A reconsideration of the reptilian relationships of *Archaeopteryx*. *Zool. J. Linn. Soc.* **69**, 149-182.

Thulborn, R. and Hamley, T. (1982). The reptilian relationships of *Archaeopteryx*. *Aust. J. Zool.* **23**, 611-634.

Van Valen, L. (1982). Homology and causes. *J. Morphol.* **173**, 305-312.

Wagner, G. P. (1989a). The origin of morphological characters and the biological meaning of homology. *Evolution (Lawrence, Kans.)* **43**, 1157-1171.

Wagner, G. P. (1989b). The biological homology concept. *Annu. Rev. Ecol. Syst.* **20**, 51-69.

Zehntner, L. (1890). Beitrage zur Entwicklung von *Cypselus melba*. *Arch. Naturgesch., Jahrg.* **56**, 189-220.

8

HOMOLOGY AND THE

MECHANISMS OF DEVELOPMENT

Günter P. Wagner

Department of Biology
Yale University
New Haven, CT 06511

Homology: The Hierarchical Basis of Comparative Biology
Copyright © 1994 by Academic Press, Inc.

I. INTRODUCTION

One of the fundamental properties of all life forms is that for each part or organ in one species there is another species with a corresponding part. Already in every day language these corresponding parts are called by the same name. We call the eye of a shark by the same name as the eye of a cow. Calling two things by the same name implies that we suppose they are "the same." Homology is a scientific conceptualization of this perception of "sameness."

With growing knowledge the idea of sameness became more refined by the realization that there are various aspects to it: the same structural organization (e.g., the "law of connections" by Geoffroy, see Hall, 1992), the same developmental origin (Roth, 1984), the same developmental constraints (Wagner, 1986), the same (genetic) information (Osche, 1973; Van Valen, 1982; Minelli and Peruffo, 1991), and common phylogenetic origin (Darwin, 1859). Most difficulties with the homology concept are because these various aspects of sameness are not congruent (Roth, 1988). There are organs with the same structural organization but radically different developmental pathways; there are structurally identical body parts which use different genetic information for their development (DeSalle and Grimaldi, 1992), and structurally identical organs do not need to have a common phylogenetic origin (homoplasia).

A necessary consequence of the discrepancy between various aspects of sameness is that not all these aspects can be included in the definition of homology [for excellent reviews of the different definitions of homology see Donoghue (1992) and Haszprunar (1992)]. Different homology concepts can be classified according to the aspect of sameness considered most fundamental. The *historical homology concept* assumes that common phylogenetic origin is more fundamental than the other aspects. The *morphological homology concept* refers to structural identity for its definition, and the *biological homology concept* refers to shared developmental constraints. Whichever definition one accepts, there is always a mix of advantages and disadvantages associated with the choice. To decide about a definition of homology is not a matter of right or wrong, but a matter of conceptual engineering. Given a certain scientific goal, and given what we know about evolution, morphology, and development an adequate homology definition can be constructed, but may differ from a definition utilized to meet another scientific goal.

The present discussion of development and homology is based on the "biological homology" concept:

> Structures from two individuals or from the same individual are homologous if they share a set of developmental constraints, caused by locally acting self-regulatory mechanisms of organ differentiation. These structures are thus developmentally individualized parts of the phenotype. (Wagner, 1989, p 62)

To justify its use I want to define the scientific goal for which this concept has been formulated. The goal is to explain patterns in the origin (Müller and Wagner, 1991) and evolution of morphological characters. It is thus intended to be a part of evolutionary biology, i.e., it should provide the conceptual framework for explaining patterns of intra- and interspecific variation. Therefore the primary emphasis is on constraints which cause the conservation of features characteristic for homologous characters. The purpose is not to provide criteria for recognizing phylogenetic relationships. Therefore criticisms that blame this approach as inadequate because it does not imply operational criteria for recognizing homology miss the point (Hall, 1992, Rieppel, 1992). But is it justified to talk about homology in this context when common ancestry has been the accepted definition of homology for over a century? Is it desirable to have a biological homology concept when the historical concept has been so valuable in guiding phylogenetic research?

These question have two aspects. One is a pragmatic and the other a conceptual/logical one. I deal with the conceptual one in the next section, in order to justify the approach to the problem of development and homology explained in this chapter. The pragmatic one I mention here briefly; it may help to avoid misunderstandings.

Until the second half of this century, phylogenetic research was based on the historical interpretation of homology defined as identity due to derivation from a common ancestor. Given this definition and independent criteria to recognize homology (Remane, 1952; Riedl, 1978; Rieger and Tyler, 1979; Rieppel, 1988), homologies can be understood as indications of phylogenetic relatedness. For this purpose the historical homology concept is an absolute necessity. However, with the advent of cladistics, a much more differentiated conceptual framework for phylogenetic reconstruction has emerged (synapomorphy, plesiomorphy, homoplasia, etc.) which functionally has taken over the role of the homology concept (Ax, 1987; Haszprunar, 1992; Patterson, 1982; and see Chapters 3 and 4). The classical homology concept does not play an indispensable role in cladistics any more. In fact, the homology concept is not required at all to pursue a cladistic analysis (Chapters 3-5). Only a shadow of it still hovers over the intuitive interpretation of cladistic methodology. Hence, reconstructing the homology concept for the purposes of an evolutionary biology of organismic design, as intended by the biological homology concept, cannot be problematic for comparative biology; its role there has already been taken over by the much better cladistic concepts.

II. WHY IS STRUCTURAL IDENTITY MORE FUNDAMENTAL FOR THE HOMOLOGY CONCEPT THAN COMMON ANCESTRY?

From the three classical components of sameness, structure, developmental and phylogenetic origin, identical developmental origin has been rejected on the basis of overwhelming embryological evidence (Hall, 1992, 1994). Too often do we find substantial developmental variation among structurally, and presumably phylogenetically identical parts. Hence, the question is where to start with a theory of homology — structural identity or phylogenetic origin? Here I will show that structural identity is conceptually required for the historical homology concept to make sense and thus must be the staring point of a theory of homology.

The original definition of homology by Owen does not refer to phylogeny at all. A "homologue is the *same* organ in different animals under every variety of form and function," (Owen 1843 cited after Hall 1992, p 56). Hence the question raised but not answered by this definition is, what does it mean to call two organs from different species "the

same," regardless of form and function? If Darwin's reinterpretation of homology would be more fundamental than
Owen's, it would have to provide an answer to the question
left open by Owen's definition on the nature of sameness.
However, this is not the case.

Taking Mayr's definition as a reference point, we realize
that the historical homology concept only adds a temporal
dimension to homology but in so doing relies also on an unspecified concept of "sameness." Mayr writes:

> A feature in two or more taxa is homologous when it is derived
> from the same (or corresponding) feature of their common ancestor
> (Mayr, 1982, p 45).

Hence the "sameness" of characters in two recent
species is "explained" by the sameness of each of these
features with the same character in a common ancestor (Fig.
1). The question raised by Owens definition of homology,
"What does it mean to call two organs in different species

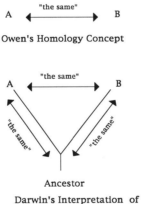

Fig. 1.   Comparison of the logical structure of Owen's
and Darwin's homology concepts. As a preevolutionary concept Owen's concept refers only to structural correspondence among recent species. Darwin's concept relates the
structures of recent species to corresponding ones in the
common ancestor of the two recent species. In neither
concept is the nature of the correspondence addressed.

the same," is not answered by the historical homology concept, but is used in the same unexplained manner as in Owens definition. Certainly, Darwin's reinterpretation does provide something not in Owen's concept.

It has long been known that the distribution of homologues among recent species follows a pattern of nested sets. The species sharing the characters of the Tetrapoda are a proper subset of the species sharing the characters of the gnathostomes, and so on. This pattern is neither explained nor addressed by Owen's definition. However, it is explained by Darwin's assumption that homologous characters are inherited from the common ancestor (Fig. 2). The historical homology concept explains why the distribution of homologous characters among recent species forms a set of nested sets, but relies as much on an unexplained notion of sameness as did Owen's original definition of structural identity.

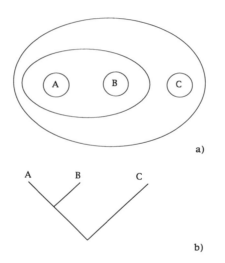

a)

b)

Fig. 2. Reinterpretation of the nested hierarchy of taxonomic groups as a tree of phylogenetic relatedness. (a) Three taxa, A, B, and C, grouped in higher taxa comprising (A & B) and [(A & B) and C] respectively. (b) This taxonomic hierarchy can be reinterpreted as a tree of genealogical relationships, such that A has a more recent common ancestor with B than C has with A & B.

The discussion in this chapter of the relationship between homology and development presumes that the main goal of a biological homology concept is to explain why certain parts of the body are passed on from generation to generation for millions of years as coherent units of evolutionary change (Roth, 1991). Hence the problem is to explain why there are individualized parts of the body that behave as units which retain their structural identity despite variation in form and function. In this chapter I concentrate on the developmental aspects of this problem.

III. DEVELOPMENT AND STRUCTURAL IDENTITY

Development is the generation of a structured material entity from a less structured precursor during ontogeny of an individual.

A. Developmental Processes

Three types of processes are known to contribute to the transformation of a zygote into a differentiated organism; self-assembly, fixation of a temporal pattern of activity into a spatial pattern, and spontaneous pattern formation (Gierer, 1985).

Self-assembly is the spontaneous aggregation of macromolecules into supramolecular structures provided the substances come together and physicochemical conditions are adequate. This process is largely guided by the principle of minimal enthalpy, just as is the generation of salt crystals out of a saturated solution of sodium chloride. Many subcellular structures (e.g., microtubules) are likely to be generated in this way.

The paradigm of a structure that is generated by fixation of a temporal pattern of cellular activity into a spatial one is the mollusc shell. A mollusc shell is generated by the deposition of skeletal material at the rim of the shell (Hyman, 1967; Jones, 1983; Kniprath, 1981). Hence, the temporal sequence of the activities of the cells depositing the material are frozen into structures of the shell. Temporally successive events lead to spatially neighboring features. These characters are usually thermodynamically metastable, passive (dead) structures. Another example is bird feathers (Sengel, 1976).

Spontaneous pattern formation is a process that requires thermodynamic disequilibrium and leads to a dissipative structure, generated and maintained at the expense of constant energy consumption. For a discussion of the physical principles underlying the formation of dissipative structures see Haken (1978) and Meinhard (1982).

This classification of developmental mechanisms is interesting in the present context because it suggests a classification of morphological structures form the thermodynamic point of view. On the one hand, organisms possess passive structures like microtubules, the mollusc shell or feathers. On the other hand there are activity-maintained, dissipative structures — most "living" structures.

B. Developmental Constraints

These different types of structures differ also in the causal factors that determine their structural identity. For instance, microtubule structure is mainly determined by the physicochemical properties of the tubulin molecule, which in turn is directly specified by genetic information. There is little if any influence of developmental processes on the structural identity of characters such as microtubules, except splicing and posttranslational modification. On the other hand the structure of a mollusc shell is completely determined by the temporal process of its development. The mollusc shell can be seen as a record of its own growth history (Jones, 1983); its structure is completely determined by its developmental pathway.

In both cases of passive structures, the identity of the structure is congruent with continuity of its material constituents. Mollusc shells do not show a turnover of material[1], and in each individual the older parts of the shell will still consist of the same atoms as they did when they were first

[1] Two notable exceptions exist in the gastropod families Neritidae and Muricidae. In the Family Neritidae, the inner whorls of the shell are remodelled (Vermeij, 1987, p. 244), while some muricid gastropods remove their spines during shell growth (Carriker, 1972).

formed. This is not the case in those organs and structures that consist of living cells, in particular those living structures that undergo massive changes in size during ontogeny of the individual.

A classical example of complete replacement of the material during growth of a structure is the bones of the skull in mammals. Figure 3 shows a comparison of the bend in the parietal bone of sheep of different ages. By comparing the bone of a 3-week-old sheep with the fetal bone, it becomes clear that at least the peripheral material of the fetal parietal bone has been removed and replaced by material at another location (Lacoste, 1923, cited after Starck, 1975). Most probably, however, all the bone has been replaced, given the constant breakdown of bone substance by osteoclasts and the buildup of bone material that takes place in lamellar bone. However, most (all?) of the acellular bones of bony fish are static structures, where material deposited early is not removed.

Structures that show a turnover of their constituents maintain structural identity in spite of noncontinuity of the material. The molecules and cells that made up the heart of a fetus are not the same ones that represent the same organ 10 years later. Hence, structural identity cannot be guaranteed by the morphogenetic process that set up the organ in the first place. Structural identity is continuously challenged during the period of organ growth, which requires the complete replacement of constituents in the transformation from a two-mm long structure of the embryo into a 10-cm piece of matter still having the same structural features as the small one. This challenge continuous into the adult stage as cell turnover and environmental and functional perturbations tend to disintegrate the organ. Maintenance of structural identity despite size transformation and cell turnover requires that the organ be actively maintained by cell interactions once the organ has been set up during morphogenesis.

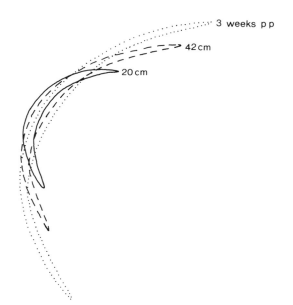

Fig. 3. Schematic cross section through the parietal bone of sheep in different ages. Note that the bend of the bone in younger individuals is much higher than in older ones. It follows that the material of the fetal bone has been replaced during growth of the animal to accommodate the shape change. (After Starck, 1975.)

C. Maintenance of Structural Integrity

Mechanisms that maintain the structural integrity of organs have to my knowledge not been a major focus of developmental biology, which is more concerned with the transformational aspects of development than with stability of structures. Knowledge about cellular functions maintaining an organ is mainly derived from pathology and neurobiology, where neuroplasticity has been a focus of research because of its possible implications for learning theory.

Given that structural integrity of a living organ cannot be guaranteed by the morphogenetic processes, one is forced to accept that regulatory mechanisms maintaining structures are a major factor in determining the structural identity of this class of morphological characters. Only those structural features that can be maintained in spite of perturbations after morphogenesis (e.g., growth) can be part of the "normal anatomy" of the character. In order to emphasize this difference the cellular mechanisms that act to maintain a structure have been called *morphostatic mechanisms* in contrast to *morphogenetic mechanisms*, which generate structure (Wagner and Misof, 1993).

As such the distinction between morphogenetic and morphostatic mechanisms indicates a difference only in the biological role of cellular activity and does not automatically imply a difference in the kind of processes involved. For instance, in the vertebrate brain cytological and biochemical evidence indicates a dynamic equilibrium of synapse formation and dissociation (Jacobson, 1978). This synapse turnover is considered to be a continuation of synapse formation during development in a dynamic equilibrium (Cotman, et al., 1981). However, the fact that there are two different biological roles for cellular activity holds the potential that morphostatic and morphogenetic processes can be mechanistically decoupled from each other. In Section IV some anecdotal evidence for the noncongruence of morphogenetic and morphostatic mechanisms is reviewed.

D. Summary

In summary, structural identity of actively maintained characters is not only determined by the generative processes transforming the anlage of the character into the character itself, but resides to a large degree in those processes that maintain the structural integrity of characters in spite of growth and other perturbations. This deemphasizes the role of generative developmental mechanisms which has important implications for the relationship between homology, in terms of structural identity, and developmental variation between species [Wagner and Misof (1993) and see below].

IV. AN ANECDOTAL REVIEW OF MORPHOSTATIC
DEVELOPMENTAL MECHANISMS

There is little systematic knowledge about morphostatic mechanisms in the literature of developmental biology. A few examples that served as a reference point in
formulating the hypothesis of morphostatic mechanisms are
introduced in the following section.

A. Innervation and Structural Maintenance

Research on vertebrate appendage regeneration revealed an interesting discrepancy between the role of
peripheral nerves on regeneration and development
[reviewed in Goss (1969) and Wallace (1981)].

It has long been known that denervated salamander
extremities cannot regenerate. Most likely this effect has to
do with a glial growth factor required for the proliferation of
some blastema cells (Brockes, 1984; Brockes and Kintner,
1986; Fekete and Brockes, 1988). However, it became clear
that structurally intact limbs can develop in the complete
absence of nerves (aneurogenic extremities) (Yntema,
1959). Moreover, such aneurogenic extremities are able to
regenerate.

The inability of denervated extremities to regenerate is
paralleled by an inability to compensate for minor injuries.
Denervated Amblystoma (=Ambystoma) limbs regress completely after minor lesions have been inflicted which include
the mesodermal tissues (Thornton and Kraemer, 1951).
Hence, not only is large-scale replacement impossible, but
maintenance of structural integrity is impaired in
denervated limbs, while limb morphogenesis is nerve independent.

A role of nerves in structural maintenance is not restricted to salamander limbs. Similar regressive phenomena
caused by denervation have been observed in fish fins and
catfish barbels (Goss, 1969). In the latter system structural
regression after denervation occurs even without additional
lesions (it is not yet known whether nerves are required for
development of the barbels).

This obvious role of nerves in structural maintenance and their minimal role during limb development may indicate a role of peripheral innervation as a generalized morphostatic factor. Once a structure has been setup and innervated the nerves play a role in maintaining what is there. Innervation may either lock-in the given structure (C. Vogl and B. Misof, personal communication), or activate structure-specific regulation mechanisms. Removing innervation from an adult structure may undercut the compensatory mechanisms necessary to maintain and repair the structure.

B. Neuromuscular Connections

Association of particular motor nerves and specific skeletal muscles is a paradigm of a phylogenetically conserved anatomical feature.

Innervation is considered to be the most reliable criterion for homology of muscles (Remane, 1952). The adult neuromuscular innervation pattern is set up by a multitude of developmental mechanisms (Holliday and Grobstein, 1981). These include axon pathway selection, target recognition, competition, and selective motor neuron death. Two of these factors, pathway selection and target recognition, are strictly generative, directing the set-up of connections. Synapse competition during development can be seen as a fine adjustment of the neuromuscular map. However, competition is also associated with synapse turnover during adulthood. Competition seems to play a major role in active maintenance of the innervation pattern during growth and addition of muscle fibers. Interestingly, the relative importance of synapse competition is correlated with the extent of growth and muscle fiber addition during postembryonic life.

Fish and amphibians show continuous growth with addition of muscle fibers. In those species competition has a strong influence on the neuromuscular projection pattern. However, in birds and mammals, which do not show continuous growth, evidence for synapse competition is weaker than in fish and amphibians. In amniotes the innervation pattern is primarily determined by pathway selection and target recognition. This indicates that continuous pattern regulation by competition contributes to the active maintenance of the innervation patterns in spite of growth of the target tissue.

C. Pectoral Fin Hooks

Pectoral fin hooks are anatomical specializations that are adaptations to the bottom dwelling life style of blennies (Brandstätter, et al., 1990; Zander, 1973). Briefly, empirical work has established three facts:

1. Fin hooks are specializations of the lower fin rays, consisting of three characters [an asymmetric regression of the fin web, a connective tissue pad associated with the bony fin ray, called a lepidotrichal cord, and a cuticular specialization of the epidermis covering the lepidotrichal cord] (Brandstätter, et al., 1990; see Fig. 4). The expression of these characters is spatially correlated. The lepidotrichal cord is most fully developed in the region of fin web regression and the cuticle is found only over the lepidotrichal cord.

2. Regeneration experiments show that expression of these characters is only loosely correlated, if at all, during histogenesis (Misof and Wagner, 1992). The sequence with which the characters appear during histogenesis is variable. For instance, there are cases where the cuticle comes later than the lc, but there are also cases where a cuticle is found

---

Fig. 4. (opposite). Schematic drawing of the first three fin rays of the pectoral fin, showing the characters that constitute a pectoral fin hook in *Salaria pavo* (Blenniidae, Perciformes). The fin is viewed form the outer side, i.e., the surface of the fin viewed from the outside when the fin is in its resting position. phr, bony fin rays; fw, fin web, lc, lepidotrichal cord (a connective tissue associated with the bony fin ray); cu, region where the epidermis is expressing a cuticle. Note the correlated expression between fin web regression, lepidotrichal cord, and cuticle. (Drawing by B. Y. Misof.)

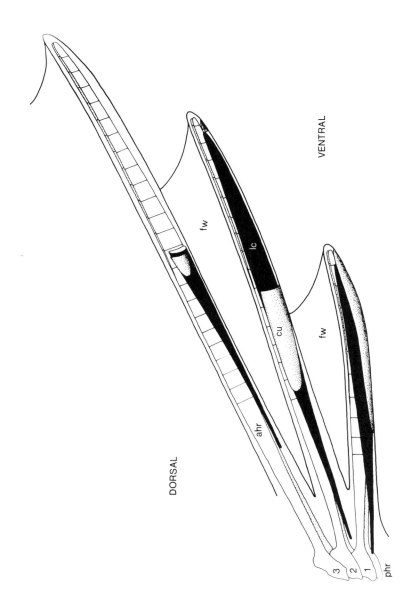

DORSAL

VENTRAL

fw

lc

cu

ahr

fw

3
2
1
phr

without a local contact to an lc. Similarly, cases have been found where the lc and the cuticle fully developed without any sign of fin web regression and *vice versa*. Such variation in the histogenetic sequence is incompatible with the view that fin hook characters are developmentally coupled. The stereotypic adult pattern appears only later in regeneration.

3. Perturbation experiments with adult fins show that the characters are mutually modulated and in part dependent on each others' presence for their maintenance; see Fig. 5 for a summary. For instance, amputation of the tip of a fin ray at the site of fin web attachment leads to a reactive fin web growth during regeneration of the fin ray tip (Wagner and Almeder, 1991). This is taken as evidence that the lc and/or the cuticle inhibits fin web growth, because during early stages of fin ray regeneration only bony fin rays are formed; the lepidotrichal cord and cuticle come later.

In another series of experiments it has been shown that resection of the lepidotrichal cord leads to loss of the cuticle (Wagner, unpublished observations), which implies that the lepidotrichal cord is necessary for maintenance of the cuticle. Transplantation of the lepidotrichal cord to non-

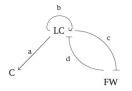

Fig. 5. Simplified scheme of tissue interactions maintaining the integrity of the fin hook. (a) The presence of the lepidotrichal cord (LC) is necessary to maintain the cuticle (C) of the epidermis, as shown by resection experiments. (b) The LC is autonomous in its maintenance, as shown by transplantation experiments. (c) The lc inhibits fin web (FW) growth, as shown by fin tip amputation experiments (Wagner and Almeder, 1991). This is the most plausible interpretation, although the possibility that the cuticle is the inhibitory agent has not been ruled out experimentally. (d) FW inhibits secretion of the glycoprotein matrix of the LC, as shown by FW resection experiments.

cuticular parts of the fin never led to an ectopic cuticular-
ization, even if the blennies are able to differentiate new
hooks as adults.

Finally, transplantation of the lepidotrichal cord to
ectopic locations leaves the cord intact for up to 8 weeks.
This shows that the lc is autonomous and does not depend
on the presence of the other characters in the fin hook,
while the cuticle and the fin web regression do. Thus the
adult pattern is maintained by constitutive epigenetic tissue
interactions.

Together these three facts imply that the correlated
expression of the characters in the adult is not due to
morphogenetic, generative (e.g., inductive) interdependen-
cies among them. Morphogenesis and histogenesis of the
subcharacters are largely independent. The correlated
expression is finally achieved by a secondary adjustment
after the subcharacters are established.

D. Conclusions

From the few examples discussed above it is clear that
there are cell-cell interactions that maintain the anatomical
structure in the adult, but it is also clear that the generative
and maintenance functions overlap to some degree. It is also
clear that there are certain structural features of actively-
maintained characters that are completely determined by
the generative processes. For instance, even if the neuromuscular innervation pattern is actively maintained in fishes
and amphibians, the structure of a brachial plexus is not
regulated by morphostatic mechanisms, but completely
determined by path-finding processes during embryonic
development.

Finally, it is not yet known to what degree the adult
pattern is specified by the morphostatic mechanisms and to
what degree by generative mechanisms. It could be that
morphostatic mechanisms are unspecific, freezing whatever
the morphogenetic process has produced. Or they could
have their own goal function which is actively sought against
accidental and developmental variation. In particular, it
needs to be determined whether those characters known to
be conserved in spite of developmental variation between

species are maintained by specific morphostatic mechanisms.

V. CONSTRAINTS AND STRUCTURAL IDENTITY

In the previous two sections the question of structural identity of morphological characters has been discussed in the context of ontogenetic development. However, this can only be a prelude to the real question of homology, which includes maintenance of structural identity in the course of phylogenesis. In particular we will ask what the impact may be of developmental mechanisms on the evolutionary modification of morphological characters and how and why structural features have been conserved to a degree that one still can talk of the same character, even if the lineages have been separated for many millions of years.

If one modifies Owen's definition to "A homologue is the structurally identical organ in two species regardless of form and function" then the task of a biological homology concept becomes clear: "How can a part of the body maintain its structural identity during the course of phylogeny, regardless of changes in form and function?" This question is far from trivial, since virtually all morphological characters are build anew in each generation. There is no continuity of descent as for a DNA molecule which is copied in a template fashion. Given further the high degrees of heritable variation measured for all aspects of the phenotype (Mousseau and Roff, 1987), one wonders why some aspects of the phenotype should be maintained despite changes in size, form and function.

Assuming the general correctness of the synthetic theory, then there are basically two explanations for conservative features of the phenotype: stabilizing selection, and lack of heritable variation. Let us follow these leads one by one.

A. Stabilizing Selection

To explain the maintenance of a character state by stabilizing selection simply means that any deviation from the character state is penalized by lower fitness. If the character is very old one has to infer that the causes for stabilizing selection are also very old and persistent. This can be due to persisting environmental features or inherited features of organismal design. The latter is the case when different parts of the organisms interact to determine the fitness of the organism, as for instance the different parts of the vertebrate eye (Riedl, 1978).

Conservativeness of a character caused by functional dependency of other characters is called *burden* (Riedl, 1978), Stearns, 1992). Burden explains the maintenance of developmental characters, even if their adult function has long been taken over by other characters (Riedl, 1978). Hence a character that plays an indispensable role in the development of functionally important characters will be phylogenetically conserved due to "burden," or stabilizing selection.

B. Developmental Constraints

Lack of heritable variation for a character has been called *developmental constraint,* to emphasize that organization of the developmental process can be responsible for biases and restrictions of phenotypic variation (Maynard Smith, *et al.,* 1985). Consequently it has been proposed that the conservation of structural identity of morphological characters (homology) can be due to the effect of developmental constraints (Wagner, 1986). However, an important qualification is necessary to this hypothesis, since homologous characters can have substantial differences in their developmental pathways

The developmental variation of homologous characters has been documented and reviewed many times (see, for example, Hall, 1992, 1994); I will not duplicate these presentations here. However, an eclectic list of 10 examples from Wagner and Misof (1993) is included on the next page. These represent variable development of homolgous characters where the variant developmental pathway does not have any substantial consequences for the adult pattern.

1. Meckel's cartilage is induced by different tissues in amphibians, birds and mammals (Hall, 1984).

2. The orbitophenoid bone develops as a membrane bone in the Amphisbaenia (Amphibia) but as a replacement bone in other vertebrates (Bellairs and Gans, 1983).

3. The columella of the chick can be derived from ectomesenchyme or mesodermal mesenchyme (Noden, 1983)

4. The sequences of chondrogenic condensations in the limbs of *Ambystoma mexicanum* and *Triturus marmoratus* are different (Blanco and Alberch, 1992)

5. Lenses develop from ectoderm but regenerate from the iris in salamanders (references in Goss, 1969)

6. The body of insects develops in two very different modes in different insect orders: short germ and long germ development (references in Raff and Kauffman, 1983).

7. The role of *even-skipped* genes is different in segment development of flies (*Drosophila*) and grasshoppers (*Schistocera americana*; Patel *et al.*, (1992).

8. Directly developing anurans (*Eleutherodactylus*) differ from indirectly developing ones in their morphogenetic sequences but not in their adult morphology (reviewed in Raff and Kaufman, 1983).

9. Directly developing sea urchins (*Heliocidaris*) differ from indirectly developing ones in their morphogenetic sequences but not in their adult morphology (Wray and Raff, 1989, 1990, 1991).

10. Cell autonomous determination of the D quadrant cell in spiralian coelomates is derived from inductive modes of development without consequences for adult morphology (Freeman and Lundelius, 1992).

None of the examples just listed refers to a morphological character that belongs to the class of passive structures. This is in accordance with the observation that these are either completely independent of developmental processes, like the self-assembly of subcellular structures, or that their structure is completely determined by the developmental process, like the mollusc shell or the feather; see Sections III, A and B. All the examples are structures that are either due to pattern formation and/or actively maintained. Note also that the skeleton of a sea urchin is a living tissue which is highly plastic and thus actively maintained in the adult.

The hypothesis to reconcile developmental constraints with variable developmental pathways is a corollary of the experimental results that suggest a distinction between generative morphogenetic and maintaining morphostatic processes (Wagner and Misof, 1993). If the maintaining mechanisms are affected by genes other than the generative ones, then developmental pathways can vary while the adult pattern is kept the same by compensatory action of the morphostatic mechanisms. The adult pattern would then be determined by the stabilizing mechanisms as long as variation of the developmental process provides the necessary material for stabilizing the adult pattern.

Of course, the stabilizing influence of morphostatic mechanisms does not allow boundless variation of the developmental pathway. Only those variations of development would be permitted which provide the necessary cellular material at the right spot and at the right time during ontogeny. Within this range, the developmental pathways can have many degrees of freedom. For instance, it does not matter where the cells originated as long as they are able to interact with other cells to build and stabilize the particular character (e.g., the cells that come from the iris to regenerate a lens in a salamander or the cells of the columnella of the bird ear). The prediction is that the developmental variation observed among species covers the range of variation that is compatible with a conserved system of morphostatic interactions.

The relationship between constrained adult variation and broad developmental variation becomes clearer if two kinds of developmental constraints are distinguished:

Generative constraints (= developmental constraints in the narrow sense) are limitations to phenotypic variation caused by the properties of the generative process of development.

Morphostatic constraints are limitations to phenotypic variation caused by the stabilization of particular patterns rather than the inability of the generative processes of development to produce them in the first place.

Both are developmental constraints in the sense of Maynard Smith and co-workers (1985); they cause biases in the production of phenotypic variation, but they differ in the ontogenetic stage at which the relevant epigenetic interactions occur. Generative constraints arise from interactions during morphogenesis; morphostatic constraints arise from regulatory interactions maintaining the characters later on.

VI. CONCLUSIONS

The task of a developmental approach to the homology problem is to identify the mechanistic underpinnings of the structural identity of homologous characters. This task cannot be fulfilled by Darwin's evolutionary reinterpretation of Owen's homology concept, but it can account for the taxonomic distribution of homologous characters.

Ever since the influential paper by Spemann (1915) on the history and critique of the homology concept, research on the mechanistic basis of structural identity of homologous characters has been paralyzed by the discovery of different developmental mechanisms for homologous characters. This impasse is mainly caused by the assumption that generative developmental mechanisms determine the structural identity of morphological characters. Under this assumption, discrepancy between structural identity and variable development cannot be reconciled. The assumption of developmental determination is, of course, largely but not entirely true.

It is proposed that the structural identity of at least some morphological characters is determined by the stabilizing (morphostatic) mechanisms which maintain the characters, rather than by the generative processes. This possibility implies a new research program in experimental morphology: to determine the contribution of generative and maintaining mechanisms to the structural organization of a character. On the basis of this information it will be possible to predict the limits of developmental variation compatible with the structural identity of the character and to predict constraints on the evolution of the character. These predictions are testable by cladistic methods, using molecular phylogenies and cladistic reconstruction methods, just as it is possible to test adaptational hypothesis by the comparative method (Baum and Larson, 1991; Harvey and Pagel, 1991).

In summary, there are three components to a mechanistic understanding of homology: (1) The *causes* of structural identity in ontogenetic development (contributions of morphogenetic and morphostatic mechanisms); (2) the *constraints* which maintain the structural identity of a morphological character during phylogeny; (3) the *evolutionary factors* that lead to the establishment of a developmentally individualized structure. The prospect that such a research program can succeed is better than ever, since some of the genes involved in determining the structural identity of a morphological character are now known; see Tabin (1992) and Chapter 6 for a review on *Hox*-gene involvement in tetrapod limb development.

ACKNOWLEDGMENTS

The author wishes to thank the following individuals for reading this chapter and for numerous discussions on the subject: L. Buss, B. K. Hall, M. Laubichler, B. Misof, G. Müller, C. Pazamandi, O. Rieppel, L. Roth, R. Sattler, C. Vogl, A. Wagner, and D. Wake. I am indebted to David Lindberg for a discussion on gastropod shell development.

REFERENCES

Ax, P. (1987). "The Phylogenetic System." J. Wiley and Sons, Chichester and New York.

Baum, D. A., and Larson, A.(1991). Adaptation reviewed: a phylogenetic methodology for studying character macroevolution. *Syst. Zool.* **40**, 1-18.

Bellairs, A., d'A., and Gans, C.(1983). A reinterpretation of the amphisbaenian orbitosphenoid. *Nature (London)* **302**, 243-244.

Blanco, M. J., and Alberch, P. (1992). Caenogenesis, developmental variability, and evolution in the carpus and tarsus of the marbled newt *Triturus marmoratus*. *Evolution (Lawrence, Kans.)* **46**, 677-687.

Brandstätter, R., Misof, B., Pazmandi, C., and Wagner, G. P. (1990). Micro-anatomy of the pectoral fin in blennies (Blenniini, Blennioidea, Teleostei). *J. Fish Biol.* **37**, 729-743.

Brockes, J. P. (1984). Mitotic growth factors and nerve dependence of limb regeneration. *Science* **225**, 1280-1287.

Brockes, J. P. and Kintner, C. R. (1986). Glial growth factor and nerve-dependent proliferation in the regeneration blastema of urodele amphibians. *Cell (Cambridge, Mass.)* **45**, 301-306.

Carriker, M. R. (1972). Observations on the removal of spines by muricid gastropods during shell growth. *Veliger* **15**, 69-74.

Cotman, C. W., Nieto-Sampedro, M., and Harris, E. W. (1981). Synapse replacement in the nervous system of adult vertebrates. *Physiol. Rev.* **61**, 684-784.

Darwin, C. (1859). "On the Origin of Species by Means of Natural Selection or the Preservation of Favored Races in the Struggle for Life." John Murray, London.

DeSalle, R., and Grimaldi, D. (1992). Characters and the systematics of Drosophilidae. *J. Hered.* **83**, 182-188.

Donoghue, M. J. (1992). Homology. *In* "Keywords in Evolutionary Biology." (E. F. Keller and E. A. Lloyd, eds.), pp. 171-179. Harvard Univ. Press, Cambridge, MA.

Fekete, D. M., and Brockes, J. P. (1988). Evidence that the nerve controles molecular identity of progenitor cells for limb regeneration. *Development (Cambridge, UK)* **103**, 567-573.

Freeman, G., and Lundelius, J. W. (1992). Evolutionary implications of the mode of D quadrant specification in coelomates with spiral cleavage. *J. Evol. Biol.* **5**, 205-247.

Gierer, A. (1985). "Die Physik, das Leben und die Seele." Piper, München.

Goss, R. J. (1969). "Principles of Regeneration." Academic Press, New York.

Haken, H. (1978). "Synergetics." Springer-Verlag, Berlin and New York.

Hall, B. K. (1984). Developmental processes underlying heterochrony as an evolutionary mechanism. *Can. J. Zool.* **62**, 1-7.

Hall, B. K. (1992). "Evolutionary Developmental Biology." Chapman & Hall, London and New York.

Hall, B. K. (1994). Homology and embryonic development. *Evol. Biol.* **28** (in press).

Harvey, P. H., and Pagel, A. D. (1991). "The Comparative Method in Evolutionary Biology." Oxford Univ. Press, Oxford.

Haszprunar, G. (1992). The types of homology and their significance for evolutionary biology and phylogenetics. *J. Evol. Biol.* **5**, 13-25.

Holliday, M., and Grobstein, P. (1981). Of limbs and eyes and neuronal connectivity. *In* "Studies in Developmental Neurobiology" (W. M. Cowan, ed), pp. 188-217. Oxford Univ. Press, Oxford and New York.

Hyman, L. H. (1967). "The Invertebrates." McGraw-Hill, New York.

Jacobson, M. (1978). "Developmental Neurobiology." Plenum, New York.

Jones, D. S. (1983). Sclerochronology: reading the record of the molluscan shell. *Am. Sci.* **71**, 384-391.

Kniprath, E. (1981). Ontogeny of the molluscan shell field. *Zool. Scr.* **10**, 61-79.

Maynard Smith, J., Burian, R., Kauffman, S., Alberch, P., Campbell, J., Goodwin, B., Lande, R., Raup, D., and Wolpert, L. (1985). Developmental constraints and evolution. *Q. Rev. Biol.* **60**, 265-287.

Mayr, E. (1982). "The Growth of Biological Thought." The Belknap Press of Harvard Univ. Press, Cambridge, MA

Meinhard, H. (1982). "Models of Biological Pattern Formation." Academic Press, London and New York.

Minelli, A., and Peruffo, B. (1991). Developmental pathways, homology and homonomy in metameric animals. *J. Evol. Biol.* **4**, 429-445.

Misof, B. Y., and Wagner, G. P. (1992). Regeneration in *Salaria pavo* (Blenniidae, Teleostei): histogenesis of the regenerating pectoral fin suggests different mechanisms for morphogenesis and structural maintenance. *Anat. Embryol.* **186**, 153-165.

Mousseau, T. A., and Roff, D. A. (1987). Natural selection and the heritability of fitness components. *Heredity* **59**, 181-197.

Müller, G. B. and Wagner, G. P. (1991). Novelty in evolution: restructuring the concept. *Annu. Rev. Ecol. Syst.* **22**, 229-256.

Noden, D. M. (1983). The role of the neural crest in patterning of avian cranial skeletal, connective and muscle tissue. *Dev. Biol.* **96**, 144-165.

Osche, G. (1973). Das Homologisieren als eine grundlegende Methode der Phylogenetik. *Aufsätze Reden Senckenberg. Naturforsch. Ges.* **24**, 155-166.

Patel, N. H., Ball, E. E., and Goodman, C. S. (1992). Changing role of *even-skipped* during the evolution of insect pattern formation. *Nature (London)* **357**, 339-342.

Patterson, C. (1982). Morphological characters and homology. In "Problems of Phylogenetic Reconstruction" (K. A. Joysey, and A. E. Friday, eds.), pp. 21-74. Academic Press, London and New York.

Raff, R. A., and Kaufman, T. C. (1983). "Embryos, Genes, and Evolution." Macmillan, New York.

Remane, A. (1952). "Die Grundlagen des natürlichen Systems, der vergleichenden Anatomie und der Phylogenetik." Geest and Portig, Leipzig

Riedl, R. (1978). "Order in living organisms: A systems analysis of evolution." Wiley, New York.

Rieger, R., and Tyler, S. (1979). The homology theorem in ultrastructural research. *Am. Zool.* **19**, 655-664.

Rieppel, O. (1988). "Fundamentals of Comparative Biology." Birkhäuser, Basel and Boston.

Rieppel, O. (1992). Homology and logical fallacy. *J. Evol. Biol.* **5**, 701-705.

Roth, V. L. (1984). On homology. *Biol. J. Linn. Soc.* **22**, 13-29.

Roth, V. L. (1988). The biological basis of homology. In "Ontogeny and Systematics" (C. J. Humphries, ed.), pp. 1-26. Columbia Univ. Press, New York.

Roth, V. L. (1991). Homology and hierarchies: problems solved and unresolved. *J. Evol. Biol.* **4**, 167-194.

Sengel, P. (1976). "Morphogenesis of Skin." Cambridge Univ. Press, Cambridge.

Spemann, H. (1915). Zur Geschichte und Kritik des Begriffs der Homologie. In "Allgemeine Biologie" (C. Chun and W. Johannsen, eds.), pp. 63-86. Teubner, Leipzig.

Starck, D. (1975). "Embryologie." Thieme, Stuttgart.

Stearns, S. C. (1992). "The Evolution of Life Histories." Oxford Univ. Press, Oxford.

Tabin, C. J. (1992). Why we have (only) five fingers per hand: *Hox* genes and the evolution of paired limbs. *Development (Cambridge, UK)* **116**, 289-296.

Thornton, C. S., and Kraemer, D. W. (1951). The effect of injury on denervated un-amputated fore limbs of *Amblystoma* larvae. *J. Exp. Zool.* **117**, 415-439.

Van Valen, L. (1982). Homology and causes. *J. Morphol.* **173**, 305-312.

Vermeij, G. (1987). "Evolution and Escalation." Princeton University Press, Princeton, NJ.

Wagner, G. P. (1986). The systems approach: An interface between development and population genetic aspects of evolution. *In* "Patterns and Processes in the History of Life" (D. M. Raup., and D. Jablonski, eds.), pp. 149-165. Springer-Verlag, Berlin.

Wagner, G. P. (1989). The biological homology concept. *Annu. Rev. Ecol. Syst.* **20**, 51-69.

Wagner, G. P., and Almeder, M. (1991). Epigenetic control of pectoral fin web shape in *Lipophrys canevai* (Blenniidae, Teleostei). *Experientia* **47**, 737-738.

Wagner, G. P., and Misof, B. Y. (1993). How can a character be developmentally constrained despite variation in developmental pathways? *J. Evol. Biol.* **6**, 449-455.

Wallace, H. (1981). "Vertebrate Limb Regeneration." Wiley, Chichester and New York.

Wray, G. A., and Raff, R. A. (1989). Evolutionary modification of cell lineage in the direct-developing sea urchin *Heliocidaris erythrogramma*. *Dev. Biol.* **132**, 458-470.

Wray, G. A., and Raff, R. A. (1990). Novel origins of lineage founder cells in the direct-developing sea urchin *Heliocidaris erythrogramma*. *Dev. Biol.* **141**, 41-54.

Wray, G. A., and Raff, R. A. (1991). Rapid evolution of gastrulation mechanisms in a sea urchin with lecithotrophic larvae. *Evolution (Lawrence, Kans.)* **45**, 1741-1750.

Yntema, C. L. (1959). Regeneration in sparsely innervated aneurogenic forelimbs of *Amblystoma* larvae. *J. Exp. Zool.* **140**, 101-123.

Zander, C. D. (1973). Zur Morphologie der Flossen von Blenniidae (Pisces) des Mittelmeeres. *J. Ichthyol.* **22**, 93-96.

9

WITHIN AND BETWEEN ORGANISMS: REPLICATORS, LINEAGES, AND HOMOLOGUES

V. Louise Roth

Department of Zoology
Duke University
Durham, NC 27708-0325

I. INTRODUCTION

Since the time of Owen, the term homology has been used in comparisons of features between as well as within individual organisms. Owen (1843, p. 379) defined homologue as "the same organ in different animals under every variety of form and function"; while serial homology, as he subsequently recognized it (Owen 1848, p. 164), is manifest by repeated structures he referred to as "homotypes," arranged on a linear axis within a single organism. Application of the term homology to comparisons within an organism, while widely considered useful (Ghiselin, 1976; Van Valen, 1982; Roth, 1984; Wagner, 1989a,b), has also been strongly criticized (Bock, 1989; de Beer, 1971; Rieppel, 1992; Schmitt, 1989). This chapter examines the justification for applying the term *homology* to cases of serial, or more broadly, iterative (*sensu* Ghiselin, 1976, as modified by Roth, 1984) and other types (Haszprunar, 1991) of homology. It will do so in an attempt to integrate what have been termed (Eldredge, 1979) taxic and transformational approaches to the study of evolution.

Conceptual discussions are commonly dichotomized, or framed as a conflict of opposing viewpoints. Eldredge (1979) and Schoch (1986), for example, saw the concept of homology as being viewed, by different workers, from predominantly a *taxic* or a *transformational* perspective. Wagner (1989b) differentiated between *historical* and *biological* concepts of homology. Rieppel (1992) saw conflict between a primary interest in *phylogenetic pattern* reconstruction and the consideration of *biological processes*. Donoghue (1992) distinguished between researchers interested in connecting homology and *ancestry*, and those connecting homology with similarity. All of these authors expand their views in chapters in this volume. I have contrasted *restrictive* (phylogenetic) definitions with more *inclusive* (phylogenetic and iterative) ones (Roth, 1991).

These dichotomies are largely congruent, with the alternative viewpoints tending to diagnose two distinct groups of researchers. A taxic approach, and a historical or phylogenetic conceptualization of homology, are often advocated by workers who identify themselves primarily as systematists. Advocates of a more transformationist perspective tend to be associated with other disciplines. Thus dichotomies can be useful, both in highlighting conceptual distinctions, and in promoting particular research agendas (see Donoghue, 1992, p. 179). Yet their utility is also limited by the extent to which they obscure conceptual links or form barriers to illumination — where the two parts of a dichotomy are simply alternative views of the same phenomenon, or where a connection allows the fuller understanding of a subject in its entirety. This chapter, while unabashedly transformationist in its approach, is intended not to negate, but to integrate with certain elements of the taxic tradition.

II. PATTERN "VERSUS" PROCESS

Some workers (e.g., Rieppel, 1992; Platnick, 1979) have asserted the primacy of pattern over process in biology. To be sure, science as the study of the natural world depends upon data. Without observations we would not bother formulating descriptions or explanations. At the same time, at a very primary level, the patterns we discern depend fundamentally on our processes of perception, and the perception of patterns is a consequence of selectively emphasizing some types of information and disregarding others. Our own visual systems, for example, differ from photographic film in responding readily to edges and contrasts but not to uniform color or illumination (Hubel, 1988). Another property of our visual systems and not film is color constancy — constancy in the perception of colors of an object despite shifts in spectral content of the light source. But the patterns different individuals see — for example, a human with normal vision versus a color-blind individual — depends on the spectral sensitivity of pigments in our retinal cones. Most of us are blind to the ultraviolet reflectance "targets" flowers display to their arthropodan pollinators.

Recognition of a pattern is a matter of selection of information also at a conscious level. The features we delineate in a pattern are influenced by what processes we conceive as producing it. Lacking correct "object-hypotheses", both Galileo and Huygens initially failed to interpret the shape of the planet Saturn correctly. Instead of a sphere encircled by a separate flattened ring, they represented it variously as a triple object, or in the shape of an anchor-chain link (Gregory, 1970, pp. 119-120). The delimitation of characters in *Heliconius* butterfly wing patterns seemed fairly straightforward and obvious, until it was discovered that genetic and developmental explanations could be greatly simplified by a reversal of what was considered figure and ground (Nijhout, 1991).

The living world is now generally represented as a hierarchical nesting of taxa, but for the centuries before Linnaeus, a ladder or chain was the chosen analogy (e.g., Mayr, 1982, p. 207). By the early 19th century, webs, nets, and maps were used in botanical systematics (Stevens, 1984b), and many 19th-century biologists, including, for a time, Charles Darwin, subscribed to W. S. MacLeay's quinarian view of higher taxa as osculating groups arrayed in sets of fives (Stresemann, 1975; Hennig, 1966).

Our current preference for the nested hierarchy, and its persistence as a model of patterned variation in the living world, is due not to any inherent "logic" intrinsic to this type of pattern, but to its isomorphism with a process of branching and divergence (Darwin, 1859; Hennig, 1966). We admit exceptions to this tree-shaped pattern, in the form of reticulations, if and when we are willing to consider a role for processes other than persistence (or extinction) and divergent branching: interbreeding (within sexual populations), hybridization (between closely related taxa), and endosymbiosis or various kinds of gene transfer (between taxa). We are willing to believe such processes occur either (1) because our experiments suggest that they do; (2) because we can observe, more or less directly, interbreeding or exchange of genetic material, or cells engulfing other cells or invading the tissues of other organisms; or (3) because we can document patterns such as incongruence in character distributions. Pattern and process are interdependent components and equally vital subjects of scientific investigation.[1]

---

[1] Rieppel's (1992) heated attack on conceptual studies of the biological basis of homology seems to arise from a confusion of ontological and epistemological issues, and an inability to allow hypothetical questions of the sort "suppose x occurred; how could that affect the pattern we would observe?"

Among the most important issues to advance in the study of systematics are the definition of criteria and the construction of tests of homology. Prominent among these are Patterson's (1982) three tests of homology, the criteria of similarity, conjunction, and congruence. Patterson's tests (although they did not originate with Patterson, he provided them a succinct and cogent presentation, just as Owen's definition of homology did not originate with Owen; see Chapter 1) have been commonly invoked and successfully applied in a large number and wide variety of problems of phylogenetic reconstruction. Within the context of cladistic analysis, applying a historical, taxic, phylogenetic conceptualization of homology, these criteria are well justified. (This is so despite the fact that the three differ in status: *similarity* leads us to postulate homology, *conjunction* indicates the presence of nonhomology, and *congruence* is the ultimate arbiter or test; see, e.g., Patterson, 1988; de Pinna, 1991). If a unified perspective on the concept of homology is to be achieved it will be important to examine these criteria also from a transformationist perspective. That is one objective of this chapter, to be undertaken after (1) presenting and justifying a flexible but comprehensive definition of the concept of homology, and (2) examining homology as it is manifest (and explained) at different levels in the hierarchy of biological organization. Language and lines of reasoning developed for the hierarchical expansion of the concept of natural selection prove useful in this analysis.

III. DEFINITION

The definition of homology I favor is Van Valen's (1982): "correspondence caused by continuity of information" (see also Osche, 1973[2]). The *correspondence* between two features referred to here may be of any of a variety of types, and may involve similarity of position, shape, material, structure, chemical composition, color, connection with other parts, etc. — Remane's criteria of position and special quality (see Riedl, 1978; and Ax, 1987). *Continuity,* for the biologi-

---

[2] "Homolog sind demnach Strukturen, deren nicht zufällige Übereinstimmung auf gemeinsamer Information beruht" (Osche, 1973, p. 156).

cal examples on which we concentrate (putting aside, in this discussion, culturally transmitted phenomena), is typically genetic or genealogical, and is a consequence of replicative processes within an organism or through generations. *Information* is that which is conveyed, directly or indirectly, by replication between generations, or reiterated within a single organism, causing the correspondence.

In discussing units of selection, Williams (1992) distinguished between two mutually exclusive domains: the codical, which deals with information, and the material, which is the material embodiment of that information. "Information" must be invoked with reference to continuity or replication in discussions of homology because the material heritage necessary for continuity of any form of life is sufficiently modest that it can be contained within a single cell[3] (e.g., a zygote). Thus, very few morphological structures or physical processes are passed along directly through the generations. Organisms and their parts are built anew, and the persistence, development, growth, and reproduction of an organism is a "continuity...of pattern, not substance" (Williams, 1992, p. 11). But how and in what form is this information encoded? It is embodied, partially, in the sequence of base pairs of DNA, but cellular components, materials, and organelles are also an essential part of this heritage. I return to this topic in Section V.

A definition merely identifies the class of phenomena of interest. Offering a definition of homology does not tell us how we are to go about recognizing instances of it. That is one of the tasks of systematics, criteria and tests having been offered by, for example, Remane and Patterson (see Table I in Roth, 1991). Nor does defining homology as "correspondence caused by continuity of information" tell us anything about precisely how that correspondence arises or is maintained, or in what form the information is stored or transmitted. That is the question increasingly being addressed by those interested in the biological basis of homology (Roth, 1988; Wagner, 1989a,b, and chapter herein; Wray and McClay, 1989; Tabin, 1992; Wagner and Misof, 1993).

---

[3] An entire cell is required even for viral replication and continuity, so a cell is truly the minimal unit.

The virtues of Van Valen's definition are that it is comprehensive, flexible, and ideologically neutral, readily able to serve the needs of various disciplines and philosophical orientations. But broad utility is no virtue if a definition thereby obscures important distinctions or conflicting viewpoints. There are common usages of the term homology that this definition does not embrace (e.g., similarity with no implication of historical continuity; see Reeck *et al.*, 1987). Furthermore, in particular applications the concept can readily be compartmentalized (see Hall, 1992, p. 192), by appropriate specification of each of the key nouns in the definition: we may specify the particular aspects or features that are shared or retained, or consider what the particularly relevant types of information are. It need not therefore be the case that "achieving consistency with every version of homology may yield a definition that is of little use to anyone" (Donoghue, 1992, p. 179). Hall (personal communication, 1987; and Hall, 1992, p. 192) indicated that:

> I ... have trouble fitting both [homology of structure vs disparity of developmental origins in a comparison of original and regenerated lenses in vertebrates] into homology as continuity of information — the structural equivalence of continuity of information of the lenses does not reflect equivalence or continuity of information in development. For me that means that the two levels require differing definitions of homology.

But what is needed is not differing *definitions*, but rather different *explanations*.

What having a single all-embracing definition (that nonetheless may be flexibly narrowed for specific applications) provides, is a means for examining the most general features of homology — what distinguishes the various types (some of which some workers would prefer to exclude from the definition), as well as what they might share. If common ground did not exist among comparative biologists — if despite variation in research objectives and philosophies, there was not some general agreement on the kinds of characters one would call homologous — discussion would be pointless; or at best, "a means of forcing [I would substitute "attempting to force"] other scientists to pay closer attention to whatever one thinks is most important" (Donoghue, 1992, p. 179). A book such as this one would be a curiosity, or the means for promoting a disparate assortment of research agendas, but it could not be expected to contribute to any synthesis.

Noting that structures that are considered homologous can be produced by different developmental programs, Hall (1992, p. 194) further noted that "Homology is a statement about pattern, and should not be conflated with a concept about processes and mechanisms." I can agree with this statement only in part. Homology is indeed about pattern, but it is more than that. In supraspecific comparisons it is a pattern of a very special type that is relevant; one (if we follow Patterson's criteria, for example) that involves similarity, or correspondence, in combination with congruence; one that, in any event, occurs in the special sort of context that permits us (by whatever means) to infer continuity (which, for Patterson, is specifically monophyly). Even virtual identity in structural pattern does not constitute homology: in the wrong organisms, structural similarity may constitute only analogy, or evolutionary convergence (or something else: the same lens structures have been discovered in the eyes of phacopid trilobites and in the notes of Christian Huygens and Charles Descartes; Levi-Setti, 1975). Thus there is always some presumption of process embodied in an hypothesis of homology. Let us assume, though, that we are satisfied (by whatever means; Patterson's tests have received wide acceptance) that a particular example is a good candidate for homology. The open-endedness of Van Valen's definition will be dissatisfying inasmuch as it confronts us with the question of how the information embodied in our homologous patterns has been preserved — in what form — and in what manner continuity is maintained. Such dissatisfaction is productive, because these questions are ready subjects for conceptual and empirical work.

I agree that Van Valen's definition deals with two separable issues: pattern and process. *Correspondence* can occur without its having any basis in historical *continuity* — examples of this are referred to as convergence. We can also observe correspondence, and infer that it is indeed based on continuity of information, without specifying that that continuity is manifest at any particular level — see examples in de Beer (1971), discussed by Roth (1988) and subsequently by Wagner (1989b, and this volume). And there can be historical continuity or even identity of genetic material[4] (an embodiment of information) without any correspondence being expressed by its products. Within a single viral or plasmid genome, sets of protein-coding genes can physically overlap (the same base pairs being a part of different genes), and some overlapping genes are translated in different reading frames (e.g., Rak *et al.*, 1982; Fiers *et al.*, 1978). In such cases, there is correspondence (actual physical identity) between the overlapping genes in a particular set of base pairs, but no correspondence at the levels of codon or protein product. According to Williams's (1992) distinction, the codical domains of such proteins are entirely distinct, although the material domains of their genes physically overlap. Whether or not we consider the evolutionary origins and relationships of such proteins to be of interest, they would not, according to Van Valen's definition, be homologous, because they exhibit no correspondences. Correspondence and the continuity of information that can cause it are the necessary and sufficient conditions for a relationship of homology.

IV. TYPES OF HOMOLOGY

Haszprunar (1992) enumerated four kinds of comparisons, which allow identification of four types of homology:

1. Iterative homology — a correspondence between different characters (or repeated characters) in the same individual at the same time.
2. Ontogenetic homology — a correspondence between characters of the same individual at different times.[5]

---

[4] *Continuity* is not to be confused with mere *contiguity* — being linked or physically adjacent; "continuity," as it is used here, is a consequence of physical overlap and of replication.

[5] "Ontogeny" often connotes monotonic change in an individual, so homologous behaviors (e.g. actual grooming, and a grooming act ritualized and incorporated into courtship; John Rowden, personal communication) or homologous physiological processes that occur in a single individual at different times, arguably can more appropriately be considered iterative homologues. It may also make sense on mechanistic grounds to distinguish within an organism between the gradual transformation of a single structure through time, and the reiterated production of variations on a

3. Polymorphic homology — a correspondence between characters of different individuals of the same species.

4. Supraspecific homology — a correspondence between characters found in different species or in different higher taxa.

In contradiction to those who would restrict application of the term homology to supraspecific comparisons, Haszprunar (1992) pointed out how readily the origins of supraspecific homology may be traced to the other three types. Through heterochrony, *ontogenetic* homology becomes supraspecific. (For example, the relationship between juvenile and adult forms of a character is ontogenetic homology, but the relationship between the adult stage of a paedomorphic descendant taxon and the corresponding juvenile stage of its ancestor is supraspecific homology, paedomorphosis being a form of heterochrony.) Through fixation, an organismal trait which originated within a population as a *polymorphism* comes to characterize an entire taxon (Roth, 1991) — "homology concerns characters, whereas apomorphy concerns taxa" (Haszprunar, 1992, p. 14), so only when a character becomes fixed among individuals within a taxon does a homology become an apomorphy. Organismal individuality is a derived trait in the living world (Buss, 1987), and for the large number of clonally reproducing taxa the distinction between *iterative* and *polymorphic* homology is blurred (Minelli and Peruffo, 1991), so by extension, iterative and supraspecific homology are also linked.

Thus in regard to the "correspondence" required by Van Valen's definition of homology, the four types of homology

---

single theme. If so, heteroblastic series in plants, and other parts added in succession to a modular organism, would be classified as iterative because they occupy different positions in the organism, even though these parts are added successively, through time, and commonly show monotonic variation. Because iterative homologues may develop both in different places and at different times, the distinction between ontogenetic and iterative homology becomes blurred: One would consider the legs of juvenile and adult ametabolous insects to be ontogenetic homologues, but are the legs of adult holometabolous insects ontogenetic homologues of the larval legs they replace, or are they iterative homologues because they develop in a distinct position and from distinct tissues (imaginal disks)? The difficulty these cases present for a rigid categorization points out that the homology relationships of these different types intergrade.

seem to be interrelated. Yet we will still want to know how "continuity of information" is manifest in causing all of these types of homology.

V. SOURCES OF CONTINUITY

Continuity of information is a matter of persistence or replication. With respect to the codical domain (*sensu* Williams, 1992), it is difficult to distinguish between persistence and replication (Williams, 1992, p. 12, referred to a "lack of a conservation principle for information, analogous to the conservation of matter or energy"), but as biologists we are also keenly interested in the material domain. Disruption of the material domain — e.g., a mutation or the deletion of a base in DNA — may, after all, have consequences for what Williams calls the codex. What form does biological information take?

Both DNA — its sequence of base pairs — and the rest of the cell — its structure, components, materials, and organelles — are important. DNA is not quite the "Only...durable archive for most of the Earth's organisms" (Williams, 1992, p. 11). It is durable, but incomplete (Bonner, 1974). DNA is inert outside the context of an appropriate cell, and the environment that such a cell provides for the DNA is sufficiently delicate and complex that attempts to completely reproduce that environment artificially have so far eluded experimentalists. One may acknowledge that all components of a cell are ultimately the material products of genes (noting that some of those genes may belong to another genome, as in the case of maternal inheritance), yet one must also acknowledge the importance of *information contained in the structure of the cell and in its dynamic state. These are what assure that the appropriate physical and chemical events occur in the right place at the right time for life to be sustained* (a nontrivial task, in view of our difficulty in sustaining life cell free *in vitro*). The information embodied in a stretch of DNA is context dependent: sex determination in *Drosophila* (Baker, 1989), and the multiple isoforms of α-tropomyosin in mammalian cells

(Goodwin *et al.*, 1991), for example, arise from alternative schemes of splicing the primary transcript of single genes. Stated in strongest terms:

> Genes [including regulatory genes] are passive sources of materials upon which a cell can draw, and are part of an evolved mechanism that allows organisms, their tissues and their cells to be independent of their environment by providing the means of synthesizing, importing, or structuring the substances (not just gene products, but all substances) required for metabolism, growth and differentiation. (Nijhout, 1990, p. 444)

There need be nothing vitalistic in such a view if one is willing to recognize the importance of information that is contained in and perpetuated by the living cell, and manifest materially outside of the static array of DNA base pairs, in the appropriate spatial and temporal distribution of cellular components.

Indeed, for many kinds of organisms, especially those characterized by organismal individuality (Buss, 1987), the only single cell capable of sustaining life independently and perpetuating a lineage indefinitely is a zygote (or an otherwise activated ovum). In other cells of such organisms, a portion of the full complement of information that is necessary for organismal development and reproduction is rendered inaccessible in the course of differentiation, so that regeneration not only of entire organisms, but even of organs or other parts, may be limited and may require a combination of cells of different types. In such cases, the source of information employed for development and reproduction must be embodied, if not in a zygote, then in multiple cells and in the physical relationships among them.

Thus, biological information is sustained in a dynamic state, and there is more to life than base-sequences (Bonner, 1974). The material form of biological information varies with time, and defies simplistic characterization: the study of mechanisms of storage and elaboration of this information is the *raison d'être* for the entire fields of developmental biology and molecular genetics. Analysis of the biological basis of homology will be enriched by these fields through the empirical elucidation of the details of specific examples, but for the purposes of conceptual analysis, a coarser, more basic mode of description than this is desirable; so, we return to the more generalizable concepts of persistence and replication.

A. Arguments from Generalized Treatments of Natural Selection

The explication of homology and its manifestation at more than one level of biological organization has some features in common with recent discussions of natural selection. For selection, effort has been directed at defining the process in the most general terms, and in identifying the entities (at multiple levels of organization) that participate. For homology, we have a general definition of the concept (see above), and it remains for us to justify application of the term for comparisons other than the most widely accepted type, which is between organisms or species.

To do this, we would like to know what kinds of objects can exhibit correspondence due to continuity of information. This is unlikely to be fully satisfying to proponents of an exclusively taxic point of view, who cite the testability of taxic (supraspecific) homology as a source of its validity as a scientific concept. Thus, a second objective will be to examine Patterson's (1982) tests of homology from a transformational perspective, and to ask whether properties of supraspecific homologues that allow them to pass these tests are generalizable to other levels of biological organization.

B. Are Homologues Replicators?

A replicator, as Hull (1980) defined it, is "an entity that passes on its structure directly in replication." The concept was originally introduced (in slightly different form) by Dawkins (1978, p. 67) to serve discussions of selection, to which Hull (1980) subsequently introduced the idea of an interactor, "an entity that directly interacts as a cohesive whole with its environment in such a way that replication is differential"; selection itself he defined as "a process in which the differential extinction and proliferation of interactors cause the differential perpetuation of the replicators that produced them." These concepts allow one to consider the process of selection in an especially general way, and to ask whether processes analogous to natural selection on individuals can occur at other levels of biological organization — e.g., parts of organisms, or groups, or species.

Brandon (1990, p. 97) identified a dual hierarchy (recognized implicitly by Hull and eventually Dawkins), of interactors and their corresponding replicators that participate in various scenarios of selection (Table I). He suggested that whereas the interactor column forms a very neat hierarchy of inclusion, the replicator column does not (Brandon, 1990, p. 98). In such discussions of levels of selection, attention has naturally focused on identifying the interactors. There has been less clarity, interest, or consensus on the identity of the replicators: "The point is that when selection occurs at a given level the entities at that level must be interactors" (Brandon, 1990, p. 93). Yet if our interest is not in selection, but rather in homology and continuity of information, replication itself may be considered the interactive process. We are led to ask at what levels replication occurs.

Williams (1992, p. 12), in emphasizing the distinction between codical and material domains, proposed that the term *replicator* be reserved for situations in which a message is copied into the same medium. It is not clear to me why an object so strictly defined as this is of interest, but it is important to recognize that Williams's focus was on the process of natural selection, where the interest lies more in the fact of replication (and in the information itself — the codical domain) than in its material form. In considerations of homology and the process of replication (rather than selection), however, a different set of entities will be most relevant. Here, both the replication of information, and its embodiment in biological structure and processes, are important.

If evolutionary success is measured in terms of abundance and potential immortality, a good replicator will have longevity, fecundity, and fidelity (Dawkins, 1976). It will persist and be copied, with the copies exhibiting the same structure or bearing the same information as the original. The persistence of any entity that might be considered a replicator is, to paraphrase Williams (1992, p. 11) again, a

Table I. Hierarchies of replicators and interactors.[a]

Selection Scenario	Interactor	Replicator
"selfish genes"	lengths of DNA	lengths of DNA
developmental or somatic selection	parts of organisms	genes or genome
organismic selection		
asexual reproduction	organisms	genome (or organism?)
sexual reproduction	organism	genes
group selection		
intrademic	group	genes
interdemic	group	group
avatar selection	avatar	avatar
species selection	species	species
clade selection	clade	clade

[a] According to Brandon (1992); an abridged version of his Table 3.1, p. 97.

continuity of pattern, not substance. In a DNA molecule, "nucleotides and their component atoms can come and go freely...as long as each replacement is by the same chemical entity,"..."a whole active organism shows a constant material flux with its environment"; and an organism "is not really an object, it is a region where certain processes take place" (Williams, 1992, p. 11). In the copying process, fidelity is important, but for Hull (1980, p. 319), and for our purposes in considering homology, a replicator "need only pass on its structure largely intact;" retention of structure need no occur with the accuracy of DNA replication,[6] (for if it had to, we would be unable to apply the concept of a replicator to other levels in the hierarchy).

Descent with modification occurs in evolution. If homology is due to continuity of information, it will be important to discover where discontinuity (modification) occurs, and to identify the objects whose replication is actually responsible for (causally involved in) the continuity of pattern. Dawkins (1978) criticized Bateson's facetious suggestion that a bird is a nest's way of making new nests by pointing out that variations in the construction of a nest are not perpetuated in future "generations" of nests; however, genetic mutations do self-propagate. Thus, we must strike a balance in requiring a replicator to show a certain amount of tolerance for error in replication (it need not be strictly accurate), yet have sufficient fidelity that at least some kinds of modifications will be heritable.[7] The objects in Brandon's hierarchy of replicators (Table I) have these qualities. The available choices form a hierarchy of physical inclusion (even if the sequence of listing in this table does not), and we can resolve some ambiguities in what serves as the replicator in various scenarios if we adopt certain conventions — e.g., if we list only the most inclusive unit that serves the function. For example, if a genome is replicated intact, then so too, in general, will be its constituent genes (allowing, again, for the possibility of errors). This choice of the most inclusive unit assures that we take account of as much of the information transferred during replication as possible.

To qualify as a replicator of the special type that interests us, two requirements should be met. (1) The replicating entity should retain a certain amount of coherence through the replication process. The genome is disqualified as a replicator in sexual reproduction because of the shuffling

---

[6] Hull also stipulated that what is more important in a replicator than strict accuracy is the directness of the replication, but, again, his purpose was to differentiate between replicators and interactors in the process of selection. The main difference between replicators and interactors is in which role — replication or interaction — they participate most directly.

[7] With strict accuracy, evolution would not occur; replication coupled with modification is what generates differing states of a character, while " 'characters' and their 'states' each represent characters — but at more and less inclusive levels" (Patterson, 1988, p. 604).

and exchange of genes that occurs between individual organisms. To describe this feature another way, in the chain of replication, replicators should form treelike lineages, or patterns of divergent branching (Hull, 1980; Williams, 1992). Where lineages exhibit a reticulate pattern, this type of replicator will be found both in the less inclusive units that retain their integrity through the shuffling (nonrecombining lengths of DNA), and in the more inclusive unit — the species or species gene pool — within which all reshuffling occurs.[8] (2) As indicated above, modifications in the replicator should also be potentially heritable.

As defined by requirements 1 and 2 above, the *special replicator* for developmental or somatic selection (Table I) might therefore more appropriately be identified not as genes or a genome, but as a cell. Although replication of cells is not exact, there are important cellular materials that are passed along in cell lineages intact. Not only the genome (in the sense of DNA sequences) but also some aspects of the cell's state of determination and differentiation are transmitted. For selection among asexual organisms, an entire organism satisfies requirement 1, but identification of the special replicator that also fulfills requirement 2 depends on the particular mode of reproduction — whether the germline is sequestered, or whether a group of somatic cells participates in regenerating a new individual. The entire organism would be the special replicator where modification of multiple types of somatic tissue has consequences for the descendant organisms.

If we are to understand homology, it will be important to understand replication, but clearly the entities one recognizes as involved in a relationship of homology (homologues) do not necessarily qualify as replicators in the special sense just described. The relationship of homology applies to features of objects, rather than entire structures (Roth, 1984; Sattler, 1990),[9] but when we do speak of the material objects that manifest relationships of homology

---

[8] There will need to be tolerance for a certain amount of reticulation as well, because virtually all lineages, at any level (cells, organisms, taxa), given sufficient time, may be subject to gene transfer in one form or another.

[9] Note that specifying with sufficient precision the features that are considered homologous is equivalent to including a qualifying phrase indicating the level of the comparison (Bock, 1989).

these homologues are more likely to be found in the *interactor* column of Brandon's table (Table I). We speak of homologous lengths of DNA, or homologous parts of organisms. An entire individual organism could be regarded as the homologue of another such individual. Moreover, homologues do not necessarily form continuous lineages of their own, as do the entities in Table II, but rather they and their corresponding features are generated anew with each cycle of reproduction. Homologues are reproduced through both replication and *development*. This point is especially pertinent, and manifest in the extreme, for the types of organisms to which Weismann's doctrine applies — those whose mode of development Buss (1987) has termed "preformistic": where the germline is determined early and is derived from discrete lineages of stem cells (Fig. 1). In this extreme situation, the process of development a homologue undergoes will be very remote from the processes by which organismal reproduction or replication occurs, and requirement 2 above will not be fulfilled: no modification arising in the course of development (even if it affects the genetic material of the participating cell lineages) will be manifest in homologues in subsequent generations.

To recapitulate then, we can think of the *process of replication* (the process by which copies of characters, of organisms, etc., are made) as being very different from the *process of selection*. Both depend upon replicators of a special type — the relevant replicators have integrity and are subject to heritable modification — but unlike the objects that interact in a process of selection, characters that are copied are not necessarily involved in the process of their own replication. Homology is manifest as a correspondence of pattern, but that correspondence depends upon a continuity of information, and not a continuity of its material manifestation. If the development of a particular character is blocked by environmental factors in one generation, for example, additional copies may nonetheless reappear in later generations. Thus, development of a character may be isolated from its transmission between generations.

Table II. Some entities that form lineages and that in appropriate circumstances can serve as special replicators.[a]

Entities whose replication forms lineages	Intervening processes (in which variation may be introduced)[b]	Sources of maintenance of fidelity[c]
DNA	DNA replication (uncoiling, initiation, polymerase binding, ligasing, etc).	use of a parental DNA template in a semi-conservative process, DNA repair mechanisms
cells	processes occurring at any phase of the cell cycle	mechanisms that assure accurate replication and segregation of organelles and redistribution of cytoplasm (using, e.g., DNA repair enzymes, spindle microtubules, etc.)
organisms	development	regulatory processes of development (involving induction, hormonal control, growth factors, etc.)
populations	anagenesis (continuation of population through generations). cladogenesis (splitting into more than one population-lineage)	mate-recognition systems (may keep population intact or contribute to its propensity to speciate)

[a] The list is not exhaustive.

[b] Processes that intervene, between bouts of replication, and that are involved in the material manifestation of the information that is replicated. If the lineage-forming entity is a special replicator, the variation is potentially heritable.

[c] Mechanisms for maintaining fidelity (these might be viewed as developmental constraints); ultimately stabilizing selection, as an overarching process, will also contribute to maintaining these sources of fidelity.

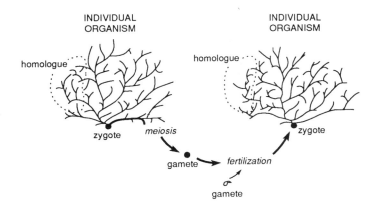

Fig. 1. Schematic diagram of the relationship between homologous features of two organisms (in this case, ancestor and descendant) of a type that reproduces sexually and sequesters germ cells early in development. Each of the two treelike objects is an individual organism — parent on the left, offspring on the right (only the gamete of the second parent is shown). Cell lineages are shown as branching lines for each organism. The formation, or development, of each homologue involves information transmitted through these finer branching lines, whereas the information transmitted between generations follows the path shown by heavy lines and heavy arrows (cf. Figs. 1.1, 1.4, and 3.14 of Buss, 1987). Note that if modifications arise in the replicators that participate in the *development* of the parent's homologue, they may not be passed along to offspring. The distinction between heavy and light lines in this diagram follows the one made by Dawkins (1982) between "germ line" and "dead end" replicators. For some other kinds of organisms — ones that are asexually reproducing and clonal, or ones without a sequestered germline — by contrast, the entire organism would be drawn in heavy lines, with all cell lineages potentially serving as germline replicators.

C. Information and Development

Figure 1 is drawn to highlight another characteristic of homologues, which is that the entity one homologizes need not be the product of a single cell lineage: it can be "polyphyletic" in the sense that it may involve interacting end-products of several such lineages, instead of comprising the single divergent branching pattern characteristic of genes or clades. Continuity of information exists between homologous features developing in different organisms ultimately by virtue of cellular and genetic replication. Interestingly, however, fidelity in the replication of the patterns we call homologues is not a strict or direct function of how literally cells or base-pair sequences are replicated (de Beer, 1971; Sander, 1983; Roth, 1988; Wagner, this volume). Homology of two features does not require identity in their genetic or developmental basis, and yet certain small genetic changes may nevertheless be capable of eliminating expression of a homologue altogether.

The reason for this is that development is not a rigid sequence of gene activity in a rigid series of cell divisions.[10] Instead, the immediate source of information employed in the development of homologues is embodied in the physical interrelationships and physiological status (including genetic activity) of participating cells (in both constitutive and inductive roles). The development of tissues or organs or structures both creates and draws upon higher levels of organization than that manifest at the level of single cells. We can use the analogy of architecture: a building of a particular architecture could be constructed by specification in three-dimensional coordinates the positions of individual bricks. The same overall form can be achieved using larger or smaller bricks, however, or irregular stones. To accommodate this a builder is guided by a blueprint; if a building were to self-assemble, however, all[11] that would be

---

[10] Even in animals with such highly deterministic development as nematodes, symmetries of gross morphology are not based on symmetrical division of cellular populations (Sulston et al., 1980).

[11] This is "all" that is required, but it's a nontrivial requirement. If such morphogenetic mechanisms were well understood, we would be well on our way to eliminating cancer or stimulating differentiated cells into regenerating lost or damaged body parts.

required would be a mechanism of regulatory feedback indicating when the position of a building block fell outside the boundaries of what was consistent with production of the particular architectural form. In biology, higher levels of organization are produced by a cascade of epigenetic processes and intercellular interactions. The particular spatial and temporal organization of genetic and other cellular activity — its cybernetic structure — rather than a narrowly specified linear sequence of events creates this organization (see Kauffman, 1983). Examples of types of processes that can produce higher level organization are cyclic regulatory feedback mechanisms, hierarchical networks, and the generative rules of pattern formation (Wagner, 1989a). Wagner has expressed particular interest in homologues of a well-individuated, developmentally highly organized type — characters that "share a set of developmental constraints, caused by locally acting self-regulatory mechanisms of organ differentiation" (Wagner, 1989b).

Where structures exhibiting such high levels of organization exist, the component processes, or the small-scale cellular or genetic participants in this organization, can undergo substitution, without any disruption of the overall pattern or the correspondence that we recognize as a relationship of homology. Systems of interaction have evolved that tend to *screen off* (i.e., make statistically irrelevant) the effects of single alleles, or the alteration of developmental processes (Brandon, 1990; Roth, 1991). Organizational information does not reside within individual codons; rather, it emerges from the interaction of their products. Indeed, Wagner and Misof (1993) provide evidence that in many cases the biological mechanisms responsible for maintaining a structure ("morphostatic" mechanisms) differ from those ("morphogenetic" mechanisms) involved in generating it during its development; the "causes" of a particular pattern may be attributed to multiple, disparate sources. What Buss (1987) has pointed out for the evolution of individuality in organisms applies equally well to the individuation of homologues: the hierarchy of biological organization is not only one of physical inclusion, but also one of developmental contingency. "Traits expressed in the higher unit [of organization] now act as selective agents on the variation arising in the lower unit. The organization of the higher unit is, however, a function of prior variation in the lower unit"

(Buss, 1987, p. 184; italics omitted); "...the lower unit may replicate, but only within bounds set by the influence of this replication on the higher unit's effectiveness in its interactions with its external environment" (Buss, 1987, pp. 171 and 172). Thus, mechanisms have evolved in which regulation at the lowest level of organization is dependent on higher level processes; gene expression, when influenced by a process such as induction, has become dependent on cellular interactions made possible by the positioning of differing tissues in gross morphology.

We can return, then, to the various types of homology, and consider the way in which continuity of information is manifest in each. (1) The correspondence observed between *iterative* homologues is produced, one postulates, by the same genetic and developmental information; the information is simply manifest in different places within the same organism. Here we call information the same if it is historically connected within the time course of the ontogeny of a single organism; if it is transmitted through the replication of cells with the same genome which are derived from a single zygote or otherwise totipotent cell; and if it involves either the same patterns of gene expression, or similarly structured interactive epigenetic networks that have historical continuity. Most iterative homologues have been identified just on the basis of correspondence in structure, but their dependence on the same information is an empirically testable hypothesis. (2) The initial and final states of *ontogenetic* homologues may look different, occur in different locations, and be made of different substances, but at least enough correspondence of pattern must be apparent for them to have been recognized. Their continuity — in space, time, and material — is expressed in a continual series of intermediates. It is possible that at the two ends of the temporal continuum the information involved in development and maintenance of ontogenetic homologues may be entirely different, but here too it is a testable hypothesis that continuity of information is the cause of correspondences of pattern. (3) In *polymorphic* homologues, identical character states in two individuals in a population are homologous to one another, and differ from alternative character states, if they are genealogically related — i.e., identical by descent. Alternative character-states of the same character are homologous at another level (that of the character). As alternatives they are evidence of change or discontinuity, but continuity is maintained (at the more general, "plesiomorphic," level in the character-tree lineage) in the context for their expression. The genital ducts of male sharks, for example, differ in morphology and function from the excretory ducts of females (as character states), but are considered homologous to them (as characters) on account of their topographic and developmental similarities (i.e., their ontogenetic homologues in the embryo are identical) (4) *Supraspecific* homologues are in principle essentially the same as polymorphic homologues, except that they are found in organisms in distinct population-lineages, and, at their most informative, are supposed to characterize not just single organisms, but entire taxa.

D. Patterson's Tests

Why should similarity, congruence, and conjunction be effective tests of phylogenetic (supraspecific) homology? For Patterson and some others of the taxic tradition, the only interesting homologues are those that characterize a monophyletic group. These, however, are but a subset of the supraspecific homologues that can be considered identical by descent.

In the evolutionary process of descent with modification, individual organisms reproduce, their progeny develop, and in the course of this reproduction and development their characteristics are also replicated. Phylogenetic reconstruction depends on an assumption that characters do change (are manifest in distinct states) but are retained for sufficient duration — i.e., replicated with sufficient fidelity or modified sufficiently slowly — that the sharing of corresponding characters can provide evidence of genealogical relationship. "Without some similarity, we should not even dream of homology," (Stevens, 1984a, p. 403), but as just implied, along with similarity, we must infer heritability of distinctive states, and independence of a character from others (see below) for that character to be putatively useful in phylogenetic inference (Mishler, 1993). Again, we consider homology both at the level of character state (in which a similarity is conserved), and at the more general level of character (which includes variations on a single theme; for homologous characters the context for a particular trait is held constant, but substitution of various states is permitted.)

But characters are features observed in individual organisms, whereas a major objective of phylogenetic reconstruction is the interrelationship of taxa. Here it becomes essential to examine the notion of a lineage, which Hull (1980, p. 327) defined as "an entity that changes indefinitely through time as a result of replication and interaction." Lineages are observed at multiple levels of biological organization — genes, cells, organisms, etc. Lineages of special replicators (defined above) have the form of a continually branching tree; replication within a lineage allows it to persist in time, or to produce coexisting duplicate lineages that can subsequently proceed on independent evolutionary courses: through duplication and divergence, distinctive multiple lineages are formed. If we ignore the organisms that contain them, or if we consider only organisms that reproduce asexually, we observe genes forming treelike lineages.

In the context of sexually reproducing organisms, however (or other types of organisms that at any point in their history participate in gene exchange or endosymbiosis), lineages of genes are woven into a network. For as long as they do not participate in gene exchange, asexual organisms form branching lineages; the lineages of sexual organisms anastomose.

Through the processes of replication, a sexually reproducing population will persist as a lineage through time as a cohesive whole. Its coherence is maintained by gene flow between individuals in sexual reproduction, while gene-lineages maintain their independence from one another through recombination and meiosis. A population in this way forms a higher-level sort of lineage. Where durable barriers to gene flow arise, a split occurs in this higher-order lineage, thereby producing coexisting duplicate population-lineages, which may diverge on independent evolutionary courses. Taxic homologies — corresponding features observed on divergent branches of these population-lineages — manifest continuity of information by virtue of common ancestry. I stated in Section V, B that homologous characters *per se* are not manifest as lineages; however, a lineage of organisms sampled only at the particular stage of their life cycle when they bear the character will give the appearance of such a lineage, and this is what I shall refer to as a lineage of characters. *In treelike lineages of characters in which a character is always overtly manifest when present, it is reasonable to look for reflections of the common ancestry of two entities in their sharing characters.* Not all lineages have this form, however, and not all characters behave in this way.

One can think in terms of the simplest sort of character — a particular allele at a locus (or even a single mutant base). On average only half the offspring of any given diploid sexual individual will inherit one of its pair of alleles at a locus, so even such close relatives as parent and offspring do not share all characters. Even if it is passed along, and replicated with great fidelity, a recessive allele will not necessarily be expressed in any of the offspring of a single ancestor, so, in their overt manifestations, characters (whether they are as simple as genetic alleles or more complicated and highly organized as Wagner's (1989b) homologues) do not always form treelike lineages.

Inferences at the level of population lineages and higher in turn face additional complications. The topology of a gene-tree is constrained only partially by the topology of the species-tree whose lineage contains it. The lineage of a gene will not cross into a neighboring divergent branch of the species (unless there is introgression), but neither need it be transmitted into all descendant branches after every bifurcation, if alternative alleles are present in each nodal population. Because of lineage sorting, *the presence of homologous alleles in individuals of two closely related taxa does not necessarily signal a sister-group relationship between the taxa* (for further explanation, see, e.g., Tajima, 1983; Pamilo and Nei, 1988; Doyle, 1992; Roth, 1991). The best-behaved characters for the purposes of phylogenetic inference, then, are those (1) that are reliably expressed in all circumstances (i.e., whenever their codical basis is present), (2) that do not change so rapidly within a lineage-segment that phylogenetic information is lost, yet (3) that originate and become fixed within a single lineage-segment (so they come to characterize not just a few individuals, but an entire population-lineage and all of its immediate descendants).

Examined character by character (without the supplemental information provided by character congruence), two populations sampled at a single time-horizon present no indication of whether the characters they share both originated *and became fixed* within a common lineage-segment in their past. We may judge independent *origin* of the characters in the two populations unlikely on the basis of detailed similarity, but independent events of *fixation* leave no trace on the character itself — even if it is fixed in both populations at the time they are sampled. Such origination and fixation in different lineage segments in the past produces distortions in present-day character distributions, even if those characters are identical by descent, because the relevant character *at the level of the population* for inferring that population's history is the *origination and fixation* of the character, not simply the presence of that character or even its presence in a fixed state. However, the probability of obtaining a character distribution in terminal taxa that actually conflicts with the population genealogy is not high (Tajima, 1983; Pamilo and Nei, 1988; Neigel and Avise, 1986). For example, in a symmetrical four-taxon situation, with one character and two states, for which the probability of remaining polymorphic on any given lineage segment is 1/3 and fixation and loss are equally likely, the probability of conflict is 0.22 (Roth, 1991). The independent inheritance of different characters becomes helpful here, because as genealogies are inferred for additional characters, the likelihood that more than one would specify the same topology purely by chance becomes very small. Specifying the same topology is evidence for lack of independence in the genealogies of the different characters, at either of two levels: at the level of the character-lineage (if the two characters are genetically linked, or pleiotropic), or the population-lineage (if the distributions of the two characters reflect the history of the populations that contain them). "Specifying the same topology" is alternative wording to describe the criterion of *congruence*.

Where multiple characters specify congruent topologies, there is good reason to use consensus topology as a working hypothesis for the population's phylogeny. We can in turn use this topology to make inferences about other characters. Assume we are studying a clade (a divergently branching population lineage descended from a single common ancestor) whose topology we have inferred from the congruent distributions of multiple characters. Distributed among our terminal taxa are another set of characters, referred to as x, that are putatively homologous with one another, and that are believed to manifest a derived state. In this context, a derived state can be thought of as a new modification whose origin (which is unique) occurred within the clade one is studying — i.e., no earlier than the clade's immediate common ancestor — and (for the sake of simplifying this analysis) has not subsequently been modified further. If the distribution of character x is congruent with the topology inferred from the other characters, there is good reason to suppose that the correspondence among characters x is due to genealogical continuity, i.e., homology.

The same suites of features can be employed in tests of similarity and congruence. Putative homologues that pass the similarity test are well supported if they are structurally complicated, because it would seem more plausible for them to have been maintained in all their complexity through continuity and replication, than that they could arise independently more than once by chance. Multiple origin is easier to imagine for a trait with only a few fairly simple features, in which there is less basis for discrimination between independently arising states, and any difference between them may go undetected. Viewed another way, a complicated object is more readily subdivided into a large number of distinct characteristics. Coded separately, the congruent behavior (joint occurrence in particular taxa) of these characteristics would be evidence for their monophyly. Bledsoe and Sheldon (1990) made exactly this point for molecular characters, in which *congruent* behavior of individual positions is used to assess *similarity* at the level of entire sequences. The application of either similarity or congruence may depend on how independent the behavior — in terms of origin and inheritance — the atomized characteristics are inferred or assumed to be.

With the *conjunction* test, the conjunction of two characters — their occurrence in a single organism — is taken as evidence against their being alternative states of a polymorphic homologue (see, e.g., McDade, 1990): If a is present in individuals of one taxon, and b is present in individuals of another, the presence of a and b in one or more individuals of a third taxon is evidence against a and b

being alternative states of a single homologous character (or else, it suggests that either *a* or *b* is not homologous with the similarly named state in the individual that bears *a* and *b* jointly). As de Pinna (1991) pointed out, an observation of conjunction suggests that homoplasy exists in the analysis, but it does not indicate its precise location. That depends, in part, on the phylogenetic positions of the taxa being considered.

Failure of the conjunction test (i.e., a finding of conjunction) suggests that for the purposes of resolving the phylogeny we have incorrectly delineated characters. The justification of the conjunction test in a transformational context has to do with similarities or differences in characters (or character states) as signals of intervening evolutionary events. In the example described above, possession of *a* and/or *b* cannot be discussed in terms of continuity or discontinuity of information in a single character lineage, because possession of *a* is observed to occur irrespective of the possession of *b* — the two characters show a certain amount of independence. Accounting for all manifest possibilities requires that we recognize more than a single evolutionary event in the transformation among states. Either a single character, "*a* and/or *b*," is involved, taking on any of three states (*a*, *b*, or *a and b*), or there are two separate characters, "*a*" and "*b*", each of which takes on one of two states (*a* or *not a*, *b* or *not b*, respectively). In either case, at least two evolutionary events, as opposed to a single substitution of state *a* for state *b*, are needed to account for existing observations. To put it in slightly different terms, we can consider the arms of humans, the wings of birds, and the arms and wings of angels. As de Pinna's (1991) analysis demonstrated, we are forced to deduce either two distinct origins for wings (in angels and in birds), or two origins for arms (in humans and angels), or the loss of arms in birds and the loss of wings in humans — which amounts to two evolutionary events, not just one, in all cases. (Sattler, this volume, discusses the relevant phenomenon of homeosis from a different perspective, in which change is not assumed to occur as discrete units or steps.)

The conjunction test obviously can't be used in refutation of iterative homology in the way it can for polymorphic, supraspecific, or even ontogenetic homology.[12] Within a single organism, a particular patterning mechanism (e.g., the system for forming eyespots within the wing cells of butterflies; Nijhout, 1991) may be repeated, or it may expand or extend its domain of influence to overlap others. For this reason, dependent as they are on replication and the formation of lineages of replicators, iterative homologues themselves (in their overt manifestations) do not arise within single continually branching processes, nor do they change by means of linear sequences of substitutions. For example, although the pectoral fins of the ancestors of tetrapods differed in structure and function from the fins of the pelvic girdle, fore- and hindlimbs apparently underwent reorganization, resulting in the evolution of a high degree of structural correspondence between these serial homologues in tetrapods (Rackoff, 1980). Evidence from experimental embryology on modern tetrapods led me to suggest that correspondence between fore- and hindlimbs was the result of "genetic piracy" — a single set of developmental processes taking over the production of these formerly independent systems (Roth, 1984; other examples in Roth, 1988). More recent work on homeobox genes, whose regional domains of expression in vertebrate embryos are arranged along the anteroposterior axis, and which may have a role in specifying positional information, has in fact shown that the *Hox*-4 cluster of genes, which is expressed in the posterior part of the vertebrate embryo and in the pelvic limbs, is also expressed in the forelimb (Tabin, 1992; see Goodwin's chapter for a further analysis of *Hox* genes). On this further evidence, Tabin too, suggested that genetic programs originally involved in development of the pelvic fin had in modern tetrapods become activated more anteriorly, and now secondarily are employed in forelimb development.

There are many other possible examples of extension and overlap in domains of influence (as opposed to a simple branching pattern of replication and proliferation) among

---

[12] The conjunction test works for ontogenetic homology because if *a* and *b* are both present at some ontogenetic stage of an individual, *b* cannot be a modified version of *a* unless iterative homology also has become involved.

iterative homologues; gene conversion provides other note-worthy instances. Hence, methods of inference designed for inferring continuity of information along linear segments of treelike topologies do not apply to iterative homologues, but iterative homology is nevertheless amenable to analysis: molecular genetics and analyses of development can provide evidence on whether their correspondences are indeed a manifestation of the same information.

VI. CONCLUSION

Interests in homology stem from several sources: from an interest in reconstructing phylogenetic history (of genes or of characters or of taxa), as well as from an interest in why some features vary in particular ways while others remain relatively invariant, among many taxa, and for long periods in history.

Van Valen's (1982) broad definition of homology, "correspondence caused by continuity of information," excludes sets of characters that manifest no correspon-dence, or characters whose correspondence has no basis in biological processes of replication. By embracing a wide range of phenomena that have been designated homologies between and within organisms, it (1) highlights the proper-ties these varying homology relationships share, and (2) permits us to identify the important features distinguishing them. Homology is a hierarchical concept applied at several levels of biological organization. Considering homology in this most general way enables us to discover what the most fundamental properties of biological lineages and replicators are that cause differences in their behavior at different levels. How information is conveyed (in what material form, and by what routes), and how it is manifest (in biological structure and function) at various levels in the biological hierarchy are subjects for empirical discovery.

A helpful prototype for this conceptual discussion of homology has been hierarchical expansion of the concept of natural selection. Although "natural selection" is usually applied to a process of selection among individual organ-isms, a word like "selection," which can be applied at other levels — to groups or to gametes — facilitates discussion of a more general and fundamental nature of relationships, properties, and processes. *Lineage, replicator,* and *interactor* are other terms that can be applied at multiple levels, in discussions of homology or of selection.

It is in the context of systematics that the study of homology has achieved greatest clarity, and progressed to a point that it is possible to test hypotheses in a rigorous way by application of the criteria of similarity, congruence, and conjunction. Questions arise about what properties of supraspecific comparisons make these sorts of tests informative, and whether tests of this sort can be applied to types of homology other than supraspecific homology.

We note that *similarity,* or correspondence of some sort, is the feature that causes us initially to hypothesize a relationship of homology between characters, and it is widely acknowledged to be a primary criterion in establishing any relationship of homology. *Congruence,* the consistency between patterns suggested by distributions of different characters, functions successfully as a test of the homology of derived character states of organisms, provided the character (1) is heritable, (2) is always overtly manifest when present, (3) both originates and becomes fixed in single lineage-segments in the past, and provided (4) lineages show continually divergent branching (as opposed to anastomosis). These requirements apply whether the discussion concerns lineages of characters, of organisms, or of populations. Where these conditions do not apply the test is less logically compelling; however, phylogenetic conclusions are relatively robust to the relaxation of these assumptions (McDade, 1992; Pamilo and Nei, 1988). If we assume that change occurs by linear substitution of states on a treelike topology, *conjunction* of putatively homologous characters is an indication that an error has been made in the accounting for the number of characters, their degree of independence, and the number of transformational events that separate them. The conjunction test cannot apply, however, to iterative homologues that do not show this type of behavior. Thus, tests of homology that were originally presented in a taxic context can also be analyzed and justified from a transformational perspective and, because at different levels of the organizational hierarchy different types of lineages, replicators, and homologues have different properties, some but not all of these tests will logically apply. The greatest challenges in the study of iterative homology lie in the empirical clarification of how biological information is transmitted and embodied, and reconstruction of the history of developmental programs (whose various topologies do not necessarily lend themselves readily to the best-developed methods of phylogenetic reconstruction)

Central in biology, the concept of homology both draws upon and contributes to the fields of population biology, systematics, genetics, and developmental biology. With continuing technological, empirical, and conceptual advances, both the integration of these fields, with their varying emphases on pattern and process, and a reconciliation of taxic and transformational approaches in comparative biology, come within reach.

ACKNOWLEDGMENTS

Thanks to Robert Brandon, John Mercer, Brent Mishler, and Fred Nijhout for comments and insights, and to Kathrin Stanger for help interpreting German. To a memory of E.B.R., himself a character with integrity, and a member of my lineage.

REFERENCES

Ax, P. (1987). "The Phylogenetic System." John Wiley & Sons, Chichester.

Baker, B. S. (1989). Sex in flies: the splice of life. *Nature* (*London*) **340**, 521-524.

Bledsoe, A. H., and Sheldon, F. H. (1990). Molecular homology and DNA hybridization. *J. Mol. Evol.* **30**, 425-433.

Bock, W. J. (1989). The homology concept: its philosophical foundation and practical methodology. *Zool. Beitr. N. F.* **32**, 327-353.

Bonner, J. T. (1974). "On Development." Harvard Univ. Press, Cambridge, MA.

Brandon, R. N. (1990). "Adaptation and Environment." Princeton Univ. Press, Princeton, New Jersey.

Buss, L. W. (1987). "The Evolution of Individuality." Princeton Univ. Press, Princeton, New Jersey.

Darwin, C. (1859). "On the Origin of Species." John Murray, London.

Dawkins, R. (1976). "The Selfish Gene." Oxford Univ. Press, New York.

Dawkins, R. (1978). Replicator selection and the extended phenotype. *Z. Tierpsychol.* **47**, 61-76.

Dawkins, R. (1982). Replicators and vehicles. *In* "Current Problems in Sociobiology" (King's College Sociobiology Group, ed.), pp. 45-64. Cambridge Univ. Press, Cambridge.

de Beer, G. R. (1971). "Homology, an Unsolved Problem." Oxford Biol. Readers, 11. Oxford Univ. Press, London.

de Pinna, M. C. C. (1991). Concepts and tests of homology in the cladistic paradigm. *Cladistics* **7**, 367-394.

Donoghue, M. J. (1992). Homology. *In* "Keywords in Evolutionary Biology" (E. F. Keller and E. A. Lloyd, eds.), pp. 170-179. Harvard Univ. Press, Cambridge, MA.

Doyle, J. J. (1992). Gene trees and species trees: molecular systematics as one-character taxonomy. *Syst. Bot.* **17**, 144-163.

Eldredge, N. (1979). Alternative approaches to evolutionary theory. *Bull. Carnegie Mus. Nat. Hist.* **13**, 7-19.

Fiers, W., Contreras, R., Haegeman, G., Rogiers, R., Van de Voorde, A., Van Heuverswyn, H., Van Herreweghe, J., Volckaert, G., and Ysebaert, M. (1978). Complete nucleotide sequence of SV40 DNA. *Nature (London)* **273**, 113-120.

Ghiselin, M. T. (1976). The nomenclature of correspondence: A new look at "homology" and "analogy". *In* "Evolution, Brain, and Behavior: Persistent Problems" (R. B. Masterton, W. Hodos, and H. Jerison, eds.), pp. 129-142. Lawrence Erlbaum, Hillsdale, New Jersey.

Goodwin, L. O., Lees-Miller, J. P., Leonard, M. A., Cheley, S. B., and Helfman, D. M. (1991). Four fibroblast tropomyosin isoforms are expressed from the rat alpha-tropomyosin gene via alternative RNA splicing and the use of two promoters. *J. Biol. Chem.* **266**, 8408-8415.

Gregory, R. L. (1970). "The Intelligent Eye." McGraw-Hill, New York.

Hall, B. K. (1992). "Evolutionary Developmental Biology." Chapman & Hall, London and New York.

Haszprunar, G. (1991). The types of homology and their significance for evolutionary biology and phylogenetics. *J. Evol. Biol.* **5**, 13-24.

Hennig, W. (1966). "Phylogenetic Systematics." Univ. of llinois Press, Urbana.

Hubel, D. H. (1988). "Eye, Brain, and Vision." Scientific American Library, W. H. Freeman, New York.

Hull, D. L. (1980). Individuality and selection. *Annu. Rev. Ecol. Syst.* **11**, 311-332.

Kauffman, S. A. (1983). Developmental constraints: Internal factors in evolution. In "Development and Evolution" (B. C. Goodwin, N. Holder, and C. C. Wylie, eds.), pp. 195-225. Cambridge Univ. Press, Cambridge.

Levi-Setti, R. (1975). "Trilobites." Univ. of Chicago Press, Chicago.

Mayr, E. (1982). "The Growth of Biological Thought." The Belknap Press of Harvard Univ. Press, Cambridge, MA.

McDade, L. A. (1990). Hybrids and phylogenetic systematics. I. Patterns of character expression in hybrids and their implications for cladistic analysis. *Evolution (Lawrence, Kans.)* **44**, 1685-1700.

McDade, L. A. (1992). Hybrids and phylogenetic systematics. II. The impact of hybrids on cladistic analysis. *Evolution (Lawrence, Kans.)* **46**, 1329-1346.

Minelli, A., and Peruffo, B. (1991). Developmental pathways, homology and homonomy in metameric animals. *J. Evol. Biol.* **3**, 429-445.

Mishler, B. D. (1993). The cladistic analysis of molecular and morphological data. *Am. J. Phys. Anthropol.* (in press).

Neigel, J. E. and Avise, J. C. (1986.) Phylogenetic relationships of mitochondrial DNA under various demographic models of speciation. *In* "Evolutionary Processes and Theory" (E. Nevo and S. Karlin, eds.), pp. 515-534. Academic Press, New York.

Nijhout, H. F. (1990). Metaphors and the role of genes in development. *BioEssays* **12**, 441-446.

Nijhout, H. F. (1991). "The Development and Evolution of Butterfly Wing Patterns." Smithsonian Institution Press, Washington, D.C.

Osche, G. (1973). Das Homologisieren als eine grundlegende Methode der Phylogenetic. *Aufsätze Reden senckenb. naturf. Ges.* **24**, 155-165.

Owen, R. (1843). "Lectures on Comparative Anatomy and Physiology of the Invertebrate Animals, Delivered at the Royal College of Surgeons, in 1843." Longman, Brown, Green, and Longman, London.

Owen, R. (1848). "On the Archetype and Homologies of the Vertebrate Skeleton." R. and J. E. Taylor, London.

Pamilo, P., and Nei, M. (1988). Relationships between gene trees and species trees. *Mol. Biol. Evol.* **5**, 568-583

Patterson, C. (1982). Morphological characters and homology. In "Problems of Phylogenetic Reconstruction" (K. A. Joysey and A. E. Friday, eds.), pp. 21-74. Academic Press, London.

Patterson, C. (1988). Homology in classical and molecular biology. *Mol. Biol. Evol.* **5**, 603-625.

Platnick, N. I. (1979). Philosophy and the transformation of cladistics. *Syst. Zool.* **28**, 537-546.

Rackoff, J. S. (1980). The origin of the tetrapod limb and the ancestry of tetrapods. In "The Terrestrial Environment and the Origin of Land Vertebrates" (A. L. Panchen, ed.), pp. 255-292. Academic Press, London.

Rak, B., Lusky, M., and Hable, M. (1982). Expression of two proteins from overlapping and oppositely oriented genes on transposable DNA insertion element *TS5*. *Nature (London)* **297**, 124-128.

Reeck, G. R., de Häen, C., Teller, D. C., Doolittle, R. F., Fitch, W. M., Dickerson, R. E., Chambon, P., McLachlan, A. D., Margoliash, E., Jukes, T. H., and Zuckerkandl, E. (1987). "Homology" in proteins and nucleic acids: a terminology muddle and a way out of it. *Cell* **50**, 667.

Riedl, R. (1978). "Order in Living Organisms: A Systems Analysis of Evolution." John Wiley & Sons, New York.

Rieppel, O. (1992). Homology and logical fallacy. *J. Evol. Biol.* **5**, 701-715.

Roth, V. L. (1984). On homology. *Biol. J. Linnaean Soc.* **22**, 13-29.

Roth, V. L. (1988). The biological basis of homology. In "Ontogeny and Systematics" (C. J. Humphries, ed.), pp. 1-26. Columbia Univ. Press, New York.

Roth, V. L. (1991). Homology and hierarchies: problems solved and unresolved. *J. Evol. Biol.* **4**, 167-194.

Sander, K. (1983). The evolution of patterning mechanisms: Gleanings from insect embryogenesis and spermatogenesis. In "Development and Evolution" (B. C. Goodwin,

N. Holder, and C. C. Wylie, eds.), pp. 137-160. Cambridge Univ. Press, Cambridge.

Sattler, R. (1990). Towards a more dynamic plant morphology. *Acta Biotheoretica* **38**, 303-315.

Schmitt, M. (1989). Das Homologie-Konzept in Morphologie und Phylogenetik. *Zool. Beitr. N. F.* **32**, 505-512.

Schoch, R. M. (1986). "Phylogeny Reconstruction in Paleontology." Van Nostrand-Reinhold, New York.

Stevens, P. F. (1984a). Homology and phylogeny: morphology and systematics. *Syst. Bot.* **9**, 395-409.

Stevens, P. F. (1984b). Metaphors and typology in the development of botanical systematics, or the art of putting new wine in old bottles. *Taxon* **33**, 169-211.

Stresemann, E. (1975). "Ornithology from Aristotle to the Present." Harvard Univ. Press, Cambridge, MA.

Sulston, J. E., Albertson, D. G., and Thomson, J. N. (1980). The *Caenorhabditis elegans* male: postembryonic development of nongonadal structures. *Dev. Biol.* **78**, 542-576.

Tabin, C. J. (1992). Why we have (only) five fingers per hand: *Hox* genes and the evolution of paired limbs. *Development (Cambridge, U.K)* **116**, 289-296.

Tajima, F. (1983). Evolutionary relationships of DNA sequences in finite populations. *Genetics* **105**, 437-460.

Van Valen, L. (1982). Homology and causes. *J. Morphol.* **173**, 305-312.

Wagner, G. (1989a). The origin of morphological characters and the biological basis of homology. *Evolution (Lawrence, Kans.)*, **43**, 1157-1171.

Wagner, G. (1989b). The biological homology concept. *Annu. Rev. Ecol. Syst.* **20**, 51-69.

Wagner, G. P. and Misof, B. Y.(1993). How can a character be developmentally constrained despite variation in developmental pathways? *J. Evol. Biol.* **6**, 449-455.

Williams, G.C. (1992). "Natural Selection: Domains, Levels, and Challenges." Oxford Series in Ecology and Evolution, 4. Oxford Univ. Press, New York.

Wray, G. A., and McClay, D. R. (1989). Molecular heterochronies and heterotopies in early echinoid development. *Evolution (Lawrence, Kans.)*, **43**, 803-813.

10

HOMOLOGY IN MOLECULAR BIOLOGY

David M. Hillis

Department of Zoology
The University of Texas
Austin, Texas 78712

I. INTRODUCTION

This book is evidence that concepts of homology are diverse among biological disciplines. These different concepts often lead to confusion when biologists of different flavors attempt to talk among themselves. Confusion, however, exists within specific research areas as well; molecular biologists may have done more to confound the meaning of the term homology than have any other group of scientists. In many circles of molecular biologists, homology has come to mean "similarity": a simple, quantifiable relationship, for which the word similarity adequately suffices. For instance, one often reads about a comparison made between two genes, in which the individual making the comparison describes the two genes as "50% homologous," whereas the author actually means that the two genes have a measured similarity of 50% (i.e., they share 50% of their aligned nucleotides in common). The two genes may or may not be similar as a result of common ancestry, and therefore may or may not be homologous under the evolutionary meaning of the term.

Why this confusion of terms has arisen in molecular biology is not clear; perhaps the term homology is thought to make the work sound more like science than would use of the simple and obvious word similarity. Wegnez (1987) suggested the substitute jargon *isology* to describe measured percent similarity of aligned sequences (as in "the two genes are 50% isologous"). Use of the term isology would seem to satisfy the desire of using a word that is unfamiliar (and therefore surely important and complex) to describe a familiar and simple concept, but its use has nevertheless not caught on among molecular biologists.

That is not to say that the meaning of words should never change after they are originally defined. Indeed, Richard Owen's concept and use of homology gradually expanded between 1843 and 1848, from "the same organ in different animals under every variety of form and function" (Owen, 1843) to include organs duplicated within an organism (Owen, 1848). As biologists began to explain homologies through an understanding of evolution, homology became inextricably linked with its evolutionary explanation (see Donoghue, 1992). Lankester (1870) suggested the terms *homogeny* and *homoplasy* to describe similarity that could be traced to common ancestry and similarity that resulted from independent evolution, respectively. Homogeny never enjoyed widespread use, and the concept of "biological similarity that can be traced to common ancestry" quickly became associated with homology. Although this use of homology has caused its share of confusion and difficulties (see Patterson, 1988; Donoghue, 1992), the concept is useful and distinct from related concepts such as similarity (which may be explained either through inheritance from a common ancestor or evolutionarily independent acquisition), *homoplasy* (similarity that arises through evolutionary convergence, parallelism, or reversal), and *analogy* (superficial similarity that arises through functional convergence).

Although similarity at some level seems a necessary prerequisite to recognize homology (Patterson, 1988), some authors have taken the concept of homology to its logical conclusion: homologous structures include any "parts that arise from the same source" (Ghiselin, 1976, p. 138) or "trace back to a single genealogical precursor" (Goodman *et al.*, 1987, p. 146). Under such a definition, structures may have diverged into such dissimilar parts that they are no longer recognizably similar, although are still homologous because of their common ancestral origin. Thus, the word homology is now used in molecular biology to describe everything from simple similarity (whatever its cause) to common ancestry (no matter how dissimilar the structures). I fall at the end of the continuum that relates homology to common ancestry, and use the word similarity to describe the likeness of structures (including molecular sequences).

None of the discussion above requires any consideration of molecular biology. This introduction merely serves as a point of departure for considering the interesting, and at times complex, ramifications of homologous relationships among genes, parts of genes, and the immediate products of genes. For the remainder of this article, I will consider two biological molecules to be homologues if they are descended (via imperfect replication) from a common ancestor.

II. CLASSES OF MOLECULAR HOMOLOGY

If we accept that most genes are evolutionarily related, and that extant genomes were derived by duplication, modification, and recombination of a small number (perhaps one?) of original replicating sequences, then it is also the case that most genes are at some level homologous. It is thus necessary to constrain the concept of homology for certain applications to make it useful. In the context of inferred evolutionary relationships among genes, we typically are interested in the most recent relationship shared by two given genes. Moreover, different classes of homology have been constructed to address different processes of divergence that generate homologous genes. The most obvious of these processes are

> **speciation** (the divergence of lineages of organisms)
>
> **gene duplication** (the divergence of lineages of genes within an organismal lineage)
>
> **horizontal gene transfer** (the divergence of lineages of genes by transfer across different organismal lineages)

Each of these processes is of interest to molecular evolutionary biologists, and each results in genes that "trace back to a single genealogical precursor." However, all three processes usually are not studied simultaneously, so it is necessary to distinguish among the various kinds of homology when only one of these processes is of interest in a given analysis. Fitch (1970) and Gray and Fitch (1983) proposed the following names for homologous macromolecules, based on the different generative processes:

> **orthologous** genes (or their products) are homologues that diverged as a result of a speciation event
>
> **paralogous** genes (or their products) are homologues that diverged as a result of a gene duplication event
>
> **xenologous** genes (or their products) are homologues that diverged as a result of lateral gene transfer.

If one is interested in reconstructing the phylogenetic history of taxa by inferring relationships among genes contained in those taxa, then it is usually necessary to examine orthologous genes (but see Section III below). If the history of gene duplication is of interest, then the genes examined need include paralogues. Study of lateral gene transfer obviously requires examination of xenologs. Although these points are obvious corollaries of the definitions, they sometimes are not appreciated by practicing biologists. The important distinction is whether history of the taxa, or history of the genes is of primary concern. Confusion of orthology with paralogy and xenology is likely to result in misleading inferences about organismal evolution.

The example in Figure 1 illustrates some possible effects of confusing orthologous and paralogous genes. The diagram shows a simplified representation of the evolution of some globin genes in four species of vertebrates: a lamprey, a frog, a mammal, and a snake. All of these genes can be traced back to an ancestral sequence, so at some level all are homologous. After the split (speciation event) that led to the lamprey lineage on the one hand and the three tetrapod lineages on the other, there was a gene duplication event in the lineage that led to the tetrapods: this duplication event gave rise to the α- and β-hemoglobin gene families. Therefore, all three tetrapods have both types of globin genes, whereas the lamprey has only one. If we wished to infer relationships among these species, we could do so by analyzing sequences of either β-hemoglobins or α-hemoglobins from the three tetrapods, together with the unduplicated hemoglobin sequence from the lamprey (Fig. 1b). However, if we were unaware that the gene duplication had occurred, and analyzed a mixture of α- and β-hemoglobins from the various tetrapods, the evolutionary relationships that we inferred would not be of the taxa, but of the genes (Fig. 1c). In other words, to reconstruct the speciation events, we have to select genes that are orthologous. If the gene duplication events are of interest, then the paralogues need to be examined as well.

The example presented above is greatly simplified; there have been at least three additional gene duplication events in the α-hemoglobin family and six additional gene duplication events in the β-hemoglobin family (Doolittle, 1987; Goodman et al., 1979, 1987). Some of these duplicated genes are expressed at different times in the ontogeny of various tetrapods (e.g., some are larval or fetal in expression, whereas others are expressed only in adults). There obviously exists a continuum of levels of duplication of genes, so that few genes are truly "single copy" (in the sense that there are no other close paralogues). In fact, much genic diversity of both eubacteria and eukaryotes is thought to have arisen through gene (or whole genome) duplication (Grime and Mowforth, 1982; Herdman, 1985; Rees and Jones, 1972; Sparrow and Nauman, 1976), so paralogy probably is the rule rather than the exception.

Orthologous, paralogous, and xenologous molecules are all likely to be detected by the similarity of their sequences. Paralogy may be distinguished from orthology by the test of conjunction: whether or not the two homologues are found in the same individual (Patterson, 1988). Xenology is inferred if

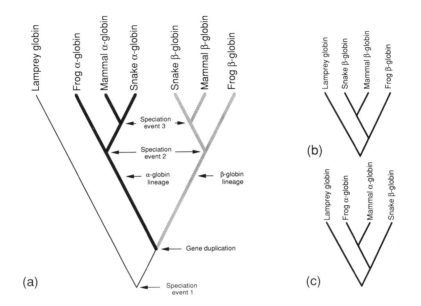

Fig. 1. (a) The origins of orthologous and paralogous globin genes in vertebrates. (b) The phylogeny of taxa can be inferred by analyzing either set of orthologous genes. (c) If a mixture of orthologous and paralogous genes is included in a phylogenetic analysis, the resulting tree will reflect gene relationships rather than relationships of taxa. (Simplified from Goodman *et al.*, 1987).

the homologues exist in distantly related species (i.e., information from the genes is highly incongruent with other information about phylogeny). Striking cases of xenology may result from retroviral transfer of genes, in which case the xenologous relationship may seem obvious. However, xenologous relationships (especially of alleles at a single locus) can also arise through hybridization, in which case xenology may be confused with convergence. In such cases, xenology of alleles may be inferred on the basis of the concentration of apparent convergence (across multiple loci) in particular branches of the inferred phylogenetic tree (see Buth, 1984; Duellman and Hillis, 1987).

III. CONCERTED EVOLUTION OF PARALOGOUS SEQUENCES

The above discussion assumes that duplicated genes will evolve independently following the duplication event. In fact, many duplicated genes continue to interact, so that evolution in the two (or more) duplicated sequences may not be independent. Sequences that are present in large numbers of tandem repeats rarely undergo independent evolution (see Arnheim, 1983; Dover, 1982, 1986; Ohta, 1980). Soon after sequences of tandem repeats were first studied, it was observed that the multiple copies of many repeated gene families were very similar within an individual and within a species, whereas the same families of repeated genes were often quite divergent among closely related species (Arnheim *et al.*, 1980; Brown *et al.*, 1972; Zimmer *et al.*, 1980). If the tandem duplications occurred in the ancestor of the two species, and the repeated sequences were evolving independently of one another, one would expect greater within-species than between-species divergence (Fig. 2). However, the observation was just the opposite: very little within-species, but great between-species divergence. It appeared that all the copies of the repeated sequences were evolving in concert; the phenomenon was termed *concerted evolution* (Zimmer *et al.*, 1980).

After concerted evolution of repeated DNA sequences was discovered, it was found to be nearly ubiquitous among mid- to highly repeated tandem gene families. In these families, concerted evolution often occurs so rapidly that relatively few differences can be detected among the copies (e.g., some nuclear ribosomal RNA genes; see Hillis and Dixon, 1991). The rate of concerted evolution among families of genes that are repeated at lower frequency is much more variable, from families that show few indications of concerted evolution (e.g., McElroy *et al.*, 1990; Shah *et al.*, 1983) to those in which the paralogous and orthologous relationships may be difficult to untangle (e.g., Doyle, 1991; Hughes and Nei, 1990; Irwin and Wilson, 1990). Furthermore, rates of concerted evolution may not be consistent within the same gene families among different taxa. In the case of ribosomal RNA genes, for instance, rates of concerted evolution may differ dramatically between tandemly repeated genes on a single chromosome versus sets

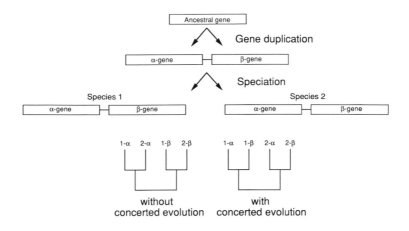

Fig. 2. The origin of a hypothetical set of paralogous sequences (α and ß) through a gene duplication event, followed by
a speciation event, producing two sets of orthologous relationships. In the absence of concerted evolution, the ortho-
logues (α in species 1 and 2, or ß in species 1 and 2) would
appear to be more closely related to each other than to the
paralogous sequences in the same species, because they share
a more recent common ancestor (the original α- and ß-genes,
respectively). However, if the paralogous sequences are undergoing concerted evolution, then the paralogues are homogenized within each species, and the α-gene in species 1 will
appear to be more closely related to the ß-gene in species 1
than to either gene in species 2.

of repeats on different chromosomes. In some species, there
may be clusters of ribosomal genes in which concerted evolution occurs, yet the different clusters may evolve independently (as in the clam genus *Corbicula*; D. M. Hillis, personal
observation). In such cases, each cluster of ribosomal RNA
repeats is evolving much like an orthologous locus, whereas
relationships among the clusters appear paralogous.

Several mechanisms have been hypothesized to explain
concerted evolution, of which two have received the most
attention: unequal crossing-over (Coen *et al.*, 1982a; Ohta,

1980, 1983; Smith, 1974; Szostak and Wu, 1980) and gene conversion (Baltimore, 1981; Nagylaki, 1984; Nagylaki and Petes, 1982; Ohta, 1984; Ohta and Dover, 1983). Both mechanisms have received some empirical support as explanations for the concerted evolution of multigene families (Hillis *et al.*, 1991; Seperack *et al.*, 1988). Unequal crossing-over usually is considered a stochastic process, in which frequency of fixation of variants is directly related to the frequency at which the relevant mutations arise. Gene conversion, on the other hand, can be biased, such that a new variant sequence is favored over an ancestral sequence, and hence can spread rapidly through all copies of the repeated sequence. Biased gene conversion is thought to be the only mechanism that can adequately explain the rapid rate of homogenization seen in many families of genes that undergo concerted evolution (Coen *et al.*, 1982a,b).

If the rate of concerted evolution is high enough, an entire family of genes will evolve almost as if they were a single sequence present in many duplicate copies. Thus, paralogous sequences that have a high rate of concerted evolution behave like orthologous sequences: they show little divergence within species, but may evolve rapidly between species. For this reason, paralogous sequences evolving under high rates of concerted evolution can be used to infer phylogenetic relationships of taxa, without fear of reconstructing gene duplication events rather than speciation events (Sanderson and Doyle, 1992). Patterson (1988) suggested the term *plerology* to describe the relationship among paralogous sequences homogenized within taxa as a result of concerted evolution.

How high does the rate of concerted evolution have to be before the distinction between orthology and paralogy becomes blurred? Sanderson and Doyle (1992) simulated the effects of concerted evolution and found that when 70% of sites underwent concerted evolution between speciation events, the inferred trees always represented the correct (simulated) relationships among the taxa rather than the genes. In order to correctly infer gene trees among paralogues, concerted evolution had to involve fewer than 10% of the sites between speciation events in the simulations. At intermediate rates of concerted evolution, the inferred trees were likely to confound paralogous and orthologous relationships. Although these values are somewhat dependent on the details of the simulations, they represent a first-order approximation of relative rates of concerted evolution that are likely to confound the inferred relationships of paralogous and orthologous genes.

IV. PARTIAL HOMOLOGY OF MOLECULES: EXON SHUFFLING

Earlier in this chapter, I was critical of those who would say that two sequences are "50% homologous" when they mean the sequences share 50% of their aligned sites. However, that does not mean that homology must be an all-or-none condition, if the units of comparison are whole genes or their products. *Partial homology* is possible because some proteins have evolved through recombination of functional modules, which often correspond to the exons of genes (Fig. 3). For instance, the gene for tissue plasminogen activator is made up of exons that appear to have been captured from the genes for plasminogen, fibronectin, and epidermal growth factor (Patthy, 1985). Therefore, the corresponding functional modules are paralogous among the proteins, and the proteins (as well as their genes) are each partially paralogous. In addition to paralogy of modules among these proteins, are several cases of within-protein module paralogy (Fig. 3). The Kringle module present in two copies in tissue plasminogen activator is present in five tandem copies in plasminogen. Likewise, the finger module of fibronectin and the growth factor module of epidermal growth factor each are repeated many times (Fig. 3).

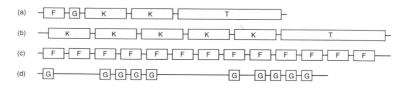

Fig. 3. Partial paralogy of some genes of proteins involved in blood coagulation and fibrinolysis. Functional modules are indicated by boxes: F, finger module; G, growth-factor module; K, Kringle module; T, trypsin-like module. (a) Tissue plasminogen activator protein. (b) plasminogen. (c) fibronectin. (d) epidermal growth factor. [Adapted from Patthy (1985) and Li and Graur (1991).]

Within-gene paralogy also can arise through slippage replication or unequal crossing-over of repeated units within genes (Hancock and Dover, 1988, 1990). Thus, genes may have regions that are internally paralogous. In the case of nuclear ribosomal RNA genes, two coevolved sets of paralogous regions may arise as a mechanism for maintaining secondary structure of the mature ribosomal RNA (Hancock and Dover, 1990).

V. POSITIONAL HOMOLOGY AND SEQUENCE ALIGNMENT

So far, I have been discussing homology of whole genes (or their products), or whole domains of genes. However, for most evolutionary analyses of sequence data, it is necessary to consider homology at a finer level: that of a single nucleotide site (or an amino acid site in the case of protein sequences). This is called *positional homology*. In a phylogenetic analysis of DNA sequences, for instance, the characters are nucleotide positions and the character states are the different nucleotides. When two orthologous genes are compared, the units of comparison typically are not the entire genes but the individual nucleotide positions. If all evolution has been through substitution, so that the two genes are exactly the same length, then it usually is a simple matter to align the homologous sites for analysis (Fig. 4a). However, when insertions or deletions have occurred in one or both sequences, gaps need to be inserted to preserve positional homology (Fig. 4b and c).

Alignment of sequences requires explicit and objective rules if inferences of positional homology are to be robust. If gaps are added without penalty to the alignment score, then unrelated (nonhomologous) sequences could be aligned without difficulty. However, if the penalty for gaps is too high, then the inferred positional homology is likely to be mistaken. Figure 5 shows a set of aligned ribosomal RNA genes in which the inferred positional homology differs depending on the weight assigned to gaps. If alternative alignments are equally good (or nearly so) for a given region, the ambiguous region should be excluded from analyses that assume accurate inference of positional homology, such as phylogenetic analyses (Swofford and Olsen, 1990).

```
      ATTCCGTAGCTGTTTCATCTTGATTGGTAACTG
(a)   ||| ||||| ||||||||| |||||||||| |||
      ATTGCGTAGATGTTTCATCATGATTGGTATCTG
```

```
      ATTCCGTAGCTGTTTCATCTTGATTGGTAACTG
(b)   |            ||||||||| |||||||||| |||
      ATTGCGTAGTGTTTCATCATGATTGGTATCTG
```

```
      ATTCCGTAGCTGTTTCATCTTGATTGGTAACTG
(c)   ||| ||||| ||||||||| |||||||||| |||
      ATTGCGTAG-TGTTTCATCATGATTGGTATCTG
```

Fig. 4. (a) Alignment of two homologous sequences is straightforward if all differences are the result of substitutions. (b) If two sequences are different in length, high apparent dissimilarity may exist if gaps are not allowed. (c) The same sequences as in (b), but with positional homology restored by the addition of a single gap (corresponding to an hypothesized deletion in the lower sequence).

Ideally, the weight of the penalty chosen for gaps should reflect the relative probability of insertion/deletion events relative to substitution events. The probability of insertion/deletion events varies greatly among different genomic regions. Such events are relatively rare in most protein-coding regions (because they usually lead to highly deleterious frameshift mutations), but are quite common within loop regions of ribosomal RNA genes and in many noncoding sequences. In regions where insertion/deletion events are common, accurate inferences of positional homology are highly unlikely in any but the most closely related sequences.

A number of secondary criteria (beyond sequence similarity) are often used to determine positional homology. In protein-coding genes, the translated amino acid sequence usually is more conserved than the encoding nucleotide sequence (because of the redundancy of the code), and often is of use for determining nucleotide alignment. Furthermore, because of the deleterious nature of frameshift mutations in protein-coding genes, it is reasonable to assign *a priori* heavy penalties (or disallow) gaps that do not correspond to multiples of codons (i.e., three nucleotides). If secondary structure is known for the gene product, alignments may be based on conserved secondary structure rather than conserved primary structure (i.e., sequence similarity).

```
            22|00              22|20              22|40
Mus         GTCAGCCAGGACTCTCTACCCGCTCACGGCAAGGCTTCCCTGCCCGCTACCGGAGGCAAC
Rattus      GTCAGCCAGGACTCTCTACCCGCTCACGGCAAGGCTTCCCTGCCCGCTACCGGAGGCAAC
Homo        GTCAGCCAGGACTCTCTACCCGCTCGCGGCAAGGCTTCCCTGCCCGCTACCGGAGGCAAC
Rhineura    GTCAGCCAGGATTCTCTATCCGCTCGCGGCAAGGCTTCCCTGCCCGCTACCGGAGGCAAC
Cacatua     GTCAGCCAGGATTCGCTATCCGCTCGCGGCAAGGCTTCCCTGCCCGCTACCGGAGGCAAC
Xenopus     GTCAGCCAGGATTCTCTACCCGCTCGCGGCAAGCCTTCCCTGCCCGCTACCGGAGGCAGC
Rhyacotriton GTCAGCCAGGATTCTCTATCCGCTCGCGGCAAGCCTTCCCTGCCCGCTACCGGAGGCAAC
Typhlonectes GTCAGCCAGGATTCTCTATCCGCTCGCGGCAAGCTTCCCTGCCCGCTACCGGAGGCAAC
Latimeria   GTCAGCCAGGATTCTCTACCCGCTTGCGGCAAGGCTTCCCTGCCCGCTACCGGAGGCAGC
Cyprinella  GTCAGTCCAGGATTCCTACCCGCTGGCGGTCAAGCCTTCCCTCCGGCTACCGGAGGCAGC
              *  **  **  **     *       **     **  **  *  *    **    *                    *
```

```
            22|00              22|20              22|40
Mus         GTCAG-CCAGGACTCTCTACCCGCTCACGG-CAAGGCTTCCCTGCCCGCTACCGGAGGCAAC
Rattus      GTCAG-CCAGGACTCTCTACCCGCTCACGG-CAAGGCTTCCCTGCCCGCTACCGGAGGCAAC
Homo        GTCAG-CCAGGACTCTCTACCCGCTCACGG-CAAGGCTTCCCTGCCCGCTACCGGAGGCAAC
Rhineura    GTCAG-CCAGGATTCTCTATCCGCTCGCGG-CAAGGCTTCCCTGCCCGCTACCGGAGGCAAC
Cacatua     GTCAG-CCAGGATTCGCTATCCGCTCGCGG-CAAGCCTTCCCTGCCCGCTACCGGAGGCAAC
Xenopus     GTCAG-CCAGGATTCTCTACCCGCTCGCGG-CAAGCCTTCCCTGCCCGCTACCGGAGGCAGC
Rhyacotriton GTCAG-CCAGGATTCTCTATCCGCTCGCGG-CAAGCCTTCCCTGCCCGCTACCGGAGGCAAC
Typhlonectes GTCAG-CCAGGATTCTCTATCCGCTCGCGG-CAAGCCTTCCCTGCCCGCTACCGGAGGCAAC
Latimeria   GTCAG-CCAGGATTCTCTACCCGCTTGCGG-CAAGGCTTCCCTGCCCGCTACCGGAGGCAGC
Cyprinella  GTCAGTCCAGGATTC-CTACCCGCTGGCGGTCAAGCCTTCCCT-CCGGCTACCGGAGGCAGC
              *      *   *   *      **    *     *        *                    *
```

Fig. 5. Alternative alignments of a segment of 28S ribosomal RNA genes in selected vertebrates. The upper set of sequences is aligned without gaps; there are 20 variable positions. Four gaps have been introduced into the lower set of sequences, thereby reducing the number of variable positions to 10 (including those with gaps). The alternative accepted depends on the weight of the penalty assigned to gaps. Sequences from Gonzales *et al.* (1985), Hadjiolov *et al.* (1984), Hassouna *et al.* (1984), Hillis and Dixon (1989), Hillis *et al.* (1990a), Larson and Wilson (1989), and Ware *et al.* (1983). (Adapted from Hillis *et al.* 1990b.)

Some authors claim that high similarity of aligned sequences never arises by convergence:

> in terms of nucleotide sequences, there seems to be no equivalent of convergence, or close similarity produced by evolution from different precursors (Goodman *et al.*, 1987, p. 147).

However, several cases of convergence of sequences are known, and more cases will undoubtedly come to light as molecular biologists begin to look for the phenomenon. For instance, the lysozymes of ruminants and colobine monkeys appear to have converged as a result of parallel development of foregut fermentation in these mammalian groups (Stewart and Wilson, 1987; Swanson *et al.*, 1991). Although these proteins are homologous as lysozymes, the high degree of sequence similarity clearly is convergent. (This convergence is analogous to the oft-cited case of convergence between bird and bat fore-limbs: the structures are homologous as forelimbs, but con-vergent as wings.) Other cases of sequence convergence are likely to arise as a result of GC or AT biases (Aoki *et al.*, 1981; Hori and Osawa, 1986; Tamura, 1992; Wilson *et al.*, 1980), the presence of simple tandem repeats (e.g., Levinson *et al.*, 1985), or because of convergent mutational spectra (Muto and Osawa, 1987; Singer and Ames, 1970).

VI. HOMOLOGY IN INDIRECT (NONSEQUENCE) MOLECULAR TECHNIQUES

A. DNA Hybridization

Homology is an important concept for nonsequence molecular data as well. In DNA-DNA hybridization, low copy portions of the complete genomes of two taxa are annealed to form hybrid duplexes, and the average similarity of the genomes is measured as a function of melting temperature of the hybrid strands. Melting temperature (temperature at which 50% of the hybrid duplexes separate into single strands, or a similar standard of comparison; see Werman *et al.*, 1990) is directly related to the number and distribution of hydrogen bonds that form between complementary base pairs, so hybrid duplexes with few mismatches melt at a higher temperature than hybrid duplexes with many mismatches. Hybrid duplexes form between any two sequences with high levels of sequence similarity.

From the description above, it should be clear that DNA hybridization provides a measure of average sequence similar-ity of cross-hybridizing sequences, which undoubtedly include paralogous as well as orthologous genes. Moreover, sequence convergence at individual sites (homoplasy) cannot be directly distinguished from sequence similarity due to common ancestry (homology) by this technique (Bledsoe and Sheldon, 1990). The melting temperature thus is confounded by similarity due to orthology, paralogy, and homoplasy. Nonetheless, some proponents of the technique claim that DNA hybridization "solves the difficulty of determining homology," although the "solution" amounts to defining homology as similarity and ignoring the distinction between orthology and paralogy:

> Fortunately, DNA-DNA hybridization data are immune to convergence, because the conditions of the experiments preclude the formation of heteroduplexes between non-homologous sequences. To form a stable duplex DNA molecule at 60°C, 80 per cent of the bases in the two strands must be correctly paired, and *only homologous sequences* have this degree of complementarity. This solves the problem of homology and thereby eliminates the possibility of convergence. (Sibley and Ahlquist, 1987, p. 100; emphasis as in original)

Other proponents of this technique (e.g., Bledsoe and Sheldon, 1990; Werman *et al.*, 1990) have a more realistic view of the problems of determining homology in DNA-DNA hybridization studies.

B. Restriction Enzyme Analysis

In restriction enzyme analyses, the targeted DNA may be entire genomes (e.g., Hillis *et al.*, 1992), an isolated organellar component (typically mitochondrial or chloroplast DNA; Moritz *et al.*, 1987; Palmer *et al.*, 1988), or a specific gene or gene region. In one common method of analyzing a specific gene, putatively homologous regions are examined by cross-hybridizing a cloned or otherwise isolated gene (called a probe) to genomic DNA that has been cleaved with a restriction enzyme, separated electrophoretically, and then attached to a support membrane (Southern, 1975). The hybrid probe–target heteroduplexes are visualized by autoradiography or its equivalent (Dowling *et al.*, 1990). As with DNA-DNA hybridization, this procedure relies on the low likelihood of hybridization between nonhomologous sequences. Hybridization between closely related paralogous sequences is commonly observed, but the presence of paralogous products usually can be detected on the basis of the presence of multiple regions of cross-hybridization. Target genes may be chosen to avoid complications of paralogy by process of elimination (Friedlander *et al.*, 1992).

Unfortunately, homology has taken on yet another meaning in the description of probes used in Southern blotting and similar procedures. A probe is often called "homologous" if it is used to study the same species from which it was derived, and "heterologous" if the probe was cloned from another species. This is somewhat confusing, since heterologous probes are used to examine presumably homologous genes in different species. To avoid confusion, the more appropriate terms *homospecific* and *heterospecific* can be used in place of *heterologous* and *homologous* to describe these relationships between probe and target.

An alternative approach for isolating a target sequence for restriction analysis is amplification by polymerase chain reaction or PCR (Kleppe *et al.*, 1971; Mullis and Faloona, 1987). This procedure involves hybridizing two short (typically 20 to 30 bases) oligonucleotide primers to opposite strands of a heat-denatured DNA target, replicating the DNA between the two primer sequences with a heat-stable DNA polymerase, denaturing the strands by heating, and repeating this cycle, usually about 30 times. The target DNA is doubled during each cycle (if all goes well), so that a large quantity of the target sequence is specifically amplified for analysis. Interpretation of the amplification products from PCR (using a given set of primers) as homologues is based on the assumption that only homologues will have the appropriate primer sequences separated by the appropriate distances in the genome; the size of the amplification product often is used to verify that the correct target was successfully amplified. However, this procedure does not distinguish orthologous from paralogous genes, both of which may retain the conserved primer sites and gene length. Worse, the amplification products may be recombinants of different alleles or even several paralogous loci — a phenomenon that produces "shuffled clones" (Saiki *et al.*, 1988; Scharf *et al.*, 1988a, b). This happens when a target sequence has not been fully replicated before the next round of denaturation and reannealing takes place. In the subsequent round, the partial product from one locus may rehybridize to a different locus to complete the product extension, thus producing a hybrid product that is partially derived from two different loci. Shuffled gene artifacts obviously are not limited to restriction analyses; any study that uses PCR should consider this process as a potential source of error that will confound assessments of homology.

Discussion so far has centered on determining homology of the portion of the genome targeted for restriction analysis. As with sequencing studies, an additional (finer) level of homology must be considered: homology of the individual characters. Data from restriction enzyme analyses typically are presented in one of two ways: restriction fragments or restriction sites (Fig. 6). Restriction fragments represent contiguous nucleotide sequences that fall between two recognition sites for a given restriction endonuclease. Such fragments may be coded as present or absent in a given individual.

Phylogenetic analysis of a presence/absence matrix of restriction fragments assumes that fragments of the same size represent homologous segments of the gene, whereas fragments of different sizes are not. There are three obvious sources of error with this assumption. The least of the three is that convergence of similarly sized but nonhomologous fragments is likely, especially if many fragments are examined and

Restriction sites

	I	II	III
One	+	+	+
Two	+	-	+
Three	-	-	+
Four	+	-	-

Restriction fragments

	1	2	3	4	7
One	+	+	-	+	-
Two	-	-	+	+	-
Three	-	-	-	+	+
Four	-	-	-	+	+

Fig. 6. Restriction site versus restriction fragment data. In this example, three restriction sites (I, II, and III) are variable in a linear segment of DNA among four individuals (One, Two, Three, and Four). The number under each restriction fragment indicates its respective length in arbitrary units. Two data matrices are shown: one of restriction sites and the other of restriction fragment size classes (in both, + indicates presence and - indicates absence). The three sites are independent characters, but the fragment size classes are not. In addition, note that the fragment patterns may be identical in species that share no restriction sites in common (e.g., individuals Three and Four).

the precision of measurement is not great. A more important problem is that insertion/deletion events will change the size of fragments and obscure homologies; this is serious because one insertion/deletion event may simultaneously affect many different but overlapping restriction fragments (each produced by a different restriction enzyme). Finally, gain of a restriction site will result in loss of one restriction fragment and gain of two smaller ones; obviously these are not independent events. The two smaller fragments are together homologous to the larger fragment. Because of all these sources of error, it is highly preferable to treat restriction sites (rather than restriction fragments) as characters in a restriction enzyme analysis (Fig. 6).

Restriction sites can be located in the target sequence by various mapping strategies (see Dowling *et al.*, 1990); the data matrix then indicates which sites are present versus absent in the studied individuals. The presence of a restriction site indicates that a specific base recognition sequence exists at a given location in the gene; a site that maps to the same position in two species is assumed to be homologous unless phylogenetic analysis suggests that the site is homoplastic. The primary source of error is related to imprecision of mapping: independent but adjacent restriction sites may be mapped to the same location. Thus, one individual may have two different restriction sites scored as one, or two individuals may have different restriction sites incorrectly assumed to be homologous. However, this source of error also affects restriction fragment data (because close sites produce small fragments that may be lost), so site data are highly preferable to fragment data when accurate assessments of homology are required.

C. Random Amplified Polymorphic DNA (RAPD)

As described above, the polymerase chain reaction can be used to amplify a specific target from a genome, based on *a priori* knowledge of the flanking regions. Welsh and McClelland (1990) and Williams *et al.* (1991) have described PCR-based methods for amplifying random polymorphic regions of the genome without any prior knowledge of specific flanking regions. The Williams *et al.* (1991) method, called RAPD analysis, utilizes a very short (typically 10 bases) oligonucleotide primer of an essentially arbitrary sequence (but with a constraint on GC content). Because of the small number of matches required for successful hybridization, the primer is likely to hybridize somewhere in the genome along opposite strands of DNA in the correct orientation (with the 3' ends facing each other), so that one or more fragments is amplified. This is especially likely at the location of inverted repeats. Some of these fragments will likely vary in length among individuals, and these variants are used as genetic markers in studies of populations and closely related species (e.g., Chapco et al., 1992; Crowhurst *et al.*, 1991; Goodwin and Annis, 1991; Hadrys *et al.*, 1992; Hunt and Page, 1992; Welsh *et al.*, 1992). Use of several different arbitrary primers can result in rapid collection of a large number of genetic markers.

Homology assignments of RAPD markers usually are based on considerations of fragment length or, less commonly, information on dominant or codominant segregation. Fragment length by itself is likely to be misleading about homology, for many of the same reasons discussed under Restriction Fragment Analysis (above). In addition, the same primer sites may exist in nonhomologous parts of the genome in two different species. Since segregation usually cannot be studied between species, assignments of homology of RAPD markers are highly tenuous in interspecific studies. If a primer site is lost (through substitution, for instance) but a second primer site exists nearby, two partially homologous fragments will be amplified that differ in length. Obviously, the presence of one such fragment is not independent of the presence of the other. Since many RAPD markers occur in regions of multiple repeats, the presence of multiple adjacent primer sites is not unlikely.

Although most of the above objections are theoretical, early empirical indications support the difficulties of using RAPD markers for phylogenetic analyses. Smith *et al.* (1993) found that loci that amplified in one strain of bacterium could be excluded from amplification in another strain because of competitive amplification of an unrelated locus. They also reported amplification of nonhomologous loci of indistinguishable size by the same primer, as well as amplification of multiple partially homologous fragments by a single primer. Smith *et al.* concluded that homology could not be reliably inferred from the RAPD fragment patterns. Furthermore,

Kambhampati *et al.* (1991) found that RAPD-based inferences of phylogeny were incongruent with established, well-supported phylogenetic relationships of mosquitoes. This suggests that shared RAPD products were not orthologous. The difficulties of assigning homology to RAPD products suggests that the technique is not appropriate for applications that assume accurate assessments of homology, such as phylogenetic analyses. Production of nongenetic artifacts in RAPD studies (e.g., Ellsworth *et al.*, 1993) suggests caution for all applications of the technique.

D. Allozyme Electrophoresis

In enzyme electrophoresis, homology among genes is determined by several functional, structural, and expressional criteria (see Murphy *et al.*, 1990). Paralogous genes are explicitly recognized as distinct loci, and usually are assumed to be evolving independently. Multiple loci that code for enzymes with the same biochemical function are thought in most cases to be paralogues of each other. In many cases, the existence of multiple loci is clear, because they may be expressed together in the same tissue at the same time in the same individual. In other cases, paralogous loci are recognized on the basis of expression in different tissue types, at different times of development, or at different seasons. Paralogy also may be determined based on multimeric structure or differential cellular location of the enzyme (e.g., whether mitochondrial or cytosolic). Therefore, for a locus to be considered orthologous in two species in an allozyme study, the two relevant enzymes typically are expected to satisfy the following conditions: (1) they have the same catalytic function; (2) they have the same multimeric structure; (3) they are expressed at the same general cellular location; (4) they are expressed at the same time in development; and (5) they are expressed in the same tissues.

In addition to homology among genes, homology among electromorphs (putative alleles) is of concern in allozyme electrophoresis. Electromorphs are defined on the basis of their distance of movement in an electric field relative to each other or to an internal standard; two electromorphs are placed in the same class if their distances of migration are indistinguishable. Convergence in electrophoretic mobility is likely because many amino acid replacements produce the same changes in net charge; this is offset somewhat by the large number of potential states of overall charge for a protein. Homology among electromorphs is determined on the basis of congruence with other characters (usually other allozyme data) in a phylogenetic analysis. Electromorphs that are phylogenetically congruent are considered to be homologous, whereas incongruent electromorphs are considered to be homoplastic.

VII. SUMMARY

Some proponents of molecular techniques have claimed that molecular biology "solves the problem of homology" (Sibley and Ahlquist, 1987). Although a better understanding of molecular biology has provided many new insights into relationships among genes and their products, there is still ample room for mistaken inferences about homologous molecular relationships. Distinguishing orthology from paralogy is the greatest obstacle for most evolutionary applications of molecular techniques. Duplicated gene loci and paralogous pseudogenes appear to be the rule rather than the exception in the eukaryote nuclear genome, so care is required if orthologous products are to be compared among taxa.

Concerted evolution among paralogous sequences helps the reconstruction of phylogenetic relationships among taxa, but hinders reconstruction of gene relationships. Some knowledge of the degree of concerted evolution for a given gene family is required to assess whether inferred relationships are those of taxa or of genes. Rates of concerted evolution appear to be high enough in some repeated gene families that paralogy is unlikely to be a confounding factor in reconstructing relationships of taxa, unless branching points in the tree are separated by very small distances. However, additional empirical studies are needed to measure rates of concerted evolution to determine its effects on phylogenetic analyses.

Relationships of paralogy extend to parts of genes as well. Functional gene modules may be duplicated in different genes or in tandem within a gene. Slippage replication can also produce paralogy of small tandem repeats.

Convergence can occur in whole genes or their parts, leading to false hypotheses of homology. In the case of whole genes, convergence may be related to positive selection, or may be secondarily related to convergence in base composition or the presence of simple tandem repeats. At a finer level, convergence at individual nucleotide sites is common, and may be facilitated by mutational biases. Inferences of positional homology requires accurate alignment criteria, which in turn require some knowledge of the relative likelihood of substitutions relative to insertions and deletions.

DNA-DNA hybridization has considerable potential for conflating paralogy and orthology, as well as homology and homoplasy. Homology and homoplasy may be factored out in analysis, but the confusion of orthology and paralogy is potentially more difficult to correct. Restriction *fragment* analyses are also likely to suffer from inaccurate homology assignments, but restriction *site* analyses are much less sensitive to this problem. Analyses of randomly amplified polymorphic DNA regions probably often involve mistaken assignments of homology, and so are best restricted to applications in which accurate assignment of homology is not critical. In allozyme electrophoresis, orthologous and paralogous proteins are identified on the basis of catalytic function, multimeric structure, location of cellular expression, timing of expression, and tissue distribution; when these criteria are applied, confusion between paralogous and orthologous loci is unlikely. However, convergence of electromorphs may occur through amino acid replacements that produce convergent changes in net charge of the protein.

As we begin to discover more about the processes that produce molecular variation among species, relationships among genes and their products look increasingly complex. The difficulties of assigning homology to molecules parallel many of the difficulties of assigning homology to morphological structures (Patterson, 1988). This may be a partial explanation for the similar levels of homoplasy that are observed in molecular and morphological phylogenetic studies (Sanderson and Donoghue, 1989). Hopefully, then, a better understanding of homologous relationships at the molecular level can lead to a better understanding of homology in general.

ACKNOWLEDGMENTS

I thank Jim Bull, David Cannatella, Paul Chippindale, Mike Donoghue, Anna Goebel, Ken Halanych, Rodney Honeycutt, John Huelsenbeck, and Barbara Mable for useful comments on the manuscript, and Jim Smith for insights on RAPD analysis. This work was supported by grants from the National Science Foundation (DEB 9106746 and DEB 9221052).

REFERENCES

Aoki, K., Tateno, Y., and Takahata, N. (1981). Estimating evolutionary distance from restriction maps of mitochondrial DNA with arbitrary G+C content. *J. Mol. Evol.* **18**, 1-8.

Arnheim, N. (1983). Concerted evolution of multigene families. *In* "Evolution of Genes and Proteins" (M. Nei and R. K. Koehn, eds.), pp. 38-61. Sinauer Associates, Sunderland, MA.

Arnheim, N., Krystal, M., Schmickel, R., Wilson, G. Ryder, O., and Zimmer, E. (1980). Molecular evidence for genetic exchanges among ribosomal genes on nonhomologous chromosomes in man and apes. *Proc. Natl. Acad. Sci. U.S.A.* **77**, 7323-7327.

Baltimore, D. (1981). Gene conversion: Some implications for immunoglobulin genes. *Cell (Cambridge, Mass.)* **24**, 592-594.

Bledsoe, A. H., and Sheldon, F. H. (1990). Molecular homology and DNA hybridization. *J. Mol. Evol.* **30**, 425-433.

Brown, D. D., Wensink, P. C., and Jordan, E. (1972). A comparison of the ribosomal DNAs of *Xenopus laevis* and *Xenopus mulleri*: Evolution of tandem genes. *J. Mol. Biol.* **63**, 57-73.

Buth, D. G. (1984). Allozymes of the cyprinid fishes: Variation and application. *In* "Evolutionary Genetics of Fishes" (B. J. Turner, ed.), pp. 561-590. Plenum, New York.

Chapco, W., Ashton, N. W., Martel, R. K. B., Antonishyn, N., and Crosby, W. L. (1992). A feasibility study of the use of random amplified polymorphic DNA in the population genetics and systematics of grasshoppers. *Genome* **35**, 569-574.

Coen , E. S., Strachan, T., and Dover, G. A. (1982a). Dynamics of concerted evolution of rDNA and histone gene families in the *melanogaster* species subgroup of *Drosophila. J. Mol. Biol.* **158**, 17-35.

Coen , E. S., Thoday, J. M., and Dover, G. A. (1982b). Rate of turnover of structural variants in the rDNA gene family of *D. melanogaster. Nature (London)* **395**, 564-568.

Crowhurst, R. N., Hawthorn, B. T., Rikkerink, E. H. A., and Templeton, M. D. (1991). Differentiation of *Fusarium solani* f. sp. *cucurbitae* races 1 and 2 by random amplification of polymorphic DNA. *Curr. Genet.* **20**, 391-396.

Donoghue, M. J. (1992). Homology. *In* "Keywords in Evolutionary Biology" (E. F. Keller and E. A. Lloyd, eds.), pp. 170-179. Harvard Univ. Press, Cambridge, MA.

Doolittle, R. F. (1987). The evolution of the vertebrate plasma proteins. *Biol. Bull. (Woods Hole, Mass.)* **172**, 269-283.

Dover, G. A. (1982). Molecular drive: A cohesive mode of species evolution. *Nature (London)* **299**, 111-117.

Dover, G. A. (1986). Molecular drive in multigene families: How biological novelties arise, spread and are assimilated. *Trends Gen.* **2**, 159-165.

Dowling, T. E., Moritz, C., and Palmer, J. D. (1990). Nucleic acids II: Restriction site analysis. *In* "Molecular Systematics" (D. M. Hillis and C. Moritz, eds.), pp. 250-317. Sinauer Associates, Sunderland, MA.

Doyle, J. J. (1991). Evolution of higher plant glutamine synthetase genes: Regulatory specificity as a criterion for predicting orthology. *Mol. Biol. Evol.* **8**, 366-377.

Duellman, W. E., and Hillis, D. M. (1987). Marsupial frogs (Anura: Hylidae: *Gastrotheca*) of the Ecuadorian Andes: Resolution of taxonomic problems and phylogenetic relationships. *Herpetologica* **43**, 141-173.

Ellsworth, D. L., Rittenhouse, K. D., and Honeycutt, R. L. (1993). Artifactual variation in randomly amplified polymorphic DNA banding patterns. *BioTechniques* **14**, 214-217.

Fitch, W. M. (1970). Distinguishing homologous from analogous proteins. *Syst. Zool.* **19**, 99-113.

Friedlander, T. P., Regier, J. C., and Mitter, C. (1992). Nuclear gene sequences for higher level phylogenetic analysis: 14 promising candidates. *Syst. Biol.* **41**, 483-490.

Ghiselin, M. T. (1976). The nomenclature of correspondence: A new look at "homology" and "analogy." *In* "Evolution, Brain and Behavior: Persistent Problems" (R. B. Masterson, W. Hodos, and H. Jerison, eds.), pp. 129-142. Erlbaum, Hillsdale, NJ.

Gonzales, I. L., Gorski, J. L., Campden, T. J., Dorney, D. J., Erickson, J. M., Sylvester, J. E., and Schmickel, R. D. (1985). Variation among human 28S ribosomal RNA genes. *Proc. Natl. Acad. Sci. U.S.A.* **82**, 7666-7670.

Goodman, M., Czelusniak, J., Moore, G. W., Romero-Herrera, A. E., and Matsuda, G. (1979). Fitting the gene lineage into its species lineage: A parsimony strategy illustrated by cladograms constructed from globin sequences. *Syst. Zool.* **28**, 132-168.

Goodman, M., Miyamoto, M. M., and Czelusniak, J. (1987). Pattern and process in vertebrate phylogeny revealed by coevolution of molecules and morphologies. *In* "Molecules and Morphology in Evolution: Conflict or Compromise?" (C. Patterson, ed.), pp. 141-176. Cambridge Univ. Press, Cambridge.

Goodwin, P. H., and Annis, S. L. (1991). Rapid identification of genetic variation and pathotype of *Leptosphaeria maculans* by random amplified polymorphic DNA assay. *Appl. Environ. Microbiol.* **57**, 2482-2486.

Gray, G. S., and Fitch, W. M. (1983). Evolution of antibiotic resistance genes: The DNA sequence of a kanamycin resistance gene from *Staphylococcus aureus. Mol. Biol. Evol.* **1**, 57-66.

Grime, J. P., and Mowforth, M. A. (1982). Variation in genome size and ecological interpretation. *Nature (London)* **299**, 151-153.

Hadjiolov, A. A., Georgiev, O. I., Nosikov, V. V., and Yavachev, L. P. (1984). Primary and secondary structure of rat 28S ribosomal RNA. *Nucleic Acids Res.* **12**, 3677-3693.

Hadrys, H., Balick, M., and Schierwater, B. (1992). Applications of random amplified polymorphic DNA (RAPD) in molecular ecology. *Mol. Ecol.* **1**, 55-63.

Hancock, J. M., and Dover, G. A. (1988). Molecular coevolution among cryptically simple expansion segments of eukaryotic 26S/28S rRNAs. *Mol. Biol. Evol.* **5**, 377-391.

Hancock, J. M., and Dover, G. A. (1990). 'Compensatory slippage' in the evolution of ribosomal RNA genes. *Nucleic Acids Res.* **18**, 5949-5954.

Hassouna, N., Michot, B., and Bachellerie, J.-P. (1984). The complete nucleotide sequence of mouse 28S rRNA gene.

Implications for the process of size increase of the large subunit rRNA in higher eukaryotes. *Nucleic Acids Res.* **12**, 3563-3583.

Herdman, M. (1985). The evolution of bacterial genomes. *In* "The Evolution of Genome Size" (T. Cavalier-Smith, ed.), pp. 37-68. Wiley, New York.

Hillis, D. M., and Dixon, M. T. (1989). Vertebrate phylogeny: Evidence from 28S ribosomal DNA sequences. *In* "The Hierarchy of Life" (B. Fernholm, K. Bremer, and H. Jörnvall, eds.), Proc. Nobel Symp. 70, pp. 355-367. Elsevier, Amsterdam.

Hillis, D. M., and Dixon, M. T. (1991). Ribosomal DNA: Molecular evolution and phylogenetic inference. *Q. Rev. Biol.* **66**, 411-453.

Hillis, D. M., Dixon, M. T., and Ammerman, L. K. (1990a). The relationships of the coelacanth *Latimeria: chalumnae:* Evidence from sequences of vertebrate 28S ribosomal RNA genes. *Environ. Biol. Fishes* **32**, 119-130.

Hillis, D. M., Larson, A., Davis, S. K., and Zimmer, E. A. (1990b). Nucleic acids III. Sequencing. *In* "Molecular Systematics" (D. M. Hillis and C. Moritz, eds.), pp. 318-370. Sinauer Associates, Sunderland, MA.

Hillis, D. M., Moritz, C., Porter, C. A., and Baker, R. J. (1991). Evidence for biased gene conversion in concerted evolution of ribosomal DNA. *Science* **251**, 308-310.

Hillis, D. M., Bull, J. J., White, M. E., Badgett, M. R., and Molineux, I. J. (1992). Experimental phylogenetics: Generation of a known phylogeny. *Science* **255**, 589-592.

Hori, H., and Osawa, S. (1986). Evolutionary change in 5S rRNA secondary structure and a phylogenetic tree of 352 rRNA species. *BioSystems* **19**, 163-172.

Hughes, A. L., and Nei, M. (1990). Nucleotide substitution at major histocompatibility complex class II loci: Evidence for overdominant selection. *Proc. Natl. Acad. Sci. U.S.A.* **86**, 958-962.

Hunt, G. J., and Page, R. E. (1992). Patterns of inheritance with RAPD molecular markers reveal novel types of polymorphisms in the honey bee. *Theor. Appl. Genet.* **85**, 15-20.

Irwin, D. M., and Wilson, A. C. (1990). Concerted evolution of ruminant stomach lysozymes. *J. Biol. Chem.* **265**, 4944-4952.

Kambhampati, S., Black, W. C., IV, and Rai, K. S. (1991). RAPD-PCR for identification and differentiation of mosquito species and populations: Techniques and statistical analysis. *J. Med. Entomol.* **29**, 939-945.

Kleppe, K., Ohtsuka, E., Kleppe, R., Molineux, I., and Khorana, H. G. (1971). Studies on polynucleotides XCVI. Repair replication of short synthetic DNAs as catalyzed by DNA polymerases. *J. Mol. Biol.* **56**, 341-361.

Lankester, E. R. (1870). On the use of the term homology in modern zoology. *Ann. Mag. Nat. Hist. Ser. [4]* **6**, 34-43.

Larson, A., and Wilson, A. C. (1989). Patterns of ribosomal RNA evolution in salamanders. *Mol. Biol. Evol.* **6**, 131-154.

Levinson, G., Marsh, J. L., Epplen, J. T., and Gutman, G. A. (1985). Cross-hybridizing snake satelite, *Drosophila*, and mouse DNA sequences may have arisen independently. *Mol. Biol. Evol.* **2**, 494-504.

Li, W.-H., and Graur, D. (1991). "Fundamentals of Molecular Evolution." Sinauer Associates, Sunderland, MA.

McElroy, D., Rothenberg, M., Reece, K. S., and Wu, R. (1990). Characterization of the rice actin gene family. *Plant Mol. Biol.* **15**, 257-268.

Moritz, C., Dowling, T. E., and Brown, W. M. (1987). Evolution of animal mitochondrial DNA: Relevance for population biology and systematics. *Annu. Rev. Ecol. Syst.* **18**, 269-292.

Mullis, K. B., and Faloona, F. A. (1987). Specific synthesis of DNA *in vitro* via a polymerase catalyzed chain reaction. In "Methods in Enzymology" (S. Fleischer and B. Fleischer, eds.), Vol. **155**, pp. 335-350. Academic Press, FL.

Murphy, R. W., Sites, J. W., Jr., Buth, D. G., and Haufler, C. H. (1990). Proteins. I. Isozyme electrophoresis. *In* "Molecular Systematics" (D. M. Hillis and C. Moritz, eds.), pp. 45-126. Sinauer Associates, Sunderland, MA.

Muto, A., and Osawa, S. (1987). The guanine and cytosine content of genomic DNA and bacterial evolution. *Proc. Natl. Acad. Sci. U.S.A.* **84**, 166-169.

Nagylaki, T. (1984). The evolution of multigene families under intrachromosomal gene conversion. *Genetics* **106**, 529-548.

Nagylaki, T., and Petes, T. D. (1982). Intrachromosomal gene conversion and the maintenance of sequence homogeneity among repeated genes. *Genetics* **100**, 315-337.

Ohta, T. (1980). "Evolution and Variation of Multigene Families." Springer-Verlag, Berlin.

Ohta, T. (1983). On the evolution of multigene families. *Theor. Popul. Biol.* **23**, 216-240.

Ohta, T. (1984). Some models of gene conversion for treating the evolution of multigene families. *Genetics* 106, 517-528.

Ohta, T., and Dover, G. A. (1983). Population genetics of multigene families that are dispersed into two or more chromosomes. *Proc. Natl. Acad. Sci. U.S.A.* **80**, 4079-4083.

Owen, R. (1843). "Lectures on the comparative anatomy and physiology of the invertebrate animals." Longman, Brown, Green, & Longmans, London.

Owen, R. (1848). "On the Archetype and Homologies of the Vertebrate Skeleton." Richard and John E. Taylor, London.

Palmer, J. D., Jansen, R. K., Michaels, H. J., Chase, M. W., and Manhart, J. R. (1988). Chloroplast DNA and plant phylogeny. *Ann. Mo. Bot. Gard.* **75**, 1180-1206.

Patterson, C. (1988). Homology in classical and molecular biology. *Mol. Biol. Evol.* **5**, 603-625.

Patthy, L. (1985). Evolution of the proteases of blood coagulation and fibrinolysis by assembly from modules. *Cell (Cambridge, Mass.)* **41**, 657-663.

Rees, H., and Jones, R. N. (1972). The origin of the wide species variation in nuclear DNA content. *Int. Rev. Cytol.* **32**, 53-92.

Saiki, R. K., Gelfand, D. H., Stoffel, S., Scharf, S. J., Higuchi, R., Horn, G. T., Mullis, K. B., and Erlich, H. A. (1988). Primer-directed enzymatic amplification of DNA with a thermostable DNA polymerase. *Science* **239**, 487-491.

Sanderson, M. J., and Donoghue, M. J. (1989). Patterns of variation in levels of homoplasy. *Evolution (Lawrence, Kans.)* **43**, 1781-1795.

Sanderson, M. J., and Doyle, J. J. (1992). Reconstruction of organismal and gene phylogenies from data on multigene families: Concerted evolution, homoplasy, and confidence. *Syst. Biol.* **41**, 4-17.

Scharf, S. J., Long, C. M., and Erlich, H. A. (1988a). Sequence analysis of the HLA-DRβ and HLA-DQβ loci from three *Pemphigus vulgaris* patients. *Human Immunol.* **22**, 61-69.

Scharf, S. J., Friedmann, A., Brautbar, C., Szafer, F., Steinman, L., Horn, G., Gyllensten, U., and Erlich, H. A. (1988b). HLA class II allelic variation and susceptibility to *Pemphigus vulgaris*. *Proc. Natl. Acad. Sci. USA* **85**, 3504-3508.

Seperack, P., Slatkin, M., and Arnheim, N. (1988). Linkage disequilibrium in human ribosomal genes: Implications for multigene family evolution. *Genetics* **119**, 943-949.

Shah, D. M., Hightower, R. C., and Meagher, R. B. (1983). Genes encoding actins in higher plants: Intron positions are highly conserved but the coding sequences are not. *J. Mol. Appl. Genet.* **2**, 111-126.

Sibley, C. G., and Ahlquist, J. E. (1987). Avian phylogeny reconstructed from comparisons of the genetic material, DNA. *In* "Molecules and Morphology in Evolution: Conflict or Compromise?" (C. Patterson, ed.), pp. 95-121. Cambridge Univ. Press, Cambridge.

Singer, C. E., and Ames, B. N. (1970). Sunlight ultraviolet and bacterial DNA base ratios. *Science* **170**, 822-826.

Smith, G. P. (1974). Unequal crossover and the evolution of multigene families. *Cold Spring Harbor Symp. Quant. Biol.* **38**, 507-513.

Smith, J. J., Scott-Craig, J. S., Leadbetter, J. R., Bush, G. L., Roberts, D. L., and Fulbright, D. W. (1993). Characterization of random amplified polymorphic DNA (RAPD) products from *Xanthomonas campestris*: Implications for the use of RAPD products in phylogenetic analysis. *Mol. Phylog. Evol.* (in press).

Southern, E. M. (1975). Detection of specific sequences among DNA fragments separated by gel electrophoresis. *J. Mol. Biol.* **98**, 503-517.

Sparrow, A. H., and Nauman, A. F. (1976). Evolution of genome size by DNA doublings. *Science* **192**, 524-529.

Stewart, C.-B., and Wilson, A. C. (1987). Sequence convergence and functional adaptation of stomach lysozymes from foregut fermenters. *Cold Spring Harbor Symp. Quant. Biol.* **52**, 891-899.

Swanson, K. W., Irwin, D. M., and Wilson, A. C. (1991). Stomach lysozyme gene of the langur monkey: Tests for convergence and positive selection. *J. Mol. Evol.* **33**, 418-425.

Swofford, D. L., and Olsen, G. J. (1990). Phylogeny reconstruction. *In* "Molecular Systematics" (D. M. Hillis and C. Moritz, eds.), pp. 411-501. Sinauer Associates, Sunderland, MA.

Szostak, J. W., and Wu, R. (1980). Unequal crossing over in the ribosomal DNA of *Saccharomyces cerevisiae*. *Nature (London)* **284**, 426-430.

Tamura, K. (1992). The rate and pattern of nucleotide substitution in *Drosophila* mitochondrial DNA. *Mol. Biol. Evol.* **9**, 814-825.

Ware, V. C., Tague, B. W., Clark, C. G., Gourse, R. L., Brand, R. C., and Gerbi, S. A. (1983). Sequence analysis of 28S ribosomal DNA from the amphibian *Xenopus laevis. Nucleic Acids Res.* **11**, 7795-7817.

Wegnez, M. (1987). Letter to the editor. *Cell (Cambridge, Mass.)* **51**, 516.

Welsh, J., and McClelland, M. (1990). Fingerprinting genomes using PCR with arbitrary primers. *Nucl. Acids Res.* **18**, 7213-7219.

Welsh, J., Pretzman, C., Postic, D., Girons, I. S., Baranton, G., and McClelland, M. (1992). Genomic fingerprinting by arbitrarily primed polymerase chain reaction resolves *Borrelia burgdorferi* into three distinct phyletic groups. *Int. J. Syst. Bacteriol.* **42**, 370-377.

Werman, S. D., Springer, M. S., and Britten, R. J. (1990). Nucleic acids I: DNA-DNA hybridization. *In* "Molecular Systematics" (D. M. Hillis and C. Moritz, eds.), pp. 204-249. Sinauer Associates, Sunderland, MA.

Williams, J. G. K., Kubelik, A. R., Livak, K. J., Rafalski, J. A., and Tingey, S. V. (1991). DNA polymorphisms amplified by arbitrary primers are useful genetic markers. *Nucleic Acids Res.* **18**, 6531-6535.

Wilson, J. T., Wilson, L. B., Reddy, V. B., Cavallesco, C., Ghosh, P. K., de Riel, J. K., Forget, B. G., and Weissman, S. M. (1980). Nucleotide sequence of the coding portion of human alpha globin messenger RNA. *J. Biol. Chem.* **255**, 2807-2815.

Zimmer, E. A., Martin, S. L., Beverley, S. M., Kan, Y. W., and Wilson, A. C. (1980). Rapid duplication and loss of genes coding for the α chains of hemoglobin. *Proc. Natl. Acad. Sci. U.S.A.* **77**, 2158-2162.

11

HOMOLOGY AND BEHAVIORAL

REPERTOIRES

Harry W. Greene

Museum of Vertebrate Zoology and
Department of Integrative Biology
University of California
Berkeley, CA 94720

I. INTRODUCTION

A unifying theme of early biological studies of behavior was that an animal's actions result in part from historical evolutionary processes. The concept of behavioral homology thus played an important, if sometimes implicit, role in the development of ethology during the early decades of this century (for a scholarly review, see Burghardt and Gittleman, 1990). Emphasis on phylogenetic studies subsequently diminished during the 1960s and 1970s, when historical approaches to the study of behavior were viewed with ambivalence or outright skepticism. Ethologists interested in "evolution" concentrated on ecological consequences of behavioral variation among individuals, rather than on similarities and differences among related species and higher taxa, as exemplified by the rise of the nominal disciplines "behavioral ecology" and "sociobiology." Atz (1970) and Brown (1975) claimed that behavioral homologies usually cannot be evaluated, while Klopfer (e.g., 1969, 1975) asserted repeatedly that behavior does not evolve. Some texts of that period discussed the behavior of extinct taxa or otherwise implied that behavior evolves (e.g., Alcock, 1975; Eibl-Eibesfeldt, 1970; Wilson, 1975), but criticisms of behavioral homology were addressed infrequently (see Greene and Burghardt, 1978; Hailman, 1976).

Although Atz's (1970) paper hardly marked "the end of attempts to homologize behavior" (Brooks and McLennan, 1991, p. 7; cf. Arnold, 1977; Böhme and Bischoff, 1976; Dewsbury, 1975; Ferguson, 1971; Greene, 1979, 1983; Greene and Burghardt, 1978; McKinney, 1975; and numerous other references in Wenzel, 1992), phylogenetic comparisons of behavior remained relatively uncommon until recently. A few pivotal papers — especially Barlow (1977) and Drummond (1981) on units of behavior, and Lauder (1981, 1982, 1986) and Gould and Vrba (1982) on the importance of historical perspectives — presaged an explosion of systematic analyses of the biology of contemporary taxa. Most of those studies depended on a phylogenetic concept of behavioral homology.

Phylogenetically defined, homologous traits are those whose presence can be traced continuously to a common ancestor. Like others (e.g., de Queiroz and Wimberger, 1993; Lauder, 1986, this volume; Wenzel, 1992), I prefer that approach to behavioral homology because it relates to important questions in evolutionary biology. Rather than reiterate conceptual problems covered in detail by those authors (see also Donoghue, Sattler, and others in this volume), the present chapter briefly presents two case studies and stresses four pragmatic issues:

1. Behavior is not necessarily more variable than morphology, nor more difficult to describe, measure, and compare among organisms. However, the former might offer special promise in elucidating truly homologous features, because experience and other epigenetic processes that complicate the analysis of behavioral similarities are relatively accessible to study.
2. The general absence of reliable fossils potentially is a serious problem for assessing behavioral homology, especially for comparisons among distantly related taxa and among taxa with markedly dissimilar characteristics.
3. Discovery of behavioral homologies has importance for biology and other human endeavors far beyond simply adding characters for phylogenetic analyses.
4. Identifying homologies and addressing broader issues will require behavioral repertoires for larger samples of taxa than are now available for most groups, and thus several logistical problems need attention.

II. TWO SERPENTINE EXAMPLES

A. *Constricting Prey*

Constricting is a behavior pattern in which prey is immobilized by two or more points on a snake's body. Alternative states of three behavioral descriptors yield 27 possible combinations (modal actions patterns, or "MAPs"; Barlow, 1977), of which at least 19 occur among more than 150 species in 60 genera and 10 families of snakes that constrict (Cundall and Irish, 1986; Greene, 1977, 1983; Greene and Burghardt, 1978; Shine and Schwaner, 1985; Willard, 1977). Except for a few highly variable species of Colubridae, constricting coil application was stereotyped within and among individuals; naive young of two genera of Boidae, one species of Cylindrophiidae, and one species of Pythonidae constricted their first prey items with behavior kinematically similar to that of adults of those taxa. The distribution of alternative MAPs among taxa implied that identical constricting patterns are homologous across eight relatively primitive families (collectively called Henophidia), and that constriction has been lost and gained several times in "advanced" snakes (Caenophidia, see Fig. 1). Homology of prey-killing tactics among henophidians was used to argue that constricting was a behavioral key innovation associated with structural modifications for swallowing large prey in the Cretaceous *Dinilysia patagonica* and other early snakes (Greene, 1983; Greene and Burghardt, 1978).

My initial assessment of constricting behavior (Greene, 1977; Greene and Burghardt, 1978) was only vaguely phylogenetic and flawed in several other respects. Scolecophidians (blindsnakes) are basal to Alethinophidia (all other living snakes) and lack constricting, but were seen as highly derived

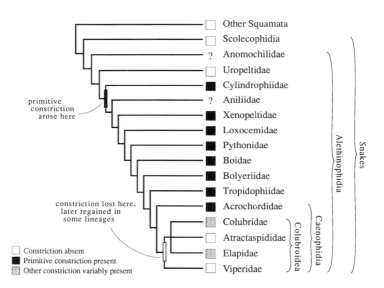

Fig. 1. Phylogenetic relationships among higher snake taxa and the origin of a widespread, homologous action pattern for applying constricting coils to prey. [Analyzed with MacClade (Maddison and Maddison, 1992), based on Cundall *et al.*, 1993; Kluge, 1991; Rieppel, 1988.]

relative to ancestral snakes and therefore ignored in my analysis (Greene, 1983). Recent work indicates that Henophidia is paraphyletic, Acrochordidae is the sister taxon of other Caenophidia rather than of all other Alethinophidia, and *Dinilysia patagonica* is perhaps outside the clade subsumed by all living snakes (Cundall *et al.*, 1993; Kluge, 1991; Rieppel, 1988). Recognition of those problems and observations on constricting by a bolyeriid (Cundall and Irish, 1986) and several elapids (Shine and Schwaner, 1985) refine my earlier conclusions (Fig. 1).

1. A particular constricting MAP, characteristic of boids and several other taxa, is indeed homologous for at least a large subset of Alethinophidia. Whether *Dinilysia* constricted prey cannot be inferred, given its controversial phylogenetic placement (Rieppel, 1988) and given uncertainties about homologizing constriction by alethinophidians with similar prey-handling behavior of certain distantly related lizards (Greene, 1977; see below).

2. Structural modifications for ingestion of large prey arose at and subsequent to the origin of Alethinophidia (Cundall and Greene, 1982; Cundall *et al.*, 1993; Rieppel, 1988). Given homology of constriction across eight lineages of "primitive" alethinophidians, that behavior accompanied or preceded at least most of those innovations. In any case, the higher snake taxa with homologous behavior shared a common ancestor no more recently than the late Cretaceous; constricting therefore arose prior to the origin of rodents — often mentioned as potentially relevant prey in discussions of early snake evolution (e.g., Rieppel, 1988).

3. Constriction is lacking or rare among most species and genera of colubrids and elapids, and lacking in atractaspidids and viperids (collectively the Colubroidea). Furthermore, constricting MAPs in colubrids and elapids usually are kinematically different from those of "primitive" snakes (Greene and Burghardt, 1978; Shine and Schwaner, 1985; Willard, 1977). The ancestor of Colubroidea thus lacked constriction, a loss perhaps concomitant there with the origin of venom-injection mechanisms (Savitzky, 1980). Subsequent gains and losses of various constricting MAPs among colubroids cannot be asssessed until relationships within that group are resolved (see Cadle, 1993).

B. Open-Mouth Threat Displays

In contrast to the widespread conservatism of constricting behavior, antipredator mechanisms are extensively convergent among and within genera and families of snakes (Greene, 1988). In one such case, independently acquired similarities defy easy correlation with habitat preferences and also imply that simply appealing to parsimony is not sufficient for understanding those phylogenetic patterns of behavioral variation.

Some snakes (mainly boids, pythonids, and viperids) strike by rapidly extending the head and foreparts from a retracted, S-shaped coil, protracting the maxillae, and thereby biting or stabbing prey. Within three of four major lineages of Viperidae, certain species respond to an adversary with open-mouth threat displays (Table I). Five species of arboreal African *Atheris* (Viperinae) gape with necks extended and fangs protracted, evidently having ritualized an intermediate stage of the strike cycle; naive young of *Atheris nietschii* gape like adults. At least 10 species in 6 genera of Old and New World pitvipers (Crotalinae) gape with necks retracted against body coils and with fangs folded, as in the initial stage of striking. Some crotaline species that gape (e.g., New World *Agkistrodon*) are congeneric with others that do not, and among 30 species of rattlesnakes only one subspecies (*Crotalus molossus nigrescens*) gapes consistently. The monotypic third lineage, *Azemiops feae*, occasionally gapes with fangs either folded or erect. Gaping threat displays are rare among other colubroids and do not involve folding or erecting the maxillary teeth.

A parsimonious interpretation is that gaping evolved once in *Azemiops feae*, once in Viperinae, and several times in Crotalinae (Fig. 2). However, the repeated origin of similar displays in the latter group is puzzling because there is no obvious correlation with ecological factors; gaping crotalines include two arboreal and at least three terrestrial origins of that behavior (assuming that the aquatic *Agkistrodon piscivorus* retained gaping from a terrestrial ancestor).

Perhaps there is something about pitvipers that constrains them to evolve a particular gaping threat display? When vipers erect their fangs the region between the eye and nostril is obscured by the rising maxillary bones, thus

Table I. Gaping Threat Displays in Vipers[a]

Taxon	Distribution	Macrohabitat	Threat display[b]
Azemiopinae			
Azemiops feae	Asia	Terrestrial	erect or folded
Viperinae			
Atheris ceratophorus	Africa	Arboreal	erect
A. chloroechis	Africa	Arboreal	erect
A. hispidus	Africa	Arboreal	erect
A. nietschii	Africa	Arboreal	erect
A. squamiger	Africa	Arboreal	erect
Crotalinae			
Agkistrodon bilineatus	M. Amer.[c]	Terrestrial	unknown
A. contortrix	N. Amer.[d]	Terrestrial	unknown
A. piscivorus	N. Amer.	Aquatic	folded
Atropoides nummifer	M. Amer.	Terrestrial	folded
A. olmec	M. Amer.	Terrestrial	folded
A. picadoi	M. Amer.	Terrestrial	folded
Bothriechis aurifer	M. Amer.	Arboreal	folded
B. lateralis	M. Amer.	Arboreal	folded
B. marchi	M. Amer.	Arboreal	folded
B. schlegelii	M. Amer.	Arboreal	folded
Crotalus molossus	N. Amer.	Terrestrial	folded
Hypnale hypnale	Asia	Terrestrial	folded
Trimeresurus wagleri	Asia	Arboreal	unknown

a Based on Campbell and Lamar (1989); Carpenter and Gillingham (1990); Greene (1988, 1992); Greene, Murphy, and Blody (unpublished data).
b Shown as whether fangs are erect or folded.
c M. Amer., Middle America.
d N. Amer., North America.

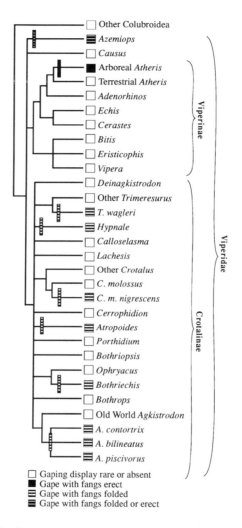

Fig. 2. Phylogenetic relationships among vipers and independent origins of open-mouth threat displays in *Azemiops*, viperines (arboreal *Atheris*), and several lineages of pitvipers (Crotalinae). [Analyzed with MacClade (Maddison and Maddison, 1992), based conservatively on Ashe and Marx, 1988; Groombridge, 1986.]

blocking sensory input to the facial pits of crotalines (A. H. Savitzky, personal communication). This fact would be of doubtful significance if pits are conceived of as primarily feeding adaptations, but has obvious relevance if pits are historically associated with evaluating and confronting adversaries (Greene, 1992). Of more importance here is the testable possibility that naive young pitvipers possess only a tendency to gape (similar to the behavior of viperines or variable, as in *A. feae*), and that they learn to inhibit fang protraction in order to maintain sensory reception by the pits during predator encounters.

Context and experience play important roles in shaping the antipredator responses of some snakes (Herzog *et al.*, 1992, and references therein), and certain pitvipers are more likely to gape when threatened at low temperatures (*Agkistrodon bilineatus* and *A. contortrix*; Carpenter and Gillingham, 1990). Perhaps the neurosensory and musculoskeletal mechanisms underlying gape displays are homologous as such across all viperids, thresholds for evoking gaping are all that evolved within Viperidae (such that many taxa rarely if ever exhibit the behavior), and particular behavioral similarities among pitvipers are constrained by morphology rather than having evolved independently. Comparative functional studies (Lauder, this volume) as well as experiments that control for context, stimulus characteristics of the simulated predator (e.g., shape and temperature), and experiential history of the vipers clearly would be of great interest.

III. CONCEPTUAL AND METHODOLOGICAL ISSUES

A. The Nature of Behavior: False Limits and Special Promise

Critics of behavioral homology often have emphasized supposedly special qualities of morphology:

> Behavior is much more difficult to treat comparatively than is structure because of its variability, continuity, extended ontogeny, and the evanescence of each behavioral act... Convergence in behavior is prevalent, probably because of intense selection pressures and

limited possible responses by the animal. (Atz, 1970, p. 69; see also, e.g., Klopfer, 1969, 1975)

A proponent of behavioral homology even conceded that:

> ...behavior has a kind of variability unknown in morphology. The length of a bone can be measured repeatedly and a frequency distribution... obtained that relates to inherent problems of measurement. The same problems apply to measuring some variable in the motion picture of a fixed action pattern. However... in repeat performances of the 'same' fixed action pattern of a given individual animal the variability in the measurements increases because there is a source attributable to differences among repeated performances. (Hailman, 1976, p. 194)

Like de Queiroz and Wimberger (1993, p. 56), my impression is that similar misconceptions may be common among biologists.

Broadly speaking, behavior is what animals do. Behavior represents fairly short-term neuromuscular, neurohormonal, and integrative responses of organisms to internal and external stimuli; useful descriptions of behavior can extend necessarily to various external factors as well (Drummond, 1981). My serpentine examples and many other phylogenetic comparisons concern MAPs: repeatable, spatiotemporal patterns of movements and postures that characterize individuals, populations, species, and higher taxa (Barlow, 1977). However defined, the distinctions between behavior and other phenotypic attributes are fuzzy. Enzyme-substrate reactions and other fast molecular events are referred to as biochemical or physiological, somewhat slower responses are called behavior, and features that appear stable over long periods are known as morphology. All these phenomena are outcomes of genetic, developmental, and control processes; all are expressed "normally" in "expected" environments.

Addressing Hailman's (1976) dichotomy, if we repeatedly measure the antlers of deer, the feathers of birds, or the uterine mucosa of mammals with estrous cycles we sample a kind of variability equivalent to repeated expression of a MAP. Even bones and other "continuous" structures of live animals would appear ever-changing if we could view them closely over time. Note also that a bone in a museum is a static collection of compounds, an inanimate shadow of the dynamic processes that maintained tissue in the living organism. It is comparable to a termite nest, a collection of nonliving materials with predictable form resulting from complex social behavior in the insects' colony. Withal, behavior should be no more difficult to homologize, *a priori*, than other organismal attributes; relative levels of variability and homoplasy therefore are questions for study on a taxon by taxon basis. Thus far, behavior does not seem to be especially variable nor especially subject to homoplasy compared to morphology (Barlow, 1977; Burghardt and Gittleman, 1990; de Queiroz and Wimberger, 1993; Greene and Burghardt, 1978; Lauder, 1986; Wensel, 1992).

In operational terms, homologies are features that define monophyletic groups. Applied strictly, this "taxic" approach equates homology with synapomorphy, to be confirmed by congruence with other characters in support of a particular phylogenetic hypothesis (e.g., Donoghue, 1992; Lauder, 1986; Wenzel, 1992). Critics often have emphasized that behavior is subject to learning and other experiential effects during the life of an organism, such that phylogenetically congruent similarities would not indicate genealogical continuity. These same critics sometimes assumed that morphological attributes are relatively free from experiential effects, despite evidence to the contrary (e.g., Hollister's (1918) dramatic comparisons of the skulls of zoo-raised and wild-shot African lions; see discussions of phenotypic plasticity by others in this volume). Lauder (1986) and Wenzel (1992) regarded this problem as one of carefully defining the level at which homologies are identified, and Wenzel (1992) noted that a distinction between intrinsic and extrinsically determined characteristics is sometimes unrealistic.

Recalling the particular gaping threat displays of pitvipers, unique to that group and seemingly independently evolved at least five times, I have a nagging feeling that we advocates of a phylogenetic concept of homology are insufficiently critical about assessing similarities across taxa. Descriptions of adult traits without regard to contextual and experiential effects on their expression, as are still routine in systematics, will not lead to carefully defining levels at which homology occurs. In this regard, Lauder (this volume) is surely right that historical patterns of concordance and divergence among classes of data, levels of organization, and diverse processes will provide exciting topics for future research. Behavior might prove especially useful for such comparisons, since experimental manipulations of ethological similarities and

dissimilarities among organisms might be more straightforward than comparable studies of morphology.

B. Fossils and Behavior

Like most evolutionary biologists until recently, early
skeptics of behavioral homology (e.g., Atz, 1970) probably
envisioned extinct taxa as actual common ancestors of living
creatures when they mentioned the absence of fossil data as a
problem in comparative studies. As such, fossils would express ancestral character states and prove the homology of
phenotypic characteristics across descendant taxa. The
absence of behavioral fossils was thus critical; note that soft
structures of most organisms are unknown as fossils either,
but the homology of vertebrate hearts and other such features
was not questioned for lack of fossil preservation. Under that
view, rare instances of animals fossilized while engaged in
some activity ("frozen behavior"; Boucot, 1990) were especially
valuable, since they could provide the only direct evidence for
behavioral homology. Likewise, paleontologists usually have
stressed fossils and inferential functional studies as keys to
the biology of extinct organisms (e.g., Boucot, 1990; Hopson
and Radinsky, 1980). "Trace fossils" (trackways, tooth marks,
bore holes, etc.) indeed can portray the activities of extinct
organisms, and relationships between form and function in
living taxa sometimes can be extrapolated reliably to fossils
(e.g., Boucot, 1990; Padian and Rayner, 1993). However, there
are three problems with fossilized behavior as proof of phylogenetic continuity and thus homology:

1. Most fossils likely represent a mosaic of primitive and
derived phenotypes (morphology, behavior, etc.) in ancient,
nevertheless descendant lineages rather than literal ancestors
of living taxa (Gauthier et al., 1988; Hailman, 1976). Fossils
thus provide additional taxa for analysis, not proof per se of
ancestral traits and thus of homology.

2. Fossilized "frozen behavior" might often be misleading.
Chance mortality and mineralization of quiescent animals
could accurately reflect normal natural history, [e.g., fossil
snake aggregations (Breithaubt and Duvall, 1986), and a fossil
snake digesting large prey (Greene, 1983)], but healthy

animals that are acting normally usually do not die and become fossils. For example, most vertebrates typically swallow prey head-first (Greene, 1976; de Queiroz and de Queiroz, 1987); rather than "ample evidence that large prey were acceptable" (Eaton, 1964, p. 2), a fossil early amniote ingesting a congeneric individual tail-first otherwise probably would not have died in the effort. Two examples of prospective fossils of inappropriate behavior include an insectivorous frog (*Rana aurora*) I found dead with a rodent stuck in its throat (deposited in the Museum of Vertebrate Zoology), and a high Andean snake (*Tachymenis peruvianus*) frozen while swallowing a relatively large lizard (J. E. Cadle, personal communication).

3. The most widely accessible and accurate window on the behavior of extinct organisms is phylogenetic study of living creatures, incorporating information from fossils whenever possible. With comparisons among recent taxa, we can infer ancestral behavioral character states and thus their likely presence in extinct species within the same clade. However, the general absence of fossils could cause difficulties for interpreting behavioral homologies among extant organisms in a way not anticipated by earlier critics. Experiments with morphological data sets show that fossils are a potential source of new taxa, new characters, and new character states; as a result, data from fossils can change the inferred polarities of character transformations and in some cases change estimates of phylogeny (Gauthier *et al.*, 1988, for amniotes; Novacek, 1992, for eutherian mammals; Wilson, 1992, for freshwater fishes). Fossils in turn might change conclusions about homology versus homoplasy as explanations for similarities across taxa.

Judging from the morphological studies cited above, errors steming from the absence of reliable fossils are most likely for comparisons involving early divergences in ancient groups and for comparisons among radically different phenotypes. In contrast to morphology, however, new fossils representing previously unkown behavioral transformations are unlikely. There is no obvious solution to this problem; investigators should simply remain aware of the potential likelihood of this bias on particular conclusions. For example, the ubiquity and essential identity of constricting behavior across eight families of snakes bolsters confidence that those

MAPs are homologous. Conversely, the ancient divergence between snakes and lizards casts doubt on the relevance of similar prey capture movements by the latter for understanding the origin of constriction.

V. PROSPECTS AND PROBLEMS

A. Homologous Behavior: So What?

A phylogenetic concept of behavioral homology has important implications for evolutionary biology and other endeavors. Two obvious applications are that behavioral characters can prove useful in reconstructing phylogenies (de Queiroz and Wimberger, 1993; McLennan et al., 1988), and that phylogenetic analyses can reconstruct the evolutionary history of particular behavioral attributes (for methods see Brooks and McLennan, 1991; Maddison and Maddison, 1992). Behavioral homologies also can be used to infer the time at which ethological character states arose. In conjunction with paleoecological information, such studies thereby extend our understanding of extinct organisms and address the role of behavioral shifts in morphological evolution (e.g., the serpentine examples discussed above; see also Losos, 1990; McLennan, 1991; Wake, 1992). Phylogenetic behavioral analyses of particular lineages can elucidate historical components of community ecology (Cadle and Greene, 1993; Losos, 1992) and help refine theories of macroevolution (e.g., Erwin, 1992; Guyer and Slowinski, 1993).

Behavioral homologies also have implications beyond basic science. Lacking contrary evidence, the biology of unstudied endangered taxa can be estimated if the behavior and ecology of representative sister taxa are known. That information in turn can be used in management programs, an important consideration when time is short or logistical factors prevent investigations of the actual endangered populations (Brooks et al., 1992; Greene, 1993). Moreover, our appreciation for nature — a prerequisite for conservation — is bolstered by historical perspectives on the behavior and ecology of living organisms (Greene and Losos, 1988). For example, educational material about Gila monsters and beaded lizards (Helodermatidae) typically is restricted to emphasizing that they are the only living venomous lizards. A phylogenetic analysis (Pregill *et al.*, 1986) revealed numerous fossil relatives, and demonstrated that the living North American relicts share several behavioral characteristics with the related Old World monitors (Varanidae). Modern species of Varanoidea (a taxon encompassing both families) are thus living reminders of ancient landscapes, of times when large carnivorous lizards were prominent predators over much of the Northern Hemisphere.

B. Looking for Repertoires: Where's "EthoBank"?

Inferring behavioral homologies requires a phylogeny for the creatures of interest and behavioral information for a representative sample of individuals, populations, and taxa (Gittleman, 1989). For robust conclusions, the behavioral sample must encompass both the "in-group" of immediate interest and appropriate "outgroups" (e.g., other colubroids for vipers, see Fig. 2). Moreover, whether a particular behavior is relevant to understanding others will be unknown in advance, so extensive behavioral inventories ("ethograms") of all taxa are desirable. Finally, observations from nature are important because they validate studies on captives, and because they expose behavior not exhibited in a laboratory context (Greene, 1986). A biologist seeking comparably detailed material for phylogenetic research on morphology, even of rare taxa, usually can turn to the literature for illustrations, or borrow museum specimens for direct examination. A molecular systematist can survey colleagues for stored tissues of needed taxa, collect new material by withdrawing a small amount of tissue from live animals, or even order DNA sequences directly from "GenBank" (Bilofsky and Burks, 1988). Unfortunately, given the importance of phylogenetic studies for broader concerns, obtaining adequate behavioral descriptions often is a much more difficult endeavor — especially for diverse and widely distributed groups.

One short-term response to the shortage of behavioral data is to settle for whatever analysis is possible with the information at hand. After all, it might prove correct, and even a shaky conclusion can stimulate additional observations. Information on some aspect of defensive behavior is available for less than 50% of the ca. 200 species of Viperidae (Greene, 1988, 1992), and only *Crotalus viridis* has been studied in reasonable detail (Duvall *et al.*, 1985). The preliminary survey of gaping threat displays (above) was insufficient to assess variability even at lower levels, and an assessment of homology for that behavior was hampered further by lack of phylogenetic resolution among major lineages of viperids (Fig. 2). Nevertheless, the analysis was sufficiently broad-based to suggest some interesting connections with morphology and to frame questions worthy of further research. In that context, the previously published observations of variable threat displays in a few captives of the exceedingly rare *Azemiops feae* were most welcome!

Other solutions to the dearth of comparative behavioral data are to scour the natural history literature for useful anecdotes (start with *Zoological Record*), to seek assistance, and to take advantages of technological innovations. My initial survey of constriction by henophidians (Greene, 1977; Greene and Burghardt, 1978) encompassed 120 individuals of 48 species in 26 genera, roughly 56 and 96%, respectively, of the taxa recognized at the time in those cosmopolitan families. Despite the fact that interactions between academic and zoo herpetologists were uncommon at the time, most of my observations were made at three U.S. zoos, and inclusion of an acrochordid was based on photographs sent by a European zoo curator. One rarely seen dwarf boa (the monotypic *Exiliboa placata*, known only from its type locality) was sent by a generous colleague. Cundall and Irish (1986) later observed the single living species of bolyeriid, found only on a small island in the Indian Ocean, thanks to hospitality of the Jersey Wildlife Preservation Trust. A few published photographs confirmed that free-living boids constricted with behavior identical to that seen in captivity. My initial findings were revised (above) to reflect improved resolution of snake phylogeny, and future observations (e.g., of anomochilids, known from six museum specimens) could further change those conclusions.

In the past 15 years, advances in radiotelemetry and night-vision devices have greatly facilitated field observations on snakes and other cryptic, previously unobservable animals (Reinert, 1992; Rodda, 1992). During that same period, research opportunities in zoological parks and aquariums have steadily increased, providing an important resource for comparative ethologists (Chiszar *et al.*, 1993). The most widespread, ongoing impediment to amassing adequate samples of behavioral data for phylogenetic analysis, however, is the general lack of appreciation for systematics and natural history by the public and even by other biologists — expressed politically as gratuitous regulation of research and poor support, especially in terms of funding. That problem turns on the widespread naivete, even among well-educated people, regarding the important roles of those disciplines in broader human endeavors (Greene, 1986, 1994; Greene and Losos, 1988; Harvey, 1991). Behavioral homology is intellectually intriguing and studying animals is personally rewarding, but our prospects for continuing these activities will improve only if we convince others of their importance.

ACKNOWLEDGMENTS

This chapter is greatly modified from part of a doctoral dissertation, supervised by G. M. Burghardt at the University of Tennessee. My research on snake behavior has been supported subsequently by the National Science Foundation; University of California, Berkeley (Annie M. Alexander Fund of the Museum of Vertebrate Zoology, Committee on Research, Distinguished Teaching Award); World Wildlife Fund — U.S.; and personal funds of H. W. Greene and D. L. Hardy, Sr. I especially appreciate working at the Dallas and Fort Worth zoos with J. B. Murphy and D. A. Blody, respectively, and discussions about behavioral homology with M. J. Donoghue, J. A. Gauthier, A. Graybeal, and D. B. Wake.

REFERENCES

Alcock, J. (1975). "Animal Behavior: An Evolutionary Approach." Sinauer, Stamford, CT.

Arnold, S. J. (1977). The evolution of courtship behavior in New World salamanders with some comments on Old World salamanders. In "The Reproductive Biology of Amphibians" (D. H. Taylor and S. I. Guttman, eds.), pp. 141-183. Plenum, New York.

Ashe, J. S. and H. Marx. (1988). Phylogeny of the viperine snakes (Viperinae): Part II. Cladistic analysis and major lineages. *Fieldiana, Zool. (New Ser.)* **52**, 1-23.

Atz, J. W. (1970). The application of the idea of homology to behavior. *In* "Development and Evolution of Behavior" (L. R. Aronson, E. Tobach, D. S. Lehrman, and J. S. Rosenblatt, eds.), pp. 53-74. Freeman, San Francisco.

Barlow, G. W. (1977). Modal action patterns. *In* "How Animals Communicate" (T. A. Sebeok, ed), pp. 98-134. Indiana Univ. Press, Bloomington, IN.

Bilofsky, H. S., and Burks, C. (1988). The GenBank (R) genetic sequence data bank. *Nucleic Acids Res.* **16**, 1861-1864.

Böhme, W., and Bischoff, W. (1976). Das Paarungsverhalten der kanarischen Eidechsen (Sauria, Lacertidae) als systematisches Merkmal. *Salamandra* **12**, 109-119.

Boucot, A. J. (1990). "Evolutionary Paleobiology of Behavior and Coevolution." Elsevier, Amsterdam.

Breithaubt, B. H. and Duvall, D. (1986). The oldest record of serpent aggregation. *Lethaia* **19**, 181-185.

Brooks, D. R. and McLennan, D. A. (1991). "Phylogeny, Ecology, and Behavior: A Research Program in Comparative Biology." Univ. of Chicago Press, Chicago.

Brooks, D. R., Mayden, R. L., and McLennan, D. A. (1992). Phylogeny and biodiversity: conserving our evolutionary legacy. *Trends Ecol. Evol.* **7**, 55-59.

Brown, J. H. (1975). "The Evolution of Behavior." W. W. Norton, New York.

Burghardt, G. M., and Gittleman, J. L. (1990). Comparative behavior and phylogenetic analyses: New wine, old bottles. *In* "Interpretation and Explanation in the Study of Animal Behavior" (M. Bekoff and D. Jamieson, eds.), pp. 192-225. Westview Press, Boulder, CO.

Cadle, J. E. (1993). The colubrid radiation in Africa (Serpentes: Colubridae): Phylogenetic relationships and evolutionary patterns based on immunological data. *Zool. J. Linn. Soc. (London)* **107** (in press).

Cadle, J. E., and Greene, H. W. (1993). Phylogenetic patterns, biogeography, and the ecological structure of neotropical snake assemblages. *In* "Species Diversity in Ecological Communities" (R. E. Ricklefs and D. Schluter, eds.), Univ. of Chicago Press, Chicago (in press).

Campbell, J. A. and Lamar, W. W. (1989). "The Venomous Reptiles of Latin America." Cornell Univ. Press, Ithaca, NY.

Carpenter, C. C., and Gillingham, J. C. (1990). Ritualized behavior in *Agkistrodon* and allied genera. *In* "Snakes of the Agkistrodon Complex: A Monographic Review" (H. K. Gloyd and R. Conant, eds.), pp. 523-531. Society for the Study of Amphibians and Reptiles, Oxford, OH.

Chiszar, D., Murphy, J. B., and Smith, H. M. (1993). In search of zoo-academic collaborations: A research agenda for the 1990s. *Herpetologica* **49** (in press).

Cundall, D., and Greene, H. W. (1982). Evolution of the feeding apparatus in alethinophidian snakes. *Am. Zool.*, **22**, 924.

Cundall, D., and Irish, F. J. (1986). Aspects of locomotor and feeding behaviour in the Round Island boa *Casarea dussumieri. Dodo* (J. Wildl. Preserv. Trust) **23**, 108-111.

Cundall, D., Wallach, V., and Rossman, D. A. (1993). The systematic status of the snake genus *Anomochilus. Zool. J. Linn. Soc.*, (in press).

de Queiroz, A., and de Queiroz, K. (1987). Prey handling behavior of *Eumeces gilberti* with comments on headfirst ingestion in squamates. *J. Herpetol.* **21**, 57-63.

de Queiroz, A., and Wimberger, P. H. (1993). The usefulness of behavior for phylogeny estimation: Levels of homoplasy in behavioral and morphological characters. *Evolution (Lawrence, Kans.)* **47**, 46-60.

Dewsbury, D. A. (1975). Diversity and adaptation in rodent copulatory behavior. *Science* **190**, 947-954.

Donoghue, M. J. (1992). Homology. *In* "Keywords in Evolutionary Biology" (E. F. Keller and E. A. Lloyd, eds.), pp. 170-179. Harvard Univ. Press, Cambridge, MA.

Drummond, H. (1981). The nature and description of behaviour patterns. *In* "Perspectives in Ethology" (P. P. G. Bateson and P. Klopfer, eds.) Vol. 4, pp. 1-33, Plenum, New York.

Duvall, D., King, M. B., and Gutzwiller, K. J. (1985). Behavioral ecology and ethology of the prairie rattlesnake. *Natl. Geogr. Res.* **1**, 80-111.

Eaton, T. H. (1964). A captorhinomorph predator and its prey (Cotylosauria). *Am. Mus. Novitates* **2169**, 1-3.

Eibl-Eibesfeldt, I. (1970). "Ethology: The Biology of Behavior." Holt, Reinhart & Winston, New York.

Erwin, D. H. (1992). A preliminary classification of evolutionary radiations. *Hist. Biol.* **6**, 133-147.

Ferguson, G. W. (1971). Variation and evolution of the push-up displays of the side-blotched lizard genus *Uta* (Iguanidae). *Syst. Zool.* **20**, 79-101.

Gauthier, J., Kluge, A. G., and Rowe, T. (1988). Amniote phylogeny and the importance of fossils. *Cladistics* **4**, 105-209.

Gittleman, J. L. (1989). The comparative approach in ethology: Aims and limitations. *In* "Perspectives in Ethology", (P. P. G. Bateson and P. H. Klopfer, eds.), Vol. 8, pp. 55-83. Plenum, New York.

Gould, S. J., and Vrba, E. (1982). Exaptation — a missing term in the science of form. *Paleobiology* **8**, 4-15.

Greene, H. W. (1976). Scale overlap, a directional sign stimulus for prey ingestion by ophiophagous snakes. *Z. Tierpsychol.* **41**, 113-120.

Greene, H. W. (1977). *Phylogeny, Convergence, and Snake Behavior.* Doctoral Dissertation, Univ. of Tennessee, Knoxville, TN.

Greene, H. W. (1979). Behavioral convergence in the defensive displays of snakes. *Experientia* **35**, 747-748.

Greene, H. W. (1983). Dietary correlates of the origin and radiation of snakes. *Am. Zool.* **23**, 431-441.

Greene, H. W. (1986). Natural history and evolutionary biology. *In* "Predator-Prey Interactions: Perspectives and Approaches from the Study of Lower Vertebrates" (M. E. Feder and G. V. Lauder, eds.), pp. 99-108. Univ. of Chicago Press, Chicago.

Greene, H. W. (1988). Antipredator mechanisms in reptiles. *In* "Biology of the Reptilia" (C. Gans and R. B. Huey, eds.), Vol. 16, pp. 1-152. Alan R. Liss, New York.

Greene, H. W. (1992). The behavioral and ecological context for pitviper evolution. *In* "Biology of the Pitvipers" (J. A. Campbell and E. D. Brodie, Jr., eds.), pp. 107-117. Selva, Tyler, TX.

Greene, H. W. (1994). Systematics and natural history, foundations for understanding and conserving biodiversity. *Am. Zool.* (in press).

Greene, H. W., and Burghardt, G. M. (1978). Behavior and phylogeny: Constriction in ancient and modern snakes. *Science* **200**, 74-77.

Greene, H. W., and Losos, J. B. (1988). Systematics, natural history, and conservation. *BioScience* **38**, 458-462.

Groombridge, B. (1986). Phyletic relationships among viperine snakes. *In* "Studies in Herpetology" (Z. Rocek, ed.), pp. 219-22. Charles Univ., Prague.

Guyer, C., and Slowinksi, J. B. (1993). Adaptive radiation and the topologies of large phylogenies. *Evolution (Lawrence, Kans.)* **47**, 253-263.

Hailman, J. P. (1976). Homology: Logic, information, and efficiency. *In* "Evolution, Brain, and Behavior: Persistent Problems" (R. B. Masterton, W. Hodos, and H. Jerison, eds.), pp. 181-198. Lawrence Erlbaum, Hillsdale, NJ.

Harvey, P. H. (1991). The state of systematics. *Trends Ecol. Evol.* **6**, 192-193.

Herzog, H. A., Jr., Bowers, B. B., and Burghardt, G. M. (1992). Development of antipredator responses in snakes. V. Species differences in ontogenetic trajectories. *Dev. Psychobiol.* **25**, 199-211.

Hollister, N. (1918). East African mammals in the United States National Museum. Part I. Insectivora, Chiroptera, and Carnivora. *Bull. U.S. Nat. Mus.* **99**, 1-194.

Hopson, J. A., and Radinsky, L. B. (1980). Vertebrate paleontology: New approaches and new insights. *Paleobiology* **6**, 250-270.

Klopfer, P. (1969). (Review of) R. F. Ewer, Ethology of Mammals. *Science* **165**, 887.

Klopfer, P. (1975). (Review of) J. Alcock, Animal Behavior: An Evolutionary Approach. *Am. Sci.* **63**, 578-579.

Kluge, A. G. (1991). Boine snakes and research cycles. *Misc. Publ. Mus. Zool. Univ. Michigan* **178**, 1-58.

Lauder, G. V. (1981). Form and function: Structural analysis in evolutionary morphology. *Paleobiology* **7**, 430-442.

Lauder, G. V. (1982). Historical biology and the problem of design. *J. Theor. Biol.* **97**, 57-67.

Lauder, G. V. (1986). Homology, analogy, and the evolution of behavior. *In* "Evolution of Animal Behavior" (M. H. Nitecki and J. A. Kitchell, eds.), pp. 9-40. Oxford Univ. Press, New York.

Losos, J. B. (1990). Concordant evolution of locomotor behaviour, display rate, and morphology in *Anolis* lizards. *Anim. Behav.* **39**, 879-890.

Losos, J. B. (1992). The evolution of convergent structure in Caribbean *Anolis* communities. *Syst. Biol.* **41**, 403-420.

Maddison, W. P., and Maddison, D. R. (1992). "MacClade: Analysis of Phylogeny and Character Evolution." Sinauer, Sunderland, MA.

McKinney, F. (1975). The evolution of duck displays. *In* "Function and Evolution of Behaviour" (G. Baerends, C. Beer, and A. Manning, eds.), pp. 331-357. Oxford Univ. Press, Oxford.

McLennan, D. A. (1991). Integrating phylogeny and experimental ethology: from pattern to process. *Evolution (Lawrence, Kans.)* **45**, 1773-1789.

McLennan, D. A., Brooks, D. R., and McPhail, J. D. (1988). The benefits of communication between comparative ethology and phylogenetic systematics: A case study using gasterosteid fishes. *Can. J. Zool.* **66**, 2177-2190.

Novacek, M. J. (1992). Fossils, topologies, missing data, and the higher phylogeny of eutherian mammals. *Syst. Biol.* **41**, 58-73.

Padian, K. and Rayner, J. V. M. (1993). The wings of pterosaurs. *Am. J. Sci.* **293-A**, 91-166.

Pregill, G. K., Gauthier, J. A., and Greene, H. W. (1986). The evolution of helodermatid squamates, with description of a new taxon and an overview of Varanoidea. *Trans. San Diego Nat. Hist. Mus.* **21**, 167-202.

Reinert, H. K. (1992). Radiotelemetric field studies of pitvipers: Data aquisition and analysis. *In* "Biology of the Pitvipers" (J. A. Campbell and E. D. Brodie, Jr., eds.), pp. 185-197. Selva, Tyler, TX.

Rieppel, O. (1988). A review of the origin of snakes. *Evol. Biol.* **22**, 37-130.

Rodda, G. H. (1992). Foraging behaviour of the brown tree snake, *Boiga irregularis. Herpetol. J.* **2**, 110-114.

Savitzky, A. H. (1980). The role of venom delivery strategies in snake evolution. *Evolution (Lawrence, Kans.)* **34**, 1194-1204.

Shine, R., and Schwaner, T. (1985). Prey constriction by venomous snakes: A review, and new data on Australian species. *Copeia* **1985**, 1067-1071.

Wake, M. H. (1992). Morphology, the study of form and function, in modern evolutionary biology. *In* "Oxford Surveys in Evolutionary Biology" (D. Futuyma and J. Antonovics, eds.), pp. 289-346. Oxford Univ. Press, New York.

Wenzel, J. W. (1992). Behavioral homology and phylogeny. *Annu.Rev. Ecol. Syst.* **23**, 361-381.

Willard, D. E. (1977). Constricting methods of snakes. *Copeia* **1977**, 379-382.

Wilson, E. O. (1975). *Sociobiology, the New Synthesis*. Harvard Univ. Press, Cambridge, MA.

Wilson, M. V. H. (1992). Importance for phylogeny of single and multiple stem-group species with examples from freshwater fishes. *Syst. Biol.* **41**, 462-470.

12

COMPLEXITY AND HOMOLOGY

IN PLANTS

Michael J. Donoghue

Department of Organismic and Evolutionary Biology
Harvard University
Cambridge, MA 02138

Michael J. Sanderson

Department of Biology
University of Nevada
Reno, NV 89557

Homology: The Hierarchical Basis of Comparative Biology
Copyright © 1994 by Academic Press, Inc.
All rights of reproduction in any form reserved.

I. INTRODUCTION

Interest in phylogenetic analysis has focused attention on the concept of homology, especially on how hypotheses of homology are tested. Our object is not to review these developments (for which see Donoghue, 1992; also de Pinna, 1991; Haszprunar, 1992; Mindell, 1991; Minelli and Peruffo, 1991; Rieppel, 1988, 1992; Roth, 1991). Instead, we analyze factors that underlie the efficacy of proposed tests of homology and attempt to develop a general conceptual framework within which to consider homology and homoplasy in plants.

Our intention is not to promote a particular definition of homology, nor to propose a new one. Because our main interest is in phylogeny, our analysis is primarily a contribution to the *historical* view of homology, broadly understood (e.g., Hennig, 1966; Wiley, 1981; Patterson, 1982), as opposed to the *biological* view (e.g., Wagner, 1989a,b; Roth, 1988, 1991; also see Chapters 8, 9, and 13 in this volume). This is not to say, however, that synapomorphy or the history of character transformations are the only legitimate concerns connected with homology (Donoghue, 1992); as we will conclude, there are obvious connections between the two approaches.

II. BACKGROUND

The view is widespread that hypotheses of homology can be tested in various ways and that some tests are more powerful than others, in the sense that they are more effective in correctly identifying instances of homology. Patterson (1982), for example, outlined a series of homology tests for morphological features, and argued that the ultimate arbiter was what he called the *congruence test*. According to Patterson's "taxic" view of homology, a character is homologous if, and only if, it marks a monophyletic group. To test whether or not this is true, a character that is hypothesized to be homologous (having passed initial similarity tests) is included in a phylogenetic analysis along with other such characters. If the character is found to be a synapomorphy then it is considered homologous. If it fails to mark a single clade, then the initial homology hypothesis is rejected and an instance of homoplasy is identified.

Patterson's (1988) analysis of homology in molecular characters led him to a different conclusion, namely that similarity is the more powerful test in this case than congruence. The reason, according to Patterson, is that it is possible to assess the probability that two molecular sequences could have achieved a certain level of similarity by chance alone. Thus, whether congruence or similarity is more powerful depends on the type of data under consideration.

We begin our analysis by taking a closer look at this view of the homology problem. Although we agree with Patterson that different tests may be more effective in different cases, we think this is not directly linked to whether characters are morphological or molecular. Instead, the critical factor, which Patterson and others before him had identified, is the relative complexity of the structures under consideration. We then bring this perspective to bear on the the value of different homology tests in plants, where much concern has centered on the supposed simplicity of plant structures and developmental systems.

III. MOLECULES AND MORPHOLOGY

Patterson's (1988) argument that similarity is the decisive test in the case of molecular data rests on an assumption about what is being homologized. He supposes that whole sequences are being compared in order to determine whether particular genes are homologous. In this case, owing to the linear, "one-dimensional" arrangement of elements (amino acids, nucleotides), it is possible to calculate the likelihood that two sequences are as similar as they are by chance alone (as opposed to reflecting common ancestry). In contrast, morphology has too many dimensions (at least three, or four, if ontogeny is included) to make such calculations feasible. In other words, the main difference that Patterson sees between morphology and molecules hinges on our ability or inability to make the appropriate probability calculations. The implication is that if we could devise methods to measure whether morphological structures matched by chance alone, then similarity would become the more decisive test in morphology as well. In fact, considering the extra dimensions associated with many morphological features, it should, theoretically, be possible to make even stronger arguments about homology, based on similarity alone.

Patterson's (1988) view that there is a difference between morphology and molecules loses force when attention is shifted, as it often is in practice, to assessment of homology at particular sites in a sequence (following an alignment procedure). In this case all we know is that some taxa share, for example, a cytosine at a particular site, while others have a guanine; there is no obvious way to make the sorts of probability calculations that apply to whole sequences. This being the case, similarity can hardly be a decisive test of whether any particular nucleotide is truly a homology. Instead, similarity (e.g., the sharing of an adenine) simply "validates it as worthy of testing" in other ways, as Patterson (1988, p. 605) argued for morphology. Congruence is the obvious test in this case, but for a somewhat different reason than in morphology. Similarity is a weak test in morphology (and congruence necessary), according to Patterson (1988), because structures are so complex that it is difficult to quantify the likelihood of homology, whereas it is weak in the case of individual nucleotides because they are too simple.

The argument above lead us to conclude that which test is most decisive is not directly linked to whether the data are morphological or molecular. Instead, decisiveness seems to vary as a function of the ability to quantify the likelihood that sharing a certain level of similarity reflects common ancestry. Even within molecular data this ability varies depending on the level of the problem at hand; for example, whether whole sequences are being compared or just nucleotides at individual sites. The same is presumably also true within morphological data, since there is variation from simpler to more complex structures.

IV. SIMPLICITY AND COMPLEXITY

The argument above shifts attention away from morphology versus molecules toward a more general factor underlying the decisiveness of different homology tests, namely the relative simplicity or complexity of the features under consideration. The notion of complexity is obviously problematical, but for our purposes it varies with the number of parts or possible correspondences, and with how irregularly the parts are arranged (McShea, 1991, 1992). As complexity in this sense increases it should be easier to weed out chance as an explanation for similarity. Whole molecular sequences happen to be complex enough for the use of statistical tests, but not so complex that it seems impossible to concoct appropriate measures. Exact measures may not yet be available for morphological characters, but the same general reasoning applies to them as well.

The relative complexity of structures figured prominently in earlier treatments of the homology problem (see Riedl, 1978). Perhaps most importantly, it is evident in Remane's (1952) division of homology criteria into "principal" and "auxiliary." Most comparative biologists are well acquainted with Remane's principal criteria — similarity in position (or topography), in special structure, and through transitional (or intermediate) forms. It is less well known that Remane equated the usefulness of these criteria with the degree of complexity. For example, in reference to the structural criterion he noted that "certainty increases with the degree of complication and of agreement in the structures compared" (in Riedl, 1978, p. 34).

Remane's three auxiliary criteria, which apply to simpler structures, are rarely mentioned. Riedl (1978, p. 36) translated these as follows:

(1) "Even simple structures can be regarded as homologous when they occur in a great number of adjacent species" (the *general conjunctional criterion*);
(2) "The probability of the homology of simple structures increases with the presence of other similarities, with the same distribution among closely similar species" (the *special conjunctional criterion*); and
(3) "The probability of the homology of features decreases with the commonness of occurrence of this feature among species which are certainly not related" (*the negative conjunctional criterion*).

While these are obviously not described in cladistic terms, they do rest squarely on the distribution of a feature in relation to other characters and on presumed phylogenetic relatedness, and therefore are similar in intent to Patterson's (1982) congruence test.

Patterson (1982), who made no mention of Remane, effectively elevated the criteria Remane intended as secondary, and for simple structures only, to his primary criterion for morphological characters (Donoghue, 1992). Many morphological characters can be evaluated using Remane's primary criteria, but Patterson (1988, p. 605) believes that these measures are too weak in morphology, which "has so far resisted quantification (and is unlikely to submit)." Thus, morphological characters are treated by Patterson as though they are simple structures — not because they really are — but because they are too complex to quantify.

We think it will clarify matters to reorient the discussion primarily around complexity, whether or not we happen to be able to perform the sorts of statistical tests that Patterson finds convincing. Under this view, complex morphological features and whole molecular sequences (both of which allow many comparisons) are situated at one end of the spectrum, where similarity can provide a more powerful preliminary test. Simple morphological structures and individual nucleotide sites are located near the other end of the continuum, where there is little else to go on but congruence with other evidence.

V. COMPLEXITY AND HOMOPLASY

It is important to remember that we are concerned here with the power of similarity tests to identify truly homologous features at the outset, before phylogenetic analysis. Because homology (at least "historical" homology) is tied to inheritance from a common ancestor, congruence testing is always necessary (de Pinna, 1991). That is, no matter how complex the structures involved, or how powerful the similarity tests may therefore be, phylogenetic analysis has at least the potential to favor homoplasy over homology. The significance of our arguments above is that they lead to a simple prediction. If the power of similarity to correctly identify homology before phylogenetic analysis varies with character complexity, then fewer instances of nonhomology should make it past the similarity filter when greater complexity makes the filter that much finer. That is, there should be fewer instances of homoplasy identified in phylogenetic analysis in the case of complex characters. This, we note, is similar to (though far less extreme than) some arguments underlying Dollo's Law (see Sanderson, 1993). On the other hand, when the filter is necessarily coarse, as in the case of simple structures, we should expect to see more homoplasy. These expectations are illustrated in Fig. 1.

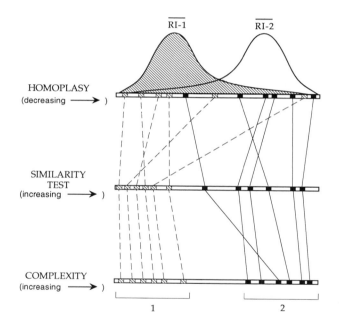

Fig. 1. Expected relationship between the relative complexity of characters, the stringency of the similarity tests applied, and levels of homoplasy observed in phylogenetic analysis. A group of less complex characters (group 1 on the lower left, each character represented by a small cross-hatched rectangle) is necessarily subjected to limited similarity tests (dotted lines connect individual characters upward through different levels in the diagram). Note that

complexity and strength of similarity test both increase to the right. Although the range of retention index (RI) values is great (some characters showing little or no homoplasy), the average level of homoplasy is relatively high. Note that the amount of homoplasy decreases to the right. The shaded curve above represents a histogram of RI values for a hypothetical large population of simple characters. In contrast, the group of more complex characters (group 2 on the lower right, represented by black rectangles) is subjected to more stringent similarity tests. On average, these characters show less homoplasy than characters of type 1. The open curve is a histogram of RI values for a large population of complex characters. The mean RI values of the two populations of characters are widely separated, although the distributions overlap.

This prediction relating complexity to homoplasy is at least potentially testable using phylogenetic studies. But is it really worth testing? We think that it is for two reasons. First, it need not be true. For example, the level of homoplasy shown by different characters might track some other factor, which may or may not covary with complexity, such as the degree of *burden* or *entrenchment* in development (Riedl, 1978; Arthur, 1988; Wimsatt and Schank, 1988). A character that is peripheral in development and of low burden might otherwise be complex in the sense of being elaborate and amenable to numerous point comparisons (e.g., the lip petal in orchid flowers). Second, attempting to test such a prediction using phylogenies will help us better understand the logic and limitations of such tests; that is, how much we can expect to learn from trees about general patterns of character evolution.

What would a proper test require? First, one would need to be able to categorize characters at the outset as to their complexity (see below). Second, phylogenies would be needed to compare levels of homoplasy in characters recognized as more and less complex. Indirectly, the proposition has perhaps already been tested in our comparison of levels of homoplasy in morphological versus molecular phylogenetic studies (Sanderson and Donoghue, 1989; also see Donoghue and Sanderson, 1992). If we could safely assume that morphological characters were prescreened by more powerful similarity tests than simpler molecular characters, we would expect to see more homoplasy in molecular data. We found, however, that there was not a significant difference between the two sorts of data. This may mean that our prediction was incorrect, but interpretation of the results is difficult. For example, there may be less prescreening of morphological characters than their complexity would allow, or there may be more prescreening of molecular characters (e.g., by elimination of segments of DNA showing ambiguity in alignment).

Givnish and Sytsma (1992) extended this kind of analysis to a comparison of chloroplast DNA restriction-site data, nucleotide sequence data, and morphology in plants. They reported significantly less homoplasy in restriction-site analyses, and concluded that such data "provide an internally more consistent basis for phylogenetic studies" (Givnish and Sytsma, 1992, p. 145). This might be interpreted as implying that restriction-site data yield a more reliable estimate of relationships. However, as we demonstrated previously (Sanderson and Donoghue, 1989; Donoghue and Sanderson, 1992) there does not appear to be a direct link between consistency index and reliability. Nevertheless, it is worth considering what might account for the observed pattern. In view of the inability to distinguish among different mutations that could lead to loss or gain of a restriction site, it is likely that some homoplasy is simply hidden. Of course, this implies that taxa in such studies are sometimes united by false synapomorphies. Another possibility is that the general pattern results from more stringent screening of restriction site data, perhaps through omission of bands that are not easily mapped. Alternatively, Givnish and Sytsma's (1992) result may reflect some other variable that was not taken into account. In this case, we suspect that differences in the level of resolution of phylogenetic relationships (not to be confused with taxonomic rank or phylogenetic depth) need to be considered. Trees based on restriction-site data may tend to show more unresolved regions owing to lack of evidence, in which case the effective number of taxa in such studies is reduced.

de Queiroz and Wimberger (1993) used a similar approach to compare behavioral and morphological characters, and concluded that there is not a significant difference in the level of homoplasy between the two. If, as is sometimes said (see de Queiroz and Wimberger, 1993), behavioral characters are subject to a limited set of similarity tests (e.g., the positional criterion might be inapplicable), then more homoplasy would be expected in the behavioral traits. That this is not observed suggests that similarity tests are actually as stringent in behavior or that less intense tests are not resulting in more mistakes in identifying homology.

VI. HOMOLOGY AND HOMOPLASY IN PLANTS

How is an expected relationship between complexity and homoplasy relevant to the homology problem in plants? The connection is that botanists have been skeptical about homology assessment in large part because plants are seen to be structurally and developmentally simpler than animals. Stebbins (1974), for example, echoes the standard view that owing to "relative indeterminism with respect to both the number and the position of plant parts, these criteria are much less diagnostic of homology in plants than in animals" (p. 142). He goes on to say that "developmental pattern is also a much less reliable criterion of homology in plants than in animals," because "patterns of development of individual organs are, in general, much simpler in plants than in animals" (p. 143). For these reasons plant morphologists have tended to shy away from the evolutionary connotations of homology, opting instead to define it in terms of similarity alone (e.g., Sattler, 1984; Kaplan, 1984; Tomlinson, 1984; but see Stevens, 1984).

Skepticism on the part of botanists is quite understandable when one lists the relevant differences between plants and animals (see Klekowski, 1988, p. 170). Plants have probably only one-tenth to one-twentieth the number of cell types, and these are distinguished mainly by the nature of the cell wall, since cell movement (as is an immune system) is basically lacking. Furthermore, plant cells tend to be totipotent and show greater plasticity than animal cells. Owing to indeterminate growth (a distinct germline is absent) and modular organization, developmental steps are generally repeated many times in different parts of the plant. Iterative (serial) homology promotes the possibility of something analogous to the paralogy problem in multigene families (Roth, 1991, and Chapter 10 in this volume). Moreover, plants may be especially prone to processes that disrupt the usual correlations seen in animals (especially vertebrates); for example, homeotic or heterotopic events that shift the position of a feature but not its structure or ontogeny (Iltis, 1983; Meyen, 1988; Sattler, 1988).

Along these same lines, Kaplan (1992) has pointed out that a fundamental distinction between multicellular construction in plants and animals stems from an underlying difference in cell division. In animals there is complete separation of protoplasts at mitosis, whereas in plants incomplete separation results from the insertion of walls with plasmodesmata. Kaplan argues that the application of an animal-biased cell theory has resulted in an inappropriate emphasis on structural qualities as a guide to homology in plants. He promotes, instead, an "organismal theory," which emphasizes positional criteria.

These observations, coupled with our arguments above, imply that similarity tests in plants might be less able to weed out nonhomology at the outset, and that plant data sets might therefore show more homoplasy than animal data sets. This difference was not observed in our earlier comparsions (Sanderson and Donoghue, 1989). In fact, we found remarkably similar levels of homoplasy in plant and animal phylogenetic studies, and we suggested that this might reflect a tendency for botanists to simply disregard more characters at the outset than zoologists. Another possible explanation is that there truly is more homoplasy in the characters included by botanists, but that phylogenies provide an underestimate because they are either false or not finely enough resolved. A third, more optimistic, explanation is that despite the relative simplicity of plant structures and development, botanists are nevertheless able to distinguish homology from nonhomology just about as effectively as zoologists.

It is difficult to choose among these possibilities, which, of course, are not mutually exclusive. The first proposition is difficult to test given the usual inattention to rejected characters. Although we are tempted to accept the third explanation, the second one remains a distinct possibility. After all, it is quite often the case that taxa added to an analysis will be positioned so as to require additional homoplasy. This is true in general (Sanderson and Donoghue, 1989) and there are many concrete examples in plants, even involving the most complex structures. Good examples are the evolution of roots and leaves in land plant sporophytes. Analysis of living plants alone implies that these organs evolved just once — that they are synapomorphies of trachaeophytes. However, analyses including fossils suggest that both organs evolved within the zosterophyte/lycophyte line and again within the trimerophyte line (e.g., Kenrick and Crane, 1991). Likewise, without "progymnosperm" fossils one would suppose that fern leaves and seed plant leaves were homologous (Doyle and Donoghue, 1986; Donoghue et al. 1989). It has not been difficult for some botanists to conclude that if we had an accurate enough phylogeny, every plant character would be seen to have evolve more than once — real synapomorphies are an illusion and parallelism must be used to infer relationships (e.g., Cronquist, 1988). Of course, a better understanding of phylogeny can also decrease homoplasy in some characters. An example is the recognition that double fertilization may be homologous in gnetales and angiosperms (see Donoghue and Scheiner, 1992).

VII. A POSSIBLE TEST

From the preceding discussion it should be clear that more direct tests are needed, designed to eliminate (or minimize) at least some of the possibly confounding variables noted above. In this spirit we have conducted a preliminary comparison of plant characters that are often viewed as being more and less complex. Although in some respects this is an improvement over the broad comparisons discussed above, it is clearly still far from ideal. Nevertheless, we think studies along these lines warrant further attention. If nothing else, they clarify what we can expect to extract from such comparisons.

Our aim was to select from the literature a set of plant phylogenetic studies containing morphological characters of varying complexity that could be compared among studies. Ideally, one would apply a specific measure of complexity to all of the characters in a number of data sets and then compare levels of homoplasy among complexity classes. Although some attempts have been made to quantify complexity (e.g., Schopf et al., 1975; Riedl, 1978; Bonner, 1988;

McShea, 1991), we know of no measures that can readily be applied to the range of morphological characters encountered in phylogenetic data sets.

Schopf and colleagues (1975) judged the complexity of organisms by the number of terms used to describe them. This suggests the possibility of quantifying the complexity of individual characters by counting the number of descriptive terms (or leads in a diagnostic key) associated with particular features. However, it is not clear to us that there is a straightforward relationship between number of descriptors and complexity. For example, although there are many terms describing pubescence in plants, botanists tend to regard hair characters as being relatively simple. Another problem is that it is unclear what phylogenetic universe to consider in counting numbers of character states. Should the number of terms be tallied for all green plants, all seed plants, all angiosperms, or perhaps only for the data sets under consideration? Bonner (1988) concentrated on the number of cell types, a measure more easily applied in broader comparisons (such as between green plants and vertebrates) than within angiosperms, especially in view of the presence in many plant structures of the same basic cell types (see above). McShea's (1992) metric uses a series of single measurements on a set of serially homologous structures, and therefore is inapplicable to most of the characters in cladistic data sets.

In view of these difficulties, we opted instead to make preliminary comparisons among sets of characters that have traditionally been viewed as more and less complex, and therefore more and less likely to provide an accurate guide to phylogeny. Furthermore, rather than attempting to categorize all characters in each data sets, we decided to focus on just a few contrasts. Initially we compared pubescence/trichome characters with flower characters (excluding pubescence, inflorescence, and fruit characters). A third class was added later, comprised of all leaf characters except pubescence and phyllotaxy. The fact that these different character types were present within the individual studies helps factor out differences between systematists in the choice and coding of characters. It also guards against artifacts that might arise if certain classes of characters were mostly used in studies at higher taxonomic levels (greater phylogenetic depth) and others at lower levels.

Regarding complexity and expected homoplasy in these three classes, we believe it is fair to say that botanists have generally considered flower characters to be more complex and more reliable, on average, than either pubescence or leaf characters. This view is expressed with little reservation in some older texts (e.g., Jeffrey, 1917), whereas in more recent discussions counter examples are usually offered to illustrate that characters of all kinds can be phylogenetically useful (e.g., Davis and Heywood, 1973; Stace, 1989; on leaves in particular, see Levin, 1986; Hershkovitz, 1993). Nevertheless, even recent treatments imply that, on average, flower characters tend to be most useful. Stebbins (1974, p. 152), for example, developed an argument similar to ours, namely that "the degree of irreversibility of a character condition depends upon the number and complexity of the separate factors that contribute to it." On this basis he concluded (pp. 148-151) that more reversibility is to be expected in characters having to do with size (e.g., leaf size) and amount (e.g., pubescence) than those reflecting the fusion and adnation of flower parts or differences in flower symmetry. In general, Stebbins (1974) adopted what we believe to be the standard view, namely that vegetative characters are more plastic and more prone to convergence and parallelism (e.g., pp. 43 and 49), whereas flower characters are more complex and therefore more conservative (e.g., pp. 100 and 125).

In selecting data sets from the literature we used the database of Sanderson *et al.* (1993) to locate 85 morphological cladistic analyses of angiosperm groups published from 1989 through 1991. These studies were scanned for data matrices that included both trichome and flower characters, and 10 were selected for further analysis (Table I). Each of these data sets was then reanalyzed using PAUP (Phylogenetic Analysis Using Parsimony, version 3.0s; Swofford, 1991). In most cases we were able to confirm the phylogenetic results presented in the original paper, but in a few cases (Anderberg and Bremer, 1991; Kron and Judd, 1990) we found additional most parsimonious trees, or (Cox and Urbatsch, 1990) even slightly more parsimonious trees. Insufficient information was given in two of the papers (Cruden, 1991; Loconte and Estes, 1989) to ascertain whether our results exactly matched those presented.

TABLE I. Summary statistics on ten numerical cladistic analyses of angiosperms and levels of homoplasy in subsets of characters.[a]

Study[b]	Taxa	Chars	Trees	Steps	CI	RI	No. Chars.			CI			RI		
							P	L	F	P	L	F	P	L	F
1	44	46	503	123	0.45	0.78	1	5	12	0.33	0.66	0.69	0.83	0.83	0.83
2	30	27	1	56	0.62	0.83	3	1	6	0.83	1.00	0.78	0.75	1.00	0.94
3	15	22	18	32	0.75	0.86	5	2	5	0.88	0.75	0.90	0.79	0.50	0.75
4	14	28	16	72	0.46	0.61	1	1	6	0.40	0.48	0.79	0.50	0.72	0.76
5	11	36	2	71	0.87	0.85	2	0	9	0.77	--	0.86	0.44	--	0.81
6	11	20	1	36	0.54	0.24	3	0	7	0.67	--	0.93	0.25	--	0.87
7	17	42	54	48	0.81	0.86	3	2	23	0.89	1.00	0.77	0.66	1.00	0.64
8	13	14	6	21	0.67	0.67	2	3	3	0.75	0.58	0.77	0.50	0.69	0.33
9	11	27	2	45	0.64	0.75	3	4	7	0.83	0.68	0.70	0.92	0.28	0.70
10	35	49	149+	124	0.48	0.78	1	4	10	0.17	0.71	0.49	0.50	0.78	0.78
Totals:							24	22	88	0.75	0.71	0.76	0.63	0.71	0.77

[a] CI, consistency index; RI, retention index; P, pubescence characters; L, leaf characters; F, flower characters. The CI for a single character is m/s, where m is the minimum number of steps a character can show on any tree, and s is the minimum number of steps required on the tree under consideration. The ensemble CI for a suite of characters is M/S, where M and S are sums of m and s over all characters in the suite. The RI for a single character is $(g - s)/(g - m)$, where m and s are as above, and g is the maximum number of steps that a character could require on any tree. The ensemble RI is analogous to the ensemble CI. For additional information see Farris (1989), Swofford (1991), and Maddison and Maddison (1992).

[b] Studies: (1) Anderberg and Bremer, 1991: *Relhania* group (Asteraceae: Gnaphalieae); (2) Axelius, 1990: *Xanthophytum* (Rubiaceae); (3) Boufford, et al., 1990: *Circaea* (Onagraceae); (4) Cox and Urbatsch, 1990: coneflower genera (Asteraceae: Heliantheae); (5) Cruden, 1991: *Isidrogalvia* (Liliaceae); (6) Gilmartin, et al., 1989: *Glomeropitcairnia* (Bromeliaceae); (7) Judd, 1989: Miconieae (Melastomataceae); (8) Kron and Judd, 1990: Rhodoreae (Ericaceae); (9) Lavin, 1990: *Sphinctospermum* (Leguminoseae); (10) Loconte and Estes, 1989: Berberidaceae and Ranunculales.

MacClade (version. 3.0; Maddison and Maddison, 1992) was used to parsimoniously optimize the selected characters on the trees obtained and to calculate the consistency index (CI) and retention index (RI) for each character, as well as ensemble CIs and RIs for suites of characters (see Table I). Because the RI is undefined (0/0) for autapomorphies, these were excluded in calculating this measure. Our attention will focus on RI comparisons, because we think this is the more appropriate statistic in this instance. The CI is affected by the distribution of character states among the taxa, whereas the stringency of similarity tests is probably independent of the number of taxa with alternative states. Where more than one most parsimonious tree was found, MacClade was used to calculate average CI and RI values over the entire set of most parsimonious trees. While this procedure effectively weights each tree equally, it is unlikely that this would bias the results and, in any case, it seems preferable to examining only a single tree.

The number of characters of the three types *within* the individual studies is so limited (Table I) that we cannot draw strong conclusions from such comparisons. For example, flower RIs exceeded pubescence RIs in six studies, pubescence RIs were higher in three, and the two were tied in one data set. Flower RIs were higher than leaf RIs in three cases, leaves were higher in three, and flowers and leaves were tied in the remaining two. None of the possible contrasts between character types was found to be significant on the basis of a nonparametric Wilcoxon signed-rank test.

Failure to find a significant difference between character types within data sets led us to pool characters from the 10 studies to increase sample sizes (Table I). The average RI is 0.63 for the 18 pubescence characters, 0.71 for the 21 leaf characters, and 0.77 for the 66 flower characters (RI calculations are based on fewer characters than shown in Table I because autapomorphies were omitted). These values are consistent with our prediction, but statistical tests (t tests and nonparametric Mann-Whitney tests) for differences among pooled character classes, as well as between vegetative characters (pubescence plus leaf characters) and flower characters, revealed no significant contrasts (although the difference between pubescence and flower RIs came closest to being significant). There may be differences in the variance of RIs among the character types, but the data are still too limited to assess this possibility. The histogram of flower RIs is skewed right, with an excess of RI = 1 and a deficit of RI = 0 characters, whereas the distribution of pubescence RIs is almost uniform from 0 to 1.

Although pooling characters increases the sample size, it also introduces a possibly confounding variable. The number of taxa included in the individual studies ranged from 11 to 44, and it is known that, on average, there is a positive correlation between the number of taxa and the amount of homoplasy, at least as measured by CI (Archie, 1989; Sanderson and Donoghue, 1989). The absolute values of CI obtained would presumably be highly influenced by the distribution of study sizes. It is unclear, however, that differences in study size would introduce a systematic error in the comparsions we have made between character classes, unless there also happens to be a correlation between the number of taxa and the number of characters of each type in the study (e.g., if the larger studies accounted for most of the pubescence characters). Ideally, one would compare studies that include approximately the same number of taxa; however, we found it was difficult enough to locate a reasonable number of studies of any size with the right kinds of characters.

These results imply that there is little difference among the character classes in levels of homoplasy. RI comparisons across data sets do suggest a tendency for pubescence (and perhaps leaf characters) to show more homoplasy than flower characters, as predicted. However, on the basis of the present analysis we cannot conclude that this is a significant difference. Our intention is to extend this analysis to more data sets to see if a clearer pattern emerges as we increase the sample size. In the meantime, it is instructive to consider possible reasons why we do not see stronger support for the predicted pattern. One possibility, of course, is that there is something fundamentally wrong with the prediction. However, before abandoning the hypothesized connection between complexity and homoplasy it will be necessary to address several limitations of the present analysis.

Perhaps the least reliable assumption of our analysis is that the character categories we established at the outset really do correspond, on average, to differences in complexity. It might be the case, as shown in Fig. 2, that within each

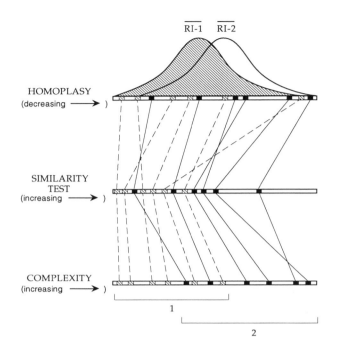

Fig. 2. Factors that may contribute to a failure to observe significant differences in homoplasy between character types. All symbols are described in Fig. 1. In contrast to the expectation shown in Fig. 1, a greater range of variation in complexity exists within each set of characters (groups 1 and 2), such that their ranges broadly overlap. Also in contrast to Fig. 1, the more complex characters are not subjected to as strong a similarity test as they could be (i.e., the black rectangles are generally shifted to the left on the axis representing the strength of similarity test). The net effect of these differences is that homoplasy levels in the two sets of characters become more broadly overlapping, and the mean RIs more similar. Under these circumstances greater sample sizes would be needed to detect a significant difference.

of these categories there is such a range of variation in complexity that the expected pattern between classes is obscured. That is, the classes we delimited, which we hoped would be distinct, are really overlapping. Our results might simply be an argument against the standard intuition about levels of homoplasy in such classes — flower characters, taken collectively, show no less homoplasy than leaf or pubescence characters.

Another possibility, shown in Fig. 2, is that there are average difference in complexity among the classes we have defined, but that these real differences are not well reflected in the actual scoring of characters for phylogenetic analysis. First, the process of prescreening characters in assembling a data matrix may tend to eliminate the very simplest morphological characters, and this might have the effect of blurring differences between character classes. Second, just because a character is potentially complex, and therefore potentially subject to stringent similarity tests, does not guarantee that such tests were actually performed. In other words, systematists may not be taking full advantage of the possibility of more detailed similarity comparisons. Instead, they may be performing about the same, relatively low, level of similarity testing regardless of the underlying complexity of different characters. In Fig. 2 this is illustrated as a general shift toward less stringent similarity tests for the more complex characters than shown in Fig. 1.

We suspect that if morphological structures were compared in greater detail (for example, by taking into account development) clearer differences among classes of characters might then emerge. A good example concerns sympetaly, or the union of petals into a floral tube. It has often been assumed that the vast majority of sympetalous dicots are related, and these are commonly united as Asteridae (e.g., Cronquist, 1988). However, phylogenies based on the chloroplast gene rbcL imply that there are several separate clades of "Asteridae" and that sympetaly must therefore be homoplastic (Donoghue et al., 1992; Olmstead et al., 1992). Recent developmental studies by Erbar (1991) have shown that there are at least two distinct developmental systems underlying sympetaly, and these appear to delimit groups that are congruent with clades, based on molecular evidence. The point is that a more detailed analysis of "sympetaly" at the outset, taking advantage of the potential to perform a more stringent similarity test using onto-genetic information, would have led to the recognition of distinct states and, consequently, to a reduction in homo-plasy. The same may also be true of such characters as "roots" (Bierhorst, 1971; Groff and Kaplan, 1988).

The two factors just discussed could be working together to blur the expected relationship between com-plexity and homoplasy. The net result is that the two RI distributions are largely overlapping and the means are much closer together than they are in Fig. 1. It may be pos-sible to establish a significant difference between classes even under these unfavorable circumstances, but it is clear that this will require much larger sample sizes.

More general problems with phylogenetic inference methods could also have influenced the outcome of our anal-ysis. Basing conclusions about patterns of character evolu-tion on trees derived in part from those same characters runs a risk of circularity. This risk may be low when a small subset of characters is of interest, as these might bias the outcome only slightly, but some of our contrasts (e.g., studies 1 and 7 in Table I) involved a sizable portion of the characters used to reconstruct the phylogeny. Circularity is only an issue, however, when the data suggest an erroneous phylogeny. If the phylogeny is correctly reconstructed it does not matter that the same characters are being sub-jected to further analysis, and the best estimate of phy-logeny may be obtained when all of the relevant data are considered (Donoghue and Sanderson, 1992). In principle, tests of robustness can be used to eliminate results based on weak or highly conflicting data. The only troublesome cases would then be results that are robust but false, which are probably obtained only under a rather limited set of condi-tions (e.g., Felsenstein, 1978). One such set of conditions is when an entire suite of functionally correlated characters has arisen several times, but, by sheer weight of numbers, these characters lead to an erroneous reconstruction sug-gesting a single origin. If that suite happens to correspond to one of the character classes involved in a contrast, then that class would have a misleadingly high retention index. However, this problem may be self-limiting inasmuch as it is unlikely that all the characters in an increasingly large sub-set of the data will be tightly correlated.

VIII. GENERAL DISCUSSION

The outlook we have developed can be viewed as an extension of Patterson's (1988) argument that the various homology tests are more or less powerful depending on the data under consideration. However, whereas Patterson saw a fundamental distinction between morphological characters (where congruence is primary) and molecular data (where similarity is decisive), we see the power of similarity tests varying as a function of the complexity of characters, of whatever type.

It is important to understand exactly where the difference arises between our outlook and Patterson's. It has nothing to do with the definition of homology, since we certainly agree that we want to identify characters that truly mark monophyletic groups. Nor is there disagreement over the types of homology tests (similarity, congruence, etc.). The difference is a more subtle one, which revolves around how the tests themselves are viewed as bearing on the evaluation of homology.

Patterson adopts the view that one or the other test will be the ultimate arbiter, depending on the type of data. He sees similarity tests as being so weak in the case of morphology that virtually the entire decision rests on congruence. In molecular data he sees similarity as such a strong test that congruence is superfluous. In any case, an hypothesis of homology is accepted or rejected on the basis of whether the decisive test is passed or failed.

In contrast, we see the power of similarity and congruence tests as varying more or less continuously, depending on the circumstances. Similarity tests are more likely to result in recognition of truly homologous structures when many points of correspondence are available for comparison. However, we would not conclude on the basis of similarity alone that structures are homologous (even in the case of molecular sequences). A congruence test is needed in all cases to evaluate whether a character evolved only once or more than once (de Pinna, 1991). However, congruence tests also vary more or less continuously in power, depending on the character analysis preceding the analysis and on the strength of support for relationships (Mindell, 1991). In

some cases a phylogenetic conclusion will be strongly supported and in other cases only weakly so. Confidence in
such conclusions varies.

Our outlook certainly does not entail the acceptance of
partial homology (e.g., Sattler, 1984; see Donoghue, 1992,
and Chapter 13 in this volume). Homology itself is generally
(and appropriately, we think) viewed as an all-or-none
proposition, in the sense that a character either is or is not
homologous in a particular group. Our discussion concerns
the bearing of homology tests on hypotheses of homology,
and it is the outcome of these tests that we see as varying
quantitatively, as opposed to qualitatively (i.e., accept or
reject; see Mindell, 1991). The strength of a particular
homology hypothesis is a complex function of the stringency
of the various tests that have been applied. On one end of
the spectrum are hypotheses supported (or rejected) on the
basis of limited similarity tests and a weak phylogenetic
hypothesis. An example would be a shared nucleotide at a
site (character) identified on the basis of dubious sequence
alignment, which also happens to be only weakly supported
as a synapomorphy by other characters in the data set. Near
the other end of the continuum are hypotheses based both
on stringent similarity tests and a strong phylogenetic
hypothesis; for example, a complex morphological character
whose status as a synapomorphy is supported by many other
characters. Most cases, of course, lie somewhere between
these extremes.

The significance of this difference in outlook is that the
view we have developed leads naturally to the sort of prediction illustrated in Fig. 1, namely that congruence testing
should uncover different levels of homoplasy depending on
differences in complexity and the strength of similarity
tests. As we have seen, phylogenetic tests of this proposition
are problematical. The main stumbling block is that we have
not found a way to readily evaluate complexity, and instead
have had to resort to using proxies based on standard intuition (molecules versus morphology, plants versus animals,
and now trichomes versus flowers). As a consequence, our
failure to find a clear difference among character classes
might only indicate that those intuitions are misguided,
rather than something more fundamental about evolution. In
view of the compromises involved, and the variety of confounding factors discussed above, we are led to the general

conclusion that tree-based studies aimed at discovering general patterns of character evolution will often require large sample sizes. Such studies will be greatly facilitated by the establishment of a database of phylogenetic evidence and trees (Sanderson *et al*, 1993).

The view we have developed helps put botanical skepticism about homology in perspective and leads to some concrete suggestions for improving the situation. Concern that the relative simplicity of plants renders similarity less reliable seems well founded in theory, but there is not a strong indication that botanists are making more mistakes about homology than zoologists. On the other hand, we suspect that botanists could do even better by taking advantage of opportunities to add dimensions to their similarity comparisons. One way to do this is to add developmental information, as in the sympetaly example (for other possibilities see Tucker, 1988; Endress, 1990; Hufford, 1990). Another very real possibility is to add information on the molecular basis of characters. This has become feasible with recent advances in understanding the genetics of flower development in *Arabidopisis* and *Antirrhinum*, and the availability of probes for some of the relevant genes (Coen and Meyerowitz, 1991). Another problem that could be addressed in this way is the homology of self-incompatibility systems in angiosperms (Weller *et al.*, 1994). Using molecular information (see Haring *et al.*, 1990) on S alleles in *Brassica* (with a sporophytic system) and *Nicotiana* (with a gametophytic system) to assess the location and sequence of S alleles in other taxa would provide the opportunity for a much finer resolution than is possible based only on scoring the presence or absence of self-compatibility.

Finally, we note that our analysis hinges on the assumption that some characters are truly more complex than others. Although this may seem reasonable, it is difficult to establish because we lack a clear concept of "character" (Fristrup, 1992) and we lack appropriate methods for quantifying complexity (McShea, 1991). These basic issues bring us directly to the work of Riedl (1978), Roth (1988, 1991), Wagner (1989a,b), (and see Chapters 8 and 9 in this volume) and others, who have tried to identify factors that promote the individuation of characters and the developmental phenomena that may maintain characters of differing complexity, such as hierarchical and cyclical ontogenetic networks. This, we think, is a critical and exceptionally promising intersection between historical and "biological" approaches to homology.

ACKNOWLEDGMENTS

We are grateful to Geeta Bharathan and to members of the Smithsonian's Laboratory of Molecular Systematics discussion group for thoughtful comments on this material, and to several contributors to the present volume for sending us their manuscripts. MJD has been supported by a grant from the NSF (BSR-8822658) and a Mellon Foundation Fellowship from the Smithsonian Institution.

REFERENCES

Anderberg, A. A., and Bremer, K. (1991). Parsimony analysis and cladistic reclassification of the *Relhania* generic group (Asteraceae-Gnaphalieae). *Ann. Mo. Bot. Gard.* **78**, 1061-1072.

Archie, J. W. (1989). A randomization test for phylogenetic information in systematic data. *Syst. Zool.* **38**, 239-252.

Arthur, W. (1988). "A Theory of the Evolution of Development." Wiley (Interscience), New York.

Axelius, B. (1990). The genus *Xanthophytum* (Rubiaceae). Taxonomy, phylogeny and biogeography. *Blumea* **34**, 425-497.

Bierhorst, D. W. (1971). "Morphology of Vascular Plants." Macmillan, New York.

Bonner, J. T. (1988). "The Evolution of Complexity." Princeton Univ. Press, Princeton, NJ.

Boufford, D. E., Crisci, J. V., Tobe, H., and Hoch, P. C. (1990). A cladistic analysis of *Circaea* (Onagraceae). *Cladistics* **6**, 171-182.

Coen, E. S., and Meyerowitz, E. M. (1991). The war of the whorles: genetic interactions controlling flower development. *Nature (London)* **353**, 31-37.

Cox, P. B., and Urbatsch, L. E. (1990). A phylogenetic analysis of the coneflower genera (Asteraceae: Heliantheae). *Syst. Bot.* **15**, 394-402.

Cronquist, A. (1988). "The Evolution and Classification of Flowering Plants." New York Botanical Garden, Bronx.

Cruden, R. W. (1991). A revision of *Isidrogalvia* (Liliaceae): Recognition for Ruiz and Pavon's genus. *Syst. Bot.* **16**, 270-282

Davis, P. H., and Heywood, V. H. (1973). "Principles of Angiosperm Taxonomy." Robert E. Krieger Co., Huntington, NY.

de Pinna, M. C. C. (1991). Concepts and tests of homology in the cladistic paradigm. *Cladistics* **7**, 367-394.

de Queiroz, A., and Wimberger, P. H. (1993). The usefulness of behavior for phylogeny estimation: Levels of homoplasy in behavioral and morphological characters. *Evolution (Lawrence, Kansas)* **47**, 46-60.

Donoghue, M. J. (1992). Homology. *In* "Keywords in Evolutionary Biology" (E. F. Keller and E. A. Lloyd, eds.), pp. 170-179. Harvard Univ. Press, Cambridge, MA.

Donoghue, M. J., and Sanderson, M. J. (1992). The suitability of molecular and morphological evidence in reconstructing plant phylogeny. *In* "Molecular Systematics in Plants" (P. S. Soltis, D. E. Soltis & J. J. Doyle, eds.), pp. 340-368. Chapman & Hall, New York and London.

Donoghue, M. J., and Scheiner, S. M. (1992). The evolution of endosperm: a phylogenetic account. *In* "Ecology and Evolution of Plant Reproduction: New Approaches" (R. Wyatt, ed.), pp. 356-389. Chapman and Hall, New York.

Donoghue, M. J., Doyle, J. A., Gauthier, J., Kluge, A., and Rowe, T. (1989). The importance of fossils in phylogeny reconstruction. *Annu. Rev. Ecol. Syst.* **20**, 431-460.

Donoghue, M. J., Olmstead, R. G., Smith, J. F., and Palmer, J. D. (1992). Phylogenetic relationships of Dipsacales based on *rbc*L sequences. *Ann. Mo. Bot. Gard.* **79**, 333-345.

Doyle, J. A., and Donoghue, M. J. (1986). Seed plant phylogeny and the origin of angiosperms: an experimental cladistic approach. *Bot. Rev.* **52**, 321-431.

Endress, P. K. (1990). Patterns of floral construction in ontogeny and phylogeny. *Biol. J. Linn. Soc.* **39**, 153-175.

Erbar, C. (1991). Sympetaly — A systematic character? *Bot. Jahrb. Syst.* **112**, 417-451.

Farris, J. S. (1989). The retention index and the rescaled consistency index. *Cladistics* **5**, 417-419.

Felsenstein, J. (1978). Cases in which parsimony or compatibility methods will be positively misleading. *Syst. Zool.* **27**,401-410.

Fristrup, K. (1992). Character: current usages. *In* "Keywords in Evolutionary Biology" (E. F. Keller and E. A. Lloyd, eds.), pp. 45-51. Harvard Univ. Press, Cambridge, MA.

Gilmartin, A. J., Brown, G. K., Varadarajan, G. S., and Neighbours, M. (1989). Status of *Glomeropitcairnia* within evolutionary history of Bromeliaceae. *Syst. Bot.* **14**, 339-348.

Givnish, T. J., and Sytsma, K. J. (1992). Chloroplast DNA restriction site data yield phylogenies with less homoplasy than analyses based on morphology or DNA sequences. *Am. J. Bot.* **79**(6), 145 (abstr).

Groff, P. A., and Kaplan, D. R. (1988). The relation of root systems to shoot systems in vascular plants. *Bot. Rev.* **54**, 387-422.

Haring, V., Gray, J. E., McClure, B. A., Anderson, M. A., and Clarke, A. E. (1990). Self-incompatibility: A self-recognition system in plants. *Science* **250**, 937-941.

Haszprunar, G. (1992). The types of homology and their significance for evolutionary biology and phylogenetics. *J. Evol. Biol.* **5**, 13-24.

Hennig, W. (1966). "Phylogenetic Systematics." Univ. of Illinois Press, Urbana, IL.

Hershkovitz, M. A. (1993). Leaf morphology of *Calindrinia* and *Montiopsis* (Portulacaceae). *Ann. Mo. Bot. Gard.* **80**, 366-396.

Hufford, L. D. (1990). Androecial development and the problem of monophyly of Loasaceae. *Can. J. Bot.* **68**, 402-419.

Iltis, H. H. (1983). From teosinte to maize: The catastrophic sexual transmutation. *Science* **222**, 886-894.

Jeffrey, E. C. (1917). "The Anatomy of Woody Plants." Univ. of Chicago Press, Chicago.

Judd, W. S. (1989). Taxonomic studies in the Miconieae (Melastomataceae). III. Cladistic analysis of axillary-flowered taxa. *Ann. Mo. Bot. Gard.* **76**, 476-495.

Kaplan, D. R. (1984). The concept of homology and its central role in the elucidation of plant systematic relationships. *In* "Cladistics: Perspectives on the Reconstruction of Evolutionary History" (T. Duncan and T.

Stuessy, eds.), pp. 51-69. Columbia Univ. Press, New York.

Kaplan, D. R. (1992). The relationship of cells to organisms in plants: problem and implications of an organismal perspective. *Int. J. Plant Sci.* **153**, 528-537.

Kenrick, P., and Crane, P. R. (1991). Water-conducting cells in early fossil land plants: implications for the early evolution of Tracheophytes. *Bot. Gaz. (Chicago)* **152**, 335-356.

Klekowski, E. J. (1988). "Mutation, Developmental Selection, and Plant Evolution." Columbia Univ. Press, New York.

Kron, K. A. and Judd, W. S. (1990). Phylogenetic relationships within the Rhodoreae (Ericaceae) with specific comments on the placement of *Ledum. Syst. Bot.* **15**, 57-68.

Lavin, M. (1990). The genus *Sphinctospermum* (Leguminoseae): taxonomy and tribal relationships as inferred from a cladistic analysis of traditional data. *Syst. Bot.* **15**, 544-559.

Levin, G. A. (1986). Systematic foliar morphology of Phyllanthoideae (Euphorbiaceae). III. Cladistic analysis. *Syst. Bot.* **11**, 515-530.

Loconte, H., and Estes, J. R. (1989). Phylogenetic systematics of Berberidaceae and Ranunculales (Magnoliidae). *Syst. Bot.* **14**, 565-579.

Maddison, W. P., and Maddison, D. R. (1992). "MacClade: Interactive Analysis of Phylogeny and Character Evolution," Vers. 3.0. Sinauer Associates, Sunderland, MA.

McShea, D. W. (1991). Complexity and evolution: What everybody knows. *Biol. Philos.* **6**, 303-324.

McShea, D. W. (1992). A metric for the study of evolutionary trends in the complexity of serial structures. *Biol. J. Linn. Soc.* **45**, 39-55.

Meyen, S. V. (1988). Origin of the angiosperm gynoecium by gamoheterotopy. *Bot. J. Linn. Soc.* **97**, 171-178.

Mindell, D. P. (1991). Aligning DNA sequences: Homology and phylogenetic weighting. *In* "Phylogenetic analysis of DNA sequences" (M. M. Miyamoto and J. Cracraft, eds.), pp. 73-89. Oxford Univ. Press, New York.

Minelli, A., and Peruffo, B. (1991). Developmental pathways, homology and homonomy in metameric animals. *J. Evol. Biol.* **4**, 429-445.

Olmstead, R. G., Michaels, H. J., Scott, K. M., and Palmer, J.
    D. (1992). Monophyly of the Asteridae and identification
    of its major lineages inferred from DNA sequences of
    rbcL. Ann. Mo. Bot. Gard. **79**, 249-265.
Patterson, C. (1982). Morphological characters and homol-
    ogy. *In* "Problems of Phylogenetic Reconstruction" (K. A.
    Joysey and A. E. Friday, eds.), pp. 21-74. Academic Press,
    London.
Patterson, C. (1988). Homology in classical and molecular
    biology. *Mol. Biol. Evol.* **5**, 603-625.
Remane, A. (1952). "Die Grundlagen des naturlichen Sys-
    tems der vergleichenden Anatomie und der Phylo-
    genetik." Geest & Portig, Leipzig.
Riedl, R. (1978). "Order in Living Organisms." Wiley, New
    York.
Rieppel, O. (1988). "Fundamentals of Comparative Biology."
    Birkhäuser Verlag, Basel.
Rieppel, O. (1992). Homology and logical fallacy. *J. Evol.
    Biol.* **5**, 701-715.
Roth, V. L. (1988). The biological basis of homology. *In*
    "Ontogeny and Systematics" (C. J. Humphries, ed.), pp. 1-
    26. Columbia Univ. Press, New York.
Roth, V. L. (1991). Homology and hierarchies: problems
    solved and unresolved. *J. Evol. Biol.* **4**, 167-194.
Sanderson, M. J. (1993). Reversibility in evolution: a maxi-
    mum likelihood approach to character gain/loss bias in
    phylogenies. *Evolution (Lawrence, Kansas)* **47**, 236-252.
Sanderson, M. J., and Donoghue, M. J. (1989). Patterns of
    variation in levels of homoplasy. *Evolution (Lawrence,
    Kansas)* **43**, 1781-1795.
Sanderson, M. J., Baldwin, B. G., Bharathan, G., Ferguson, D.,
    Porter J. M., von Dohlen, C., Wojciechowski, M. F., and
    Donoghue, M. J. (1993). The rate of growth of phylo-
    genetic information, and the need for a phylogenetic
    database. *Syst. Biol.* (in press).
Sattler, R. (1984). Homology — a continuing challenge. *Syst.
    Bot.* **9**, 382-394.
Sattler, R. (1988). Homeosis in plants. *Am. J. Bot.* **75**, 1606-
    1617.
Schopf, T. J. M., Raup, D. M., Gould, S. J., and Simberloff, D.
    S. (1975). Genomic versus morphological rates of evolu-
    tion: Influence of morphologic complexity. *Paleobiology*
    1, 63-70.

Stace, C. A. (1989). "Plant Taxonomy and Biosystematics, 2nd Edition." Edward Arnold, London.

Stebbins, G. L. (1974). "Flowering Plants: Evolution Above the Species Level." Harvard Univ. Press, Cambridge, MA.

Stevens, P. F. (1984). Homology and phylogeny: morphology and systematics. *Syst. Bot.* **9**, 395-409.

Swofford, D. L. (1991). "PAUP: Phylogenetic Analysis Using Parsimony," Vers. 3.0s. Ill. Nat. Hist. Surv., Champaign, IL.

Tomlinson, P. B. (1984). Homology: An empirical view. *Syst. Bot.* **9**, 374-381.

Tucker, S. C. (1988). Loss versus suppression of floral organs. *In* "Aspects of Floral Development" (P. Leins, S. C. Tucker, and P. Endress, eds.), pp. 69-82. Borntraeger, Berlin/Stuttgart.

Wagner, G. P. (1989a). The origin of morphological characters and the biological basis of homology. *Evolution (Lawrence, Kansas)* **43**, 1157-1171.

Wagner, G. P. (1989b). The biological homology concept. *Annu. Rev. Ecol. Syst.* **20**, 51-69.

Weller, S. G., Donoghue, M. J., and Charlesworth, D. (1994). The evolution of self-incompatibility in flowering plants: a phylogenetic approach. *In* "Experimental and Molecular Approaches to Plant Biosystematics" (P. C. Hoch, ed.). Missouri Botanical Gardens, St. Louis. (in press)

Wiley, E. O. (1981). "Phylogenetics. The theory and Practice of Phylogenetic Systematics." Wiley, New York.

Wimsatt, W. C., and Schank, J. C. (1988). Two constraints on the evolution of the complex adaptations and the means of their avoidance. In "Evolutionary Progress" (M. H. Nitecki, ed.), pp. 231-273. Univ. of Chicago Press, Chicago.

13

HOMOLOGY, HOMEOSIS, AND

PROCESS MORPHOLOGY IN

PLANTS

Rolf Sattler

Department of Biology
McGill University
Montreal, Quebec
Canada H3A 1B1

Homology: The Hierarchical Basis of Comparative Biology
Copyright © 1994 by Academic Press, Inc.
All rights of reproduction in any form reserved.

I. INTRODUCTION

The term homology refers to a great diversity of concepts (see, e.g., Donoghue, 1992), which have been grouped into classes. Wiley (1975) suggested the following four classes: classical, phenetic, evolutionary, and cladistic homology. Patterson (1982) added utilitarian homology as a fifth class. Wagner (1989b) distinguished between two types of homology (homology within and between organisms) and three concepts of homology: idealistic, historical and biological homology. These concepts differ in their explanation of homology, whereas the two types of homology contrast comparisons within one organism and among different organisms.

Other authors have suggested yet other distinctions within the broad field of homology (e.g., van der Klaauw, 1966; Sattler, 1984), some having emphasized the close relationship or overlap between different types, concepts, or classes of homology. For example, Van Valen (1982) pointed out the intergradation between homologies within and among organisms so that "historical homology cannot be separated sharply from repetitive [or iterative] homology" (Van Valen, 1982, p. 305; see also Minelli and Peruffo, 1991). Thus, despite differences between various versions of homology, there appears to be a close relationship between them (e.g., Roth, 1988, 1991). This relationship becomes most obvious when we ask how we establish homology. The first step in this endeavour appears to be similar; whether we look for historical or ahistorical homologies within or among organisms, we try to find correspondences or relations. Not all of them may eventually qualify as homologies, but nonetheless our first step in the establishment of homologies is the search for correspondences or relations (e.g., de Pinna, 1991, p. 373; see also Chapters 7 and 8). Correspondences are usually perceived as 1:1, hence, for the majority of homology concepts only total correspondence (sameness with regard to selected features) is admitted. Everything else is either ignored or arbitrarily forced into the mold of a 1:1 correspondence. This conceptual filter insisting on 1:1 correspondence impoverishes our perception of biological diversity from the start. If we want to obtain a richer representation of the comparison of organic structures, we will have to include *partial corre-*

spondence, which is the basis of *partial homology* (e.g., van der Klaauw, 1966; Sattler, 1966, 1984; Meyen, 1973, 1987; Roth, 1984; Wagner 1989b). Partial correspondence or partial homology represents the degree of similarity with regard to selected structural parameters (see Sections IV and V). These concepts apply to all organizational levels, including the molecular; see Chapter 10.

At least methodologically, the first step in the establishment of (potential) homologies cannot be phylogenetic relationship (see Chapter 8). Therefore, instead of referring to relationships that might be understood genealogically, it might be better to simply refer to the establishment of relations. An example of such relations can be seen in the following linear series of three parts, each of which is characterized by two properties indicated by letters: **AB, BC, CD** (Table I; see also Sattler 1984, p. 384). Obviously, **BC** is related to both **AC** and **CD** because it shares one property with each of them. Such a relation also can be seen as a partial correspondence since **BC** corresponds partially to both **AB** and **CD** (see, e.g., Sattler, 1984, p. 384, 1992, p. 709; Brady, 1987).

Correspondences or relations usually refer to structures, i.e., structures correspond or are related to each other. Both concepts, however, also may be applied to nonstructural correspondences and relations. For example, functions or behaviors may be related or correspond to each other, as discussed in Chapters 4 and 12. If structure itself is seen as function (in the sense of process), then morphological correspondences and relations become dynamic, i.e., the dichotomy between structure and function (in the sense of process) is superseded (Sattler, 1990, 1992; and see Section V).

Cladistic homology (or *synapomorphy*) is also grounded in correspondence (e.g., Sattler, 1984; Stevens, 1984; Crisp and Weston, 1987; Weston, 1988; Bateman et al.; and see Chapters 3 and 8 in this volume). Since it is based on 1:1 correspondence (even of rather dissimilar structures that are thought, but not actually, transformed into each other), it is more limited than phenetic homology, which may include a wide range of partial correspondences (Sneath and Sokal, 1973). In another sense, cladistic homology goes beyond phenetic homology because after the establishment of 1:1 correspondence additional steps are taken that lead to cladistic assertions (e.g., Patterson 1982; Brady, 1987; Rieppel, 1992).

Although I refer to cladistic homology and other homology concepts, the emphasis in this chapter is on the common basis of *all* versions of homology, i.e., correspondence or relation. To avoid bias toward particular notions of homology, I often refer to "*correspondence* or (potential) homology". This formulation comprises both ahistorical and historical homology concepts. For ahistorical concepts, correspondence entails homology; for historical concepts correspondence is only potential homology.

Table I.   Example of a morphocline[a]

Morphocline:

AB, BC, CD[b]

Explanation of Abbreviations:

A = radial symmetry
C = dorsiventral symmetry
B = branching
D = lack of branching

Examples of Structures:

AB = axillary branch of radial symmetry
BC = fertile phylloclade of *Semele androgyna*, which is dorsiventral, but branches
CD = phylloclade of *Danae racemosa*, which is dorsiventral and unbranched

---

[a] If the whole complexity of the above examples is taken into consideration, each letter would represent a whole set of properties and/or additional letters would have to be added (for details see Cooney-Sovetts and Sattler, 1987).

[b] Although symbols are logically more correct, letters provide greater clarity, ease of reading, and uniformity with conventions used in other chapters.

I begin the discussion of correspondence or (potential) homology with cases that allow the establishment of 1:1 correspondence. Then I proceed to increasingly complex situations and present ways of dealing with them. *Homeosis* and the notion of *morphological distance* will be illustrated in this context. Finally, I discuss *process morphology*, a more dynamic approach to the comparative morphology of plants. Since in this approach structures are seen as process combinations, process morphology deals with relations of process combinations, implying a dynamic multirelational view of plant diversity and evolution.

II. HOMOLOGY AND CORRESPONDENCE

A. *Criteria of Homology*

Regardless of how homology is defined, we begin our enquiry into homology by establishing correspondence or relation. This may be done solely through intuition, or, more commonly, by the application of specific criteria. These criteria are referred to as homology criteria (Remane, 1952; Wiley, 1981, p. 130; Sattler, 1984). According to ahistorical homology concepts, they establish homology; according to historical homology concepts, they provide only potential candidates of homology (e.g., Stevens, 1984; Chapter 3 in Rieppel, 1988; de Pinna, 1991; Donoghue, 1992, p. 176). There is no agreement on which criteria and how many criteria should be used. However, many botanists and zoologists have placed great emphasis on the criterion of relative position. Some biologists have insisted that only this criterion should be used to establish (potential) homology so that two structures are homologous (or potentially homologous) only if they occupy the same relative position within a more inclusive whole (Froebe, 1982; Rieppel 1988, p. 44, and see Chapter 2 in this volume). I show below by means of examples that in practice correspondence or relation often is established by both relative position and special quality as a second criterion, or by special quality alone (Remane, 1952; Eckardt, 1964; Sattler and Rutishauser, 1992, p. 71). The criterion of special quality is rather encompassing: it may refer to any feature or process that homologues have in common and distinguish them from other nonhomologous structures.

A third criterion often invoked is that of intermediate forms (see Remane, 1952, 1956; Sattler, 1984, p. 383). According to this criterion, two different structures are (potentially) homologous if there are intermediate forms that link them. It should be noted that, in contrast to the other two criteria, this criterion does not necessarily establish sameness or similarity with regard to a set of properties, but rather indicates a continuum or an injunction (Hassenstein, 1954, 1971; Sattler 1986, p. 85). In extreme cases homologues established through this criterion may have no defining properties in common (Sattler 1984, p. 384). For example, in the cline referred to above (namely **AB**, **BC**, **CD**) **AB** and **CD** share no common defining properties, but are considered (potentially) homologous on the basis of the third criterion because the intermediate form **BC** provides a linkage (see also Brady, 1987).

All three homology criteria mentioned so far, which were clearly distinguished by Remane (1952, 1956) and which are adhered to by many other biologists (e.g., Wiley, 1981, p. 130), may be applied to structures of mature organisms, to any developmental stage, or to the whole continuum that comprises all developmental stages, including maturity. Since an organism can be seen as a four-dimensional space time-extension (Woodger, 1929; Tomlinson, 1984, p. 380), the mature stage, like any other developmental stage, is only one time slice in this four-dimensional continuum. From this point of view, it is arbitrary to separate the mature stage from preceding development. For establishment of morphological correspondence or relation it therefore appears desirable to apply the three criteria not only to the mature stage or to any other time slice, but to the whole space-time extension of corresponding or related structures. This may lead, however, to difficulties; relative position and special quality may change during the development of certain structures (e.g., Rieppel, 1988, p. 45). Consequently, the application of these criteria may yield different correspondences or relations in different developmental stages. Thus, conflicts may arise (see, e.g., Sattler, 1984; Roth, 1988; and Chapter 10 in Hall, 1992).

B. *Conflict between Homology Criteria*

Conflicts are not restricted to discrepancies between different developmental stages, including maturity. Conflicts also occur between applications of the three criteria mentioned above. For example, the criterion of relative position may lead to a different, even contradictory, correspondence from the criterion of special quality (e.g., Sattler, 1984, p. 383). Even within the topographical criterion different correspondences can be obtained depending on which frame of reference is chosen for the analysis of positional relations (Rieppel, 1988, pp. 45-49). The same applies to the criterion of special quality: different correspondences may be obtained depending on which particular feature is selected. As a result, a considerable divergence of more or less contradictory correspondences or relations may arise at the very first step of homologization that forms the basis for all homology concepts.

C. *Resolution of the Conflict*

How can we resolve this conflict between criteria? We could decide in an *ad hoc* fashion which criterion should be considered the decisive one in each particular case. Thus, in conflict situations between the criteria of relative position and special quality we could decide in one case that position is more important and in another that special quality is decisive. Such *ad hoc* decisions appear to be arbitrary and inconsistent. A more uniform and consistent solution is to recognize the breakdown of a general concept of 1:1 correspondence that is based on all the above criteria. Each criterion then defines its own kind of correspondence: the position criterion defines *homotopy* (the same or similar relative position of structures); the criterion of special quality defines *homomorphy* (the same or similar special quality of structures; see Sattler, 1992). In cases of discrepancies between different features or processes of the special quality, homomorphy would have to be further broken down into different kinds of homomorphy. Roth (1988) recognized this. She wrote that we:

avoid some of the ambiguity and confusion biologists have encountered with the concept of homology in the past, if we cease to insist on homologizing entire structures (Roth 1984), and instead decompose structures into their individual characteristics (properties, individual features, or aspects of development). (Roth, 1988, p. 17)

Although this methodology is helpful in resolving conflicts between homology criteria, it may not be palatable to those who want to compare and homologize entire structures. I therefore want to suggest an alternative way of conflict resolution by means of a recognition of *partial correspondence* or *partial homology* (van der Klaauw, 1966; Sattler 1966, 1984; Meyen 1973, 1987; Roth, 1984; Wagner 1989b, pp. 55 and 67; Sattler and Jeune, 1992). In a partial correspondence or partial homology, a structure does not share all relevant properties with another one. Thus, instead of insisting on 1:1 correspondence, which is characteristic of a categorical world view where everything has to be pigeonholed, the concept of partial correspondence or partial homology allows for a great variety of relations: it reflects a multirelational world view. This may be illustrated by the above-mentioned theoretical example of clinal variation, namely **AB**, **BC**, **CD**. The structure that exhibits the properties **B** and **C**, partially corresponds to the structures **AB** and **CD** because it shares properties with both of them. According to the methodology proposed by Roth (1988), **BC** would correspond to **AB** with regard to **B** (let's say, its position) and to **CD** with regard to a special quality (let's say, symmetry). Thus, **BC** would be homotopous with **AB**, but homomorphous in a specific sense (e.g., sharing its symmetry) with **CD**. It is obvious that as the number of uncorrelated properties increases, the analysis becomes increasingly complex (see Sections IV and V below).

So far the two methodologies of conflict resolution between contradictory homology criteria appear to be complementary to each other (for the notion of complementarity see Rutishauser and Sattler, 1985). This means that we can compare certain structures such as **AB** and **BC** *both* in terms of 1:1 correspondence (when we deal with properties) *and* in terms of partial correspondence (when we compare entire structures). A closer analysis of properties shows, however, that partial correspondence also may apply to them. This can be illustrated by means of symmetry.

We may distinguish the attributes radial and bilateral symmetry. Although theoretically one can always distinguish between radial and bilateral symmetry, in practice this need not be so. As bilateral symmetry approaches radial symmetry (or *vice versa*), it may no longer be possible to make a clearcut distinction between the two; there are practical limits of exact measurement that prevent us from distinguishing between the two kinds of symmetry (see also Hay and Mabberley, 1993). In addition to this measurement problem, is a more fundamental issue: radial and bilateral symmetry can be seen as a continuum (Fig. 1). As the length of x approaches the length of y, bilateral symmetry, characterized by a difference between y and x, gradually turns into radial symmetry in which y equals x (Fig. 1). Hence, from a quantitative point of view, the qualitative distinction between radial and bilateral symmetry is a simplification. This simplification works in extreme cases such as Fig. 1a and c, but it cannot adequately represent cases such as Fig. 1b. Figure 1b is not exactly radial because of a slight difference between y and x. If, however, it is classified as bilateral, it is lumped together with Fig. 1c, although the quantitative difference between Fig. 1b and c is much greater than that between Fig. 1a and b. Obviously, correspondence between the symmetries of Fig. 1a, b, and c is a matter of degree. Therefore, decomposing structures into properties such as symmetry attributes does not necessarily eliminate the need of correspondence or (potential) homology as a matter of degree, as suggested by Roth (1988, p. 22, footnote 8). The same applies to other properties such as weight, length, growth rate, growth duration, or branching angle.

I therefore conclude that we cannot eliminate partial correspondence with regard to all properties. As a consequence, correspondence has to be understood in a broad sense so as to include partial correspondence, which implies a multirelational world view in contrast to the categorical world view that insists on 1:1 correspondence as a basis for all (potential) homologies.

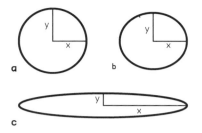

Fig. 1. Diagrammatic cross-sections of radial (a) and bilateral (b and c) structures. In the radial structure (a), $x = y$, whereas in the bilateral structures (b and c), $x > y$. In (b) the difference between x and y is minimal, hence the formally bilateral structure approaches a radial one.

D. *Examples of Correspondence or (Potential) Homology in Seed Plants*

There are different ways of decomposing seed plants into structural units. Each way constitutes a model of plant construction (see Rutishauser and Sattler, 1985, p. 422).

According to the classical model, seed plants are organized as a cormus that consists of three fundamental types of organs in characteristic relative positions (Fig. 2): roots (that are occasionally absent), caulomes, and phyllomes (e.g., Troll 1954, pp. 3-5; Hagemann, 1976, p. 253). A caulome is a vegetative stem (Fig. 2:1) or the axis of an inflorescence or flower. Phyllomes are lateral appendages such as cotyledons (seed leaves) (Fig. 2:8), various types of foliage leaves (Fig. 2:7), sepals (Fig. 2:18), petals (Fig. 2:19), stamens (Fig. 2:13), and carpels (Fig. 2:16). From this perspective, a flower is a monaxial shoot homologue because, like a shoot, it consists of a caulome with one or usually more than one phyllome (Fig. 2:11). By definition, branching does not occur in a flower. In the vegetative region branches (shoots) may be borne in the axils of phyllomes (Fig. 2:6 and 2:9). This is a positional relation to which great importance has been attached for the homologization of phyllomes and caulomes, especially in so-called difficult cases of seed plants [see also Hagemann's (1984, p. 210, 1989, p. 245) extended notion of phyllome-conjunct branching].

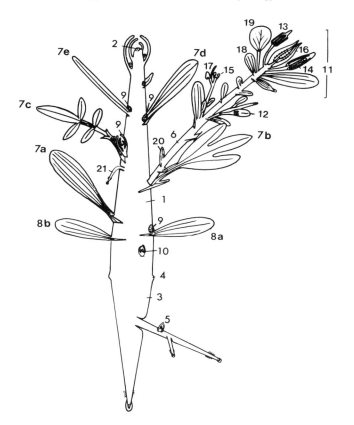

Fig. 2. A fictitious seed plant illustrating many structural relations found in different taxa. Classical terminology is used, indicating homologies of classical plant morphology. 1, Stem (caulome); 2, shoot apex; 3, root; 4, crown; 5, root bud; 6, branch of main stem; 7, foliage leaves (phyllomes): 7a, simple leaf without axillary bud; 7b, lobed leaf with axillary branch; 7c, pinnately compound leaf with axillary bud; 7d, simple leaf with axillary bud (9); 7e, narrow leaf with axillary bud (9); 8, cotyledons (seed leaves): 8a, cotyledon with axillary bud (9); 8b, cotyledon without axillary bud; 9, axillary bud; 10, hypocotyledonary bud, not subtended by a leaf; 11, flower; 12 and 13, stamen; 14, pollen sac; 15, ovule (in a gymnosperm); 16, carpel; 17, ovule bearing appendage (in a gymnosperm); 18, sepal; 19, petal; 20, prophyll; 21, adventitious root. [After Guédès (1979, Fig. 1).]

I now show by some examples from seed plants how the homology criteria discussed in Section II,A can be used to establish correspondence or (potential) homology of structures. I begin with a case where all criteria are in agreement and then proceed to more complicated cases where conflicts between criteria occur. Finally, I illustrate possible resolutions of the conflicts and conclude that partial correspondence or partial homology (i.e., homology as a matter of degree) plays a rather important role.

I begin with a 1:1 correspondence of mature organs illustrated in Fig. 2: the correspondence between the mature cotyledon or seed leaf (Fig. 2:8a) and the mature foliage leaf (Fig. 2:7d). Both structures have the same special quality and the same relative position, i.e., they subtend an axillary bud (Fig. 2:9). Thus, this correspondence is based on an agreement of all homology criteria used for the establishment of correspondence [Remane's (1952, 1956) first two criteria; his third criterion does not apply in this case. Sneath and Sokal (1973, p. 77) termed these two criteria *structural* and *compositional*. De Pinna (1991) and others simply referred to similarity].

If we now include the origin of the seed leaf (Fig. 2:8a) and the foliage leaf (Fig. 2:7d), we notice a discrepancy: whereas the foliage leaf arises laterally at the shoot apex, the seed leaf does not because the shoot apex is formed after the inception of the seed leaf. Hence, at their origin the seed leaf and foliage leaf do not have a comparable position. This illustrates a conflict in the application of the position criterion at different developmental stages.

Probably no one would question the correspondence of two seed leaves at the same plant, provided they either both subtend an axillary bud (as in Fig. 2:8a) or both lack an axillary bud (the condition of Fig. 2:8b). If, however, we compare a seed leaf subtending an axillary bud (Fig. 2:8a) with a seed leaf of another taxon that does not subtend an axillary bud (Fig. 2:8b), then the position criterion in terms of the relation between axillary bud and its subtending phyllome cannot be applied: correspondence or (potential) homology is established only on the basis of the special quality of the structures. One might argue that the two structures nonetheless share a lateral position. The question is whether "lateral position" is sufficient to distinguish between phyllomes and caulomes or shoots. Occasionally

buds (i.e., shoots) are also formed laterally without a subtending leaf (Fig. 2:10). For example, in *Nymphaea alba* and
other water lilies, buds may occupy leaf sites in the
phyllotactic spiral (Cutter, 1966; M. Wolf, personal communication). Thus, according to their position, these buds
correspond to leaves, whereas their special quality clearly
identifies them as shoots, each consisting of a caulome and
phyllomes. Despite this conflict between criteria, everyone
recognizes these buds as shoot homologues. This shows that
at least in this case the criterion of special quality is considered more important than the position criterion. In other
cases the same observation can be made: although theoretically the importance of the position criterion is emphasized,
in practice correspondence or (potential) homology often is
established solely by the criterion of special quality (Sattler
and Rutishauser 1992, p. 71). For example, in several
species of *Utricularia* (bladderworts) such as *U. longifolia*,
the position of leaf and branch (stolon) may be inverted so
that the leaf is more distal than the phyllome-conjunct
branch (Fig. 3). If the position criterion would be enforced,
we would have to conclude that the inverted leaf is homologous with a shoot and the inverted shoot with a leaf. Nobody
seems to go to such extremes despite the theoretical
emphasis on the position criterion (e.g., Froebe, 1982).
Nonetheless, a conflict between criteria persists. We can
resolve this conflict by breaking down the general concept
of correspondence or (potential) homology into homotopy
(positional correspondence) and homomorphy (correspondence of special quality). Then the inverted leaf is homotopous with a shoot but homomorphous with a phyllome
(leaf), whereas the inverted shoot is homotopous with a leaf
but homomorphous with a shoot. To avoid a tautologous
formulation, we would have to say that the structure that
looks like a leaf is homomorphous with a leaf and the structure that looks like a shoot is homomorphous to a shoot.
Corresponding conclusions can be drawn for the phylloclades of the Asparagaceae such as those of *Danae racemosa*
(Cooney-Sovetts and Sattler, 1987; Sattler, 1984, pp. 383-
384): since they have the special quality of leaves, they are
homomorphous with leaves, but since they occupy the position of axillary branches, they are homotopous with shoots.
More complex situations arise when phylloclades such as
those of *Semele androgyna* share leaf and shoot features so

that with regard to their special quality they correspond partially to both a shoot and a leaf (Cooney-Sovetts and Sattler, 1987).

Another correspondence illustrated in Fig. 2 is that between a simple leaf (Fig. 2:7d) and a pinnately compound leaf (Fig. 2:7c). In classical plant morphology, this correspondence is considered a 1:1 correspondence. However, according to recent developmental studies, it is only a partial correspondence because the early developmental stages of the pinnate leaves investigated resemble a distichous shoot and the later stages become more leaflike (e.g., Jeune, 1984; Cusset 1991, p. 1817; Sattler and Rutishauser,

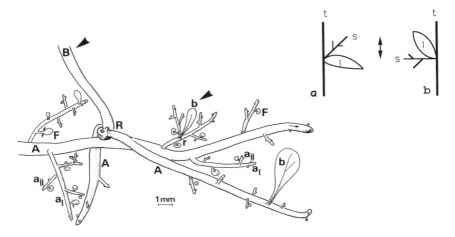

Fig. 3. <u>Left</u>: Drawing of a portion of the shoot of *Utricularia longifolia*. A, a$_1$, and a$_2$, stolons; B, basal portion of leaf; b, leaf; F, bladder. Arrow points to shoot tip. Note that two leaves (B and b, marked with an arrowhead) have an inverted position, i.e., they are more distal than the associated axillary stolon (branch) which is in opposition to the rules of axillary branching [see Fig. 2, where the axillary buds (9) or branch (6) are more distal than their subtending leaves].[After Brugger and Rutishauser (1989, Fig. 17, p. 104).] <u>Right</u>: Diagrammatic representation of the leaf-branch (stolon) inversion in *Utricularia longifolia* (b) as compared to the typical topography in seed plants (a); s, branch (stolon); l, leaf; t, tip of main stem.

1992). Thus, in their whole space-time extension, these leaves are partially homomorphous to both a shoot and a leaf. Since they subtend an axillary bud or branch, they are homotopous to a leaf. Homotopy is one of the invariant properties in this case; see Chapter 6.

A similar conclusion can be reached in a comparison of the simple leaf (Fig. 2:7d) and structure 7e in the same figure. Since both subtend an axillary bud, they are homotopous. However, if we assume (contrary to Guédès 1979, p. 14), that structure 7e is radial (i.e., circular in cross-section), its symmetry does not correspond to that of the simple leaf (Fig. 2:7d) but rather to that of a typical stem (Fig. 2:1) which is also circular in cross section. Such a situation has been found in the so-called phyllodes of *Acacia longipedunculata*, but with additional complexities (Rutishauser and Sattler, 1986; Sattler *et al.*, 1988). There are phyllodes with and without axillary buds. Both are radial from inception to maturity. However, the following qualifications have to be made (Sattler *et al.*, 1988). (1) There may be slight deviations from a perfectly circular outline in cross-section. (2) Even in those phyllode primordia whose outline appears circular there is a slight internal differentiation gradient from the adaxial to the abaxial side. (3) The phyllode primordia are acrovergent which is related to the differentiation gradient. All this means that with regard to their anatomy, the phyllodes are slightly dorsiventral, whereas with respect to their external form (i.e., cross-sectional outline), they are more or less radial. Because of these complexities, there is no complete 1:1 correspondence between the symmetry of the phyllodes and that of typical stems or leaves. Overall, the symmetry of the phyllodes comes rather close to that of typical stems, but not in all respects (see Sattler *et al.*, 1988, pp. 1273 and 1277).

The description and homologization of flowers has led to many controversies (e.g., Croizat, 1960, 1962; Stebbins, 1974; Rutishauser and Sattler, 1985; Meeuse 1986, 1990; Leins *et al.*, 1988; Cresens and Smets, 1989; Fries and Endress, 1990; Brückner, 1991; Leins and Erbar, 1991; Rutishauser, 1993a). One controversy concerns the question of whether a particular flower or flowers in general represent a monaxial or polyaxial system (e.g., Stebbins, 1974; Rutishauser and Sattler, 1985; Lyndon, 1978; Leins and Erbar, 1991; Ronse Decraene 1992; Ronse Decraene and

Smets, 1992). This question is related to the homologization of stamens and carpels. According to the classical
model, a stamen is thought to correspond to a phyllome.
Although there are taxa with stamens that lend themselves
to this interpretation, there are many other taxa whose
stamens show remarkable resemblance to caulomes (e.g.,
Sattler and Jeune, 1992). These stamens are initiated in the
corpus like axillary buds; they are more or less radial at
inception, and they may be subtended by a phyllome (e.g.,
Lacroix and Sattler, 1988). In addition, there are taxa with
stamen fascicles reminiscent of short shoots (e.g., Leins and
Erbar, 1991). These fascicles may or may not be subtended
by phyllomes. Hence, they may or may not be homotopous to
axillary branches. With regard to their special quality, they
show a range of partial correspondences to the special
quality of short shoots (see Sattler, 1988b; Sattler and
Jeune, 1992).

III. HOMEOSIS AND HOMOLOGY

Homeosis occurs in many plant taxa, both in the vegetative and reproductive region (Leavitt, 1909; Sattler,
1988a; Rutishauser, 1989, 1993a; Greyson, 1993). How
widespread it is, depends on whether it is narrowly or
broadly defined and on implied phylogenetic hypotheses
when genealogy has not been observed (Sattler 1988a, p.
1609). Bateson (1894. p. 84), who introduced the term into
biology and used it mainly with regard to animals, defined
homeosis as "the assumption by one member of a Meristic
series, of the form or characters proper to other members
of the series." This definition restricts homeosis to the
replacement of one part by a serially homologous part. In
contrast, Leavitt (1909), who first demonstrated the general
importance of homeosis in plants, implied a wider definition, according to which homeosis is "the assumption by one
part of likeness to another" (Holmes 1979). In other words
homeosis is the total or partial replacement of one part by
another of the same organism. This definition is broader in
two ways. (1) Homeosis is a matter of degree, i.e. one part
may be more or less replaced by another part. At one
extreme, there is a total replacement; at the other extreme,
only one feature is replaced either entirely or partially (see

Sattler, 1988a). (2) Parts or structures that replace each other may or may not be homologous. If nonhomologous structures replace each other, transgressions of homology occur as a result of homeosis (Leavitt, 1909; Sattler, 1988a). Such transgressions, that may be (nearly) total or partial, may be saltational, and are therefore of great importance for macroevolution (McCook and Bateman, 1990; Charlton, 1991; Hay and Mabberley, 1991; see also Stidd, 1987).

One might argue (in vain, I think) whether homeosis should be defined in the narrow or broad sense. Like so many other terms, it can be used in both senses. The broad definition is useful for a number of reasons (Sattler, 1988a, p. 1607). (1) If we defined homeosis only in the narrow sense, we would need yet another term for the phenomena of partial replacement and homology transgressions. (2) Whether a replacement is total or partial, is not always clear. What may appear to be a total replacement as long as only mature stages are compared, may turn out to be a partial replacement when development becomes known (Hill and Lord, 1989; Polowick and Sawhney, 1987; Smith-Huerta, 1992). (3) Finally, it is not always clear whether structures that replace each other are homologous, and homology may differ from one level to another (see also Sattler, 1988a, p. 1607). I illustrate all this by some examples.

The classical case of homeosis in plants is the replacement of a stamen by a petal. This case already was discussed by Bateson (1894) and now is investigated at the molecular level in homeotic mutants of *Arabidopsis* and other taxa (e.g., Bowman *et al.*, 1989, 1991; Meyerowitz *et al.*, 1989; Coen, 1991; Lord, 1991). It has been known in plant teratology for a long time (Penzig, 1921, 1922; Meyer, 1966). In horticulture it led to the cultivation of double flowers in which (besides the addition of petals) many stamens are replaced by petals, thus creating particularly showy flowers in roses and many other taxa (e.g., Reynolds and Tampion, 1983; Innes *et al.*, 1989; Lehmann and Sattler, 1993). Replacement of stamens by petals (or *vice versa*) is, however, not restricted to artificially induced mutants, breeding, and terata. Implying observed genealogy or phylogenetic hypotheses, there is evidence that it also occurs naturally and thus has played a role in floral diversification during evolution (Zimmermann, 1961; Tucker, 1989, 1992; McCook and Bateman, 1990; Kirchoff, 1991; Lehmann and

Sattler, 1992, 1993; Rutishauser, 1993a). In some cases, this evidence for homeosis persists even when phylogenetic hypotheses are contradictory. For example, it can be concluded that homeosis played a role in the evolution of the Papaveraceae regardless of whether a four-petalled or many-petalled ancestor of the Papaveraeae is postulated (Lehmann and Sattler, 1993).

According to the classical model of plant morphology, a stamen is serially homologous to a petal, at least at the level of phyllomes, i.e., both stamen and petal are interpreted as phyllomes. Therefore, replacement of a stamen by a petal is replacement by a serially homologous structure (as the replacement of one *Drosophila* segment by another is a serially homologous replacement). The replacement may be total or partial in artificially induced mutants, cultivars and naturally occurring variation (Rutishauser, 1989, 1993a; Bowman *et al.*, 1989; Schultz *et al.*, 1991; Lehmann and Sattler, 1993, and personal observation).

As pointed out in Section II,D, contrary to the classical interpretation of stamens, there is evidence that in many taxa stamens partially correspond to caulomes or branchlets and therefore are partially homologous to caulomes or branchlets (Rutishauser and Sattler, 1985). According to this evidence, the replacement of a stamen by a petal (or *vice versa*) amounts to a partial homology transgression, i.e., it is homeosis *sensu lato*, whereas according to the classical interpretation it is homeosis *sensu stricto* (if the replacement is total).

One could cite many examples of such partial homology transgressions in both the reproductive and vegetative region of plants (Leavitt, 1909; Meyen 1987; Sattler, 1988a; Rutishauser 1989, 1993 a,b; Rutishauser and Huber, 1991; but see also Jäger-Zürn, 1992). Besides these partial transgressions, there also seem to be total homology transgressions. In these cases one structure is (nearly) totally replaced by a nonhomologous structure (e.g., Sattler, 1988a). One example is the replacement of a shoot by a leaf in the axil of a scale leaf of *Danae racemosa* (see Cooney-Sovetts and Sattler, 1987). Another example is found in *Nasturtium* (watercress) (Champagnat and Blatteron, 1966). In this taxon roots occur in leaf axils. Although these roots are typical in their structure, they arise exogenously, which is typical of shoots, not roots. Therefore, if the origin of these homeotic roots is taken into consideration, the replacement is not complete: with regard to one feature, namely their origin, the roots still resemble the original shoots they replaced. In general, partial replacement appears to be much more frequent than total replacement. Negrutiu *et al.* (1991, p. 8) went as far as to claim that homeosis "is only exceptionally complete in plants."

I conclude that homeosis in the broad sense is fundamentally relevant to homology and the basis of plant morphology (as well as animal morphology). It undermines two widely accepted tenets of comparative morphology: (1) the importance of relative position in the establishment of correspondence or (potential) homology, and (2) the assumption that all correspondences between structures must be 1:1 correspondences.

1. Relative position. Since Owen's lectures (1843) and earlier (e.g., Geoffroy Saint-Hilaire (1830), quoted by Rieppel, 1988, p. 44; see also the discussion in Chapter 1 in this volume), there has been a widespread reliance on the criterion of relative position for the establishment of correspondence or (potential) homology (Woodger, 1945; Jardine, 1967, 1969; Froebe, 1982; Rieppel, 1988). However, when structures are replaced by totally or partially nonhomologous structures, this criterion becomes useless. Correspondence or (potential) homology is then solely determined by the criterion of special quality and relations may be elucidated by the criterion of transitional forms (see Sections I, and II,A). Thus, the axillary roots of *Nasturtium* are homologized as such because they have the special quality of roots (except that their origin is exogenous). Similarly, the inverted leaves and stolons (shoots) of *Utricularia longifolia* are homologized as such on the basis of their special quality (see Section II,D). Use of the criterion of relative position would lead to the conclusion that the leaves are homologous with shoots and the shoots with leaves.

One lesson we learn from homeosis *sensu lato* is that not only the special quality of structures, but also their relative position may change. Therefore, the same or similar relative position is not necessarily a better indication of homology than the same or a similar special quality. If during a particular evolutionary change the relative position of a structure has been retained and change has affected only its special quality, then the criterion of relative position can be used to establish relationships. If, however, the relative position of a structure has changed, then the criterion of relative position can no longer be used for homologization; in this case (potential) homology will have to be established by the criterion of special quality. Finally, if both the relative position and special quality of a structure have changed, then the relations will have to be determined by methods that will be discussed below in Sections IV and V.

2. 1:1 correspondence. The second widespread tenet of comparative morphology that is undermined by homeosis *sensu lato* is the exclusiveness of 1:1 correspondences, i.e., the idea that all correspondences of structures must be 1:1. This idea is at the root of the widespread notion that homology must be a qualitative concept, which means that a structure is either homologous with another structure or it is not. *Tertium non datur*, i.e., partial homologies are excluded. This rigid either-or philosophy finally is at the root of the widespread notion that homology and analogy are mutually exclusive; but see Panchen's discussion in Chapter 1.

Homeosis *sensu lato* that comprises partial replacements, and hence partial correspondences of structures, challenges all of these ideas. As partial correspondences of structures and properties of structures are recognized, homology may be seen as comparative (semiquantitative) or quantitative (see Sattler, 1966, 1984; Sneath and Sokal, 1973, p. 77; Minelli and Peruffo, 1991). Thus, the central concept of comparative morphology also becomes comparative, a matter of more or less (degree) rather than either-or (Minelli and Peruffo, 1991).

Nature is seen not only in terms of black and white, but also as the enormous multitude of nuances between black and white. In this continuum, it is difficult to maintain the mutual exclusiveness of homology and analogy (in the sense of similarity due to similar function) (Sneath and Sokal, 1973, pp. 86 and 87). This may be illustrated in cacti.

It is generally thought that the so-called stems in cacti are analogous to leaves. They function like leaves and all the morphological resemblances with leaves are thought to be related to this transfer of function; hence they are analogous, not homologous. Recently, Mauseth (1992) and Mauseth and Sajeva (1992), however, have questioned this common interpretation. They showed that the cortex of the so-called stems contains numerous features that correspond rather closely to features found in leaves. The simplest explanation of this finding is that leaf features have been expressed in the stem cortex whereby the so-called stem becomes a hybrid organ. Sachs (1982, p. 128) discussed other examples of *developmental hybridization* (Sattler, 1988a, p. 1613). Whether the terms "hybridization" and "hybrid" should be used in this context is debatable. Bowman *et al.* (1989), Rutishauser (1989a, p. 61) and others referred to mosaic organ as a synonym for hybrid or intermediate. Regardless of which term is used, the phenomenon is rather well established in natural taxa and experimentally induced mutants (e.g., Cooney-Sovetts and Sattler, 1987; Broadhvest *et al.*, 1992; Sawhney, 1992).

In general, the recognition of homeosis *sensu lato*, which includes partial homeosis based on partial correspondence, softens a rigid categorical morphology. It dissolves man-made boundaries between supposedly sharp categories and homologies and leads to what Cusset and Ferrand (1988) called *morphologie floue* which may be translated as "fuzzy morphology" (Rutishauser, 1989, p. 65, 1993b) or "open morphology" (Cusset, 1982, p. 55). Openness, like fuzziness, means there is a continuum between structures of one organism (e.g., the root and the stem) and between the so-called categories such as root, caulome, and phyllome (Veilex 1973-1976, 1979-1981). This continuum, which appears to be heterogeneous, is illustrated in Sections IV and V.

IV. MULTIVARIATE ANALYSIS AND MORPHOLOGICAL DISTANCE

As pointed out in Section I, all homology concepts are rooted in correspondence or relation. These latter concepts are understood in the wide sense so that they comprise cases of partial correspondence such as in the example of clinal variation presented in Section II,A: **AB, BC, CD**. In this very simple theoretical example, it is obvious how **BC** is related to both **AB** and **CD**. However, in the comparison of concrete structures, relations often are not so obvious. When the criterion of special quality is broken down into sub-criteria to avoid conflict between criteria (Section II), many characters are created. The states of these characters are not necessarily correlated, which renders intuitive methods of comparison difficult. Multivariate analysis such as principal component analysis (PCA) or discriminant analysis (DA) then become useful tools (concerning morphometrics see Bookstein's discussion in Chapter 5). Multivariate analysis shows the relation or correspondence *sensu lato* (which includes partial correspondence) in a quantitative manner so that the relation between any two structures can be characterized as a linear distance in a p-dimensional space. This distance may vary from zero to rather large. For example, if we compare typical simple leaves in flowering plants, the distance is zero, which means that on the basis of the selected criteria, they are the same, which means 1:1 correspondence. If, on the other hand, we compare a simple leaf with a shoot, the distance is great, which means that these two structures are very different.

It is important to note that morphological distances expressing relations or correspondences in the broad sense are relative: they reflect the methodology used to calculate them. This methodology includes the selection of characters and character states as well as the method of analysis, such as principal component analysis or discriminant analysis. Relativity is, however, not restricted to quantitative analyses. Relativity characterizes the results arrived at in whatever fashion because a methodology is always used to obtain the results. Even intuitive comparisons may imply a methodology which, however, in this case has not been objectified.

Riedl (1978, p. 130) also used the term *morphological distance*, but in a more restrictive sense because he saw homology as an all-or-none relation and homologues at different levels of a hierarchy. As I shall point out below, analyses by Sattler and Jeune (1992) invalidate the hierarchical relation of structures, if the whole spectrum of structural diversity is taken into consideration (see also Minelli and Peruffo, 1991). If, however, we limit our analyses either to one or a few correlated properties or typical forms that exhibit a set of correlated features, then structural categories and hierarchies may be appropriate notions.

I now briefly describe some results of a pilot study by Sattler and Jeune (1992) that analyzed morphological distances between a great diversity of plant structures, especially in flowering plants. In this study, principal component analysis was used, based on eight characters mostly with two states. The first two characters were position characters, with the following states: extraaxillary, axillary; axillant/ lateral, nonaxillant (axillant means subtending an axillary bud or branch). The remaining six characters referred to aspects of special quality: growth period (determinate, indeterminate), symmetry (dorsiventral, radial), construction (composite, not composite), abscission (present, absent), root cap (present, absent), vascular tissue (present, absent). Obviously, more characters could have been added. Some or all of the above have been used by many plant morphologists to distinguish structural categories, although there is no consensus on how many characters should be used and which characters are the most important ones.

Using the eight selected characters, we first established the relations between typical representatives of the following structural categories: root, caulome (stem), shoot, phyllome (leaf), and trichome (hair). As expected, the representative(s) of each category formed a distinct point or cluster (Fig. 4A), confirming the validity of the structural categories for the typical representatives. However, when we included atypical or controversial structures the picture changed drastically. These structures, whose homology and relationship has been discussed endlessly in the literature, occupy intermediate positions between the typical structures. As a result, we obtain a heterogeneous continuum in which the typical structures are linked to each other by intermediate atypical structures (Fig. 4B).

Analysis of stamen relations confirms their partial correspondence to both phyllomes and caulomes or shoots (Section II,D). The stamens of Comandra umbellata (Cs in Fig. 5) have almost the same morphological distance from the phyllome and caulome centroids; 2.43 from the phyllome centroid and 2.54 from the caulome centroid. The stamen fascicle of Hypericum hookerianum (Hsf in Fig. 5) shows morphological distances of 2.62 from the phyllome centroid, 2.77 from the caulome centroid, and 2.05 from the shoot centroid; it is more closely related to a typical shoot than to a typical caulome or phyllome (for developmental details see Leins, 1964, 1983; Leins and Erbar, 1991).

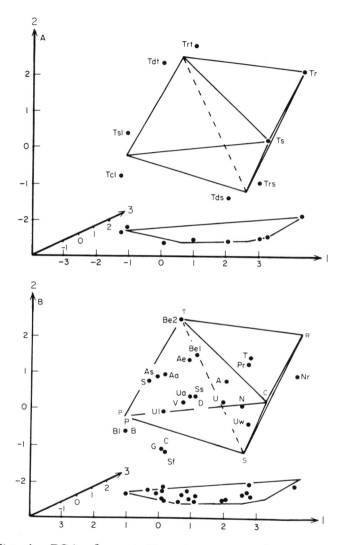

Fig. 4. PCA of vegetative structures of angiosperms. (A)
Typical structures: simple leaf (Tsl), compound leaf (Tc1),
stem (Ts), radial shoot (Trs), dorsiventral shoot (Tds), root
(Tr), radial trichome (Trt), and dorsiventral trichome (Tdt).
The correlation structure has the form of a double tetra-

hedron whose corners represent the centroids of phyllome (small P), caulome (small C), shoot (S), root (R), and trichome (small T) (see B). Note that the triangle C-S-R (in B) almost approaches a line. The lower figure is a projection of the double-tetrahedron. (B) Atypical or controversial structures placed into the correlation structure shown in (A). Where two structures coincide, their labels are placed on different sides of the dot that marks the position of the structures. A, So-called phylloclade of *Asparagus plumosus*; Aa, phyllode subtending an axillary bud of *Acacia longipedunculata*; Ae, phyllode not subtending an axillary bud of the same species; As, stipule of the same species; Bel, vascularized epiphyllous emergence of *Begonia hispida* var. *cucullifera*; Be2, nonvascularized epiphyllous emergence of the same species; B1, leaf of the same species (coincides with B, leaf of *Bryophyllum daigremontianum*, and P, leaf of *Phyllonoma integerrima*); C, indeterminate "leaf" of *Chisocheton tenuis* (coincides with G, indeterminate "leaf" of *Guarea*); D, phylloclade of *Danae racemosa* (coincides with Ss, sterile phylloclade of *Semele androgyna*); N, branch occupying a leaf site of *Nymphaea alba*; Nr, axillary root of *Nasturtium officinale*; Pr, adventitious root of *Pinguicula moranesis*; S, leaf of *Sedum dasyphyllum*; Sf, fertile *Semele* phylloclade; T, flattened root of *Taeniophyllum*; Ua, air-shoot of *Utricularia foliosa*; Ul, leaf of the same species; Uw, water-shoot of the same species; V, *Vitis riparia* tendril [After Sattler and Jeune (1992, Fig. 2, p. 253).]

These results are relevant to many fundamental issues of comparative plant morphology (see Sattler and Jeune, 1992). I mention only the following:

1. Analyses, based on widely used criteria and characters, confirm the continuum view of plant form, which states that there is a heterogeneous continuum of forms between and around the typical representatives of the structural categories. Hence, structural categories are not mutually exclusive. As pointed out above (Section III), they are fuzzy or open; they have no boundaries and in that sense do not exist as entities or essences. Correspondences of structures then must include partial correspondences. As the morphological distance approaches zero, partial corre-

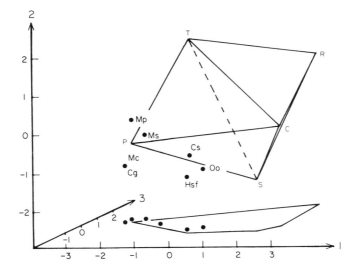

Fig. 5. Various floral structures placed into the correlation structure of Fig. 4A. Cg, Gynoecial appendage of
Comandra umbellata; Cs, stamen of the same species; Hsf,
stamen fascicle of *Hypericum hookerianum*; Mc, carpel of
Magnolia denudata; Mp, perianth member of the same
species; Ms, stamen of the same species; Oo, ovule of *Ochna
atropurpurea*. C, caulome; P, phyllome; R, root; S, shoot; T,
trichome. [After Sattler and Jeune (1992, Fig. 6, p. 255).]

spondences become increasingly complete so that a distance of zero represents a 1:1 correspondence. Obviously
1:1 correspondences do exist (within the context of a
selection of characters), but only as a special subset of all
correspondences. Unless this is taken into consideration,
homology concepts, identified with or based on 1:1 correspondence of entire structures, convey a rather distorted
view of nature. Cladistic analysis that is based on such distorted correspondences may lead to distorted cladograms
(see also Stevens, 1991).
    2. The findings of a heterogeneous morphological con-
tinuum and the entailed partial correspondences explain
why there have been endless debates on the qualitative
homology (in terms of 1:1 correspondence) of many atypical

structures (Eckardt, 1964; Cooney-Sovetts and Sattler, 1987). Obviously, these atypical structures do not fit the structural categories because they are more or less intermediate between them. Depending on which character is emphasized, they may be forced into one or the other category. For example, if in **BC** (of the cline **AB, BC, CD**) **B** is emphasized, then **BC** is categorized with **AB**; if, however, **C** is considered decisive, then **BC** is classified with **CD**. But how do we decide which property is decisive? As long as for special purposes we are interested only in **B** or **C**, there is no problem, because this selective focus decides the categorization. If, however, we are interested in the whole structure and how entire structures are related to each other, then we have to consider all properties.

Eckardt (1964) pointed out that many atypical structures are controversial. However, they are controversial only as long as we try to press them into mutually exclusive categories and thus homologize them in terms of a qualitative homology concept. In terms of a continuum morphology that recognizes partial correspondence or partial homology, these cases are no longer difficult; they simply turn out to be more or less intermediate between the typical representatives of the so-called structural categories that do not exist as distinct entities or essences.

3. It should be noted that in a continuum the distinction between typical and atypical or intermediate structures also vanishes. Reference to "typical structures" can only mean those structures that are considered typical by most morphologists. Thus, the expression "typical structure" is simply a means for communication. It does not refer to an inherent quality apart from the fact that the so-called typical structures are more common than the atypical ones.

4. In the continuum the postulated hierarchy of classical plant morphology disappears. According to this hierarchy, the plant comprises a root system consisting of root(s) and a shoot system composed of caulome(s) and phyllome(s). These units in turn comprise lower level units such as trichomes. If we want to recover this hierarchy, we have to restrict ourselves to the so-called typical structures, i.e., more or less arbitrarily selected regions of the heterogeneous continuum. In a more global view hierarchical thinking appears inappropriate (see also Sattler, 1986, pp.

96, 105, and 135; Anthony and Sattler, 1990, p. 169; Minelli and Peruffo, 1991).

In conclusion, continuum morphology that entails the recognition of partial correspondences or partial homologies not only provides a more comprehensive view of the diversity of plant form but also explains the futility of endless debates on the "true" nature or homology of structures perceived to be difficult, but only difficult because of our preconception that nature (or our approach to the study of nature) must be categorical.

V. PROCESS MORPHOLOGY AND HOMOLOGY

The analyses of morphological distances between structures by Sattler and Jeune (1992) are based on criteria that often have been used in plant morphology. These criteria refer to both dynamic characters (processes) and static characters. An example of a process is determinate or indeterminate growth; an example of a static character is radial or dorsiventral symmetry. When a structure exhibits the same kind of symmetry from its inception to maturity, a static symmetry concept may appear appropriate. However, if symmetry changes during development, symmetry must be seen as a process (symmetrization) that is a reflection of differential growth.

Now we have to go one step further. Even when symmetry does not change during development, there is an underlying dynamic. Symmetry and any other static character or character state are an abstraction from this dynamic (Hay and Mabberley, 1993). Furthermore, structure itself (not only its properties) is an abstraction from this dynamic (e.g., Chapter 7 in Woodger, 1929). Bohm (1980, pp. 10 and 30) illustrated this by a vortex. Having a shape, a vortex can be seen as a structure, but first of all it is dynamic, and second, it is totally integrated into its environment, i.e., it has no boundary. This movement within the whole — Bohm referred to "holomovement" — is primary and the so-called discretely static structure is an abstraction (see also Hay and Mabberley, 1993). As long as we see the abstraction for what it is, there is no problem. But if we mistake the abstraction for the territory from which it has been abstracted, then we delude ourselves. We become victims of the "fallacy of misplaced concreteness" (Whitehead, 1967, p. 51). In Korzybski's (1948) terms, we mistake the map for the territory or as Bateson (1972, p. 280) put it, "eating the menu card instead of the dinner" (see Hay and Mabberley, 1993).

On the "map" (i.e. our abstractions), structure and process may be different: structure tends to be considered static, whereas process is dynamic. If we mistake the map for the territory, we conclude that plants consist of structures within which processes occur. On closer inspection we learn, however, that what appears static is in fact also dynamic. The reason why something like a leaf may appear static is because, apart from occasional fast movements such as nastic movements in the sensitive plant (*Mimosa pudica*), it changes only very slowly. What we actually find in plants (and also in animals) is a superimposition of slow and fast movements. The slow movements are growth and decay, differentiation and dedifferentiation, whereas the fast movements are, for example, metabolic processes (e.g., in Chapter 7, Woodger, 1929).

In process morphology structure is seen dynamically as process or processes (Sattler, 1990, 1992). Process is not perceived as merely an attribute of structure; structure itself *is* process. Even a mature structure is process. The speed by which it changes may have slowed down enormously, but its morphological change does not come to an absolute standstill. Eventually decay occurs in continuity with growth. One process is associated with or gives way to others.

Many processes of form-making (morphogenetic processes) may be distinguished. I considered growth and decay, differentiation and dedifferentiation as the most fundamental processes or dynamic aspects of the general form process (Sattler, 1990, 1992). For each of these four aspects further, more detailed aspects may be distinguished which I called modalities or parameters. Finally, for each parameter, two (or three) values may be discerned. Table II presents parameters and values of growth (Jeune and Sattler, 1992). The values should not be considered as mutually exclusive, but rather as extremes in a continuum. A quantification of the parameters would represent this continuum more appropriately. Furthermore, other parameters could be added (Sattler, 1990), including parameters at the molecular and ecological levels (Sattler, 1992, p. 709).

Table II. Parameters (a-g) and subparameters (1, 2, 9-13) with their values[a]

Parameters		Values		
(a) Positioning	(1)	Axillary (−1)		Extra-axillary (+1)
	(2)	Non-axillant (−1)		Axillant (+1)
(b) Growth rate	(3)	low (−1)		High (+1)
(c) Growth period	(4)	Indeterminate (−1)		Determinate (+1)
(d) Orientation	(5)	Inward (−1)		Inward and outward (+1)
(e) Geotropism	(6)	Negative (−1)	Zero (0)	Positive (+1)
(f) Growth symmetry	(7)	Radial (−1)	Changing (0)	Dorsiventral (+1)
(g) Branching s. lat.	(8)	Branching (−1)	Lobing (0)	No branching (+1)
Subparameters of branching				
	(9)	Acropetal (−0.5)	Basipetal (0)	No branching (+0.5)
	(10)	Alternate (−0.5)	Opposite/ whorled (0)	No branching (+0.5)
	(11)	Extra-axillary (−0.5)	Axillary/ axillant (0)	No branching (+0.5)
	(12)	Endogenous (−0.5)	Exogenous (0)	No branching (+0.5)
	(13)	Symmetry change (−0.5)	No symmetry change (0)	No branching (+0.5)

[a] After Jeune and Sattler (1992), Table 1, p. 152).

It is important to note that each value represents a process. Thus, the values of Table II refer to growth processes such as radial growth or indeterminate growth. Any particular structure can then be seen as a particular process combination. For example, a typical simple leaf is a combination of extraaxillary, axillant positioning, a relatively high growth rate, determinate growth of inward orientation, more or less zero geotropism, and dorsiventral growth (the branching parameter and its subparameters do not apply in this case). From this dynamic perspective, morphological diversity is a diversity of process combinations and morphological evolution is a change in process combinations [see also Hinchliffe (1990) with regard to process thought in animal morphogenesis].

I now briefly illustrate process morphology with reference to a study by Jeune and Sattler (1992). In this study, only growth parameters were used because the focus was on external morphology before the onset of decay. A great diversity of growth process combinations were analyzed by principal component analysis (PCA) and discriminant analysis (DA). As in the more static analyses by Sattler and Jeune (1992) (see Section IV), we first analyzed process combinations that represent typical roots, caulomes (stems), phyllomes (leaves), and trichomes (hairs). With both methods (PCA and DA) we obtained four distinct points or clusters (Fig. 6a). When we analyzed process combinations representing atypical or controversial structures, the discreteness of the four points or clusters disappeared because the atypical process combinations formed a heterogeneous continuum with the typical ones (Fig. 6b).

Although process morphological analyses yielded a heterogeneous continuum as the more static analyses, there are differences between the two approaches. First of all, the values of the distances differ to some extent, although there are remarkable similarities. Second, in the process morphological approach, the process combinations representing caulomes (stems) and shoots form one common cluster (Fig. 6a, c and s). This contradicts classical plant morphology, according to which shoot and caulome are not only distinct categories, but in addition occupy different levels of the categorical hierarchy: the shoot category comprises caulome (stem) and phyllome (leaf) as categories at a lower level of the hierarchy. It is, then, not surprising that the distinction

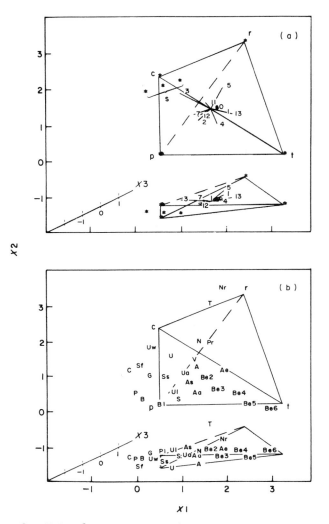

Fig. 6.   DA of process combinations corresponding with
the vegetative structures of flowering plants. The upper
representation of each figure is a three-dimensional representation. Since the depth of the third axis is not obvious,
the lower representation was added, which is a projection
of the upper one onto the plane of axis-1 and axis-3, thus
recovering the depth that is missing in the upper representation. (a) Typical structures. The corners of the tetra-

hedron represent the centroids of caulome (c), phyllome (p), root (r), and trichome (t); s = shoots. The DA norm is placed into the middle of the tetrahedron. (b) Atypical or controversial structures: A, radial *Asparagus plumosus* "phylloclade"; Aa, *Acacia longipedunculata* phyllode subtending an axillary branch; Ae, phyllode of the same species without axillary branch; As, stipule of the same species; B1, leaf of *Begonia hispida* (B, leaf of *Byrophyllum*, nearly coincides and therefore was not labelled); Be2-Be5, vascularized epiphyllous appendages of *Begonia hispida*; Be6, nonvascularized epiphyllous emergence of the latter species; G, indeterminate leaf of *Guarea* (C, indeterminate leaf of *Chisocheton*, is close and therefore was not labelled); N, branch of *Nymphaea* occupying a leaf site; Nr, axillary root of *Nasturtium*; P, leaf of *Phyllonoma integerrima*; Pr, adventitious *Pinguicula moranensis* root; S, *Sedum dasyphyllum* leaf; Sf, fertile phylloclade of *Semele*; Ss, sterile phylloclade of the same species; T, *Taeniophyllum* root; U, *Ulex* thorn; Ua, air-shoot of *Utricularia foliosa*; U1, "leaf" of the same species; Uw, water-shoot of the same species; V, tendril of *Vitis riparia*. [After Jeune and Sattler (1992, Fig. 2, p. 157).]

---

between shoot and caulome should be absent in process morphological analyses. Shoot and caulome as well as the other categories are abstractions from the underlying dynamic. The process combination that represents a stem includes processes such as branching that are also characteristic of the process combination representing a shoot. As a consequence, the two combinations form one cluster.

How is process morphology related to the notions of correspondence and relation that are at the root of homology concepts? Since process morphology leads to a heterogeneous continuum of process combinations, correspondence between the latter must again include partial correspondences and multiple relations. As in the more static analyses by Sattler and Jeune (1992), the degree of relatedness or correspondence can be quantitatively expressed as the distance between process combinations in the p-dimensional space of the analyses. If the term *morphology* is understood in the wide sense so that it refers also to morphogenesis, then the distance could again be referred to as morphological distance. However, if we want to underline the fact that this measure refers to a distance between morphogenetic process combinations, we could call it *morphogenetic distance*.

In any case, the morphogenetic distance, as the morphological distance, is a measure of relation or correspondence in the broad sense. As the distance approaches zero, the partial correspondences approximate 1:1 correspondence and a distance of zero represents exactly a 1:1 correspondence. Thus, again, 1:1 correspondence is only a subset or special case of the wider range of morphological relations. Insistence on 1:1 correspondence as the basis for homology of entire structures (process combinations) means that we omit a large part of the spectrum of morphological diversity and/or force partial correspondences into the mold of 1:1 correspondences and thus violate and distort nature.

VI. PROCESS MORPHOLOGY AND EVOLUTION

From the dynamic point of view, any particular plant is a particular sequential change of process combinations. In a plant population or species a certain amount of variation of this dynamics occurs. However, certain sequences of process combinations are more frequent than others (Meyen, 1987, pp. 360 and 361). As a species evolves, its particular frequency of process combinations and their sequences change. Furthermore, new combinations and sequences may appear.

This dynamics can be described in terms of process morphology, although so far process morphology has not yet been applied to whole plants, but only to parts of plants (Sattler and Rutishauser, 1990; Jeune and Sattler, 1992). A limited subset of the dynamics can be dealt with in terms of 1:1 correspondence and qualitative homology concepts that are based on this simple type of correspondence. For example, certain changes of phyllotaxis (such as from alternate to whorled) can be represented in the terminology of classical plant morphology: it can be said that the phyllotaxis of the shoot has changed. In this statement, it is implied that the shoot as shoot category has not changed. This is all right since a change in phyllotaxis need not affect the definition of a shoot. The situation changes, however, when other processes related to the definition of the shoot undergo change. For example, if the symmetry changes from radial to dorsiventral, problems arise. Classically, it would still be claimed that a radial shoot changed to a dorsiventral "shoot." A dorsiventral "shoot," however, no longer totally corresponds to the shoot category. Since it shares one process (namely, dorsiventral growth) with the processes that define a typical leaf, it corresponds partially to both the shoot category and the leaf category. Thus, it evolved toward the leaf category. This change is, however, obscured if we refer to this structure as a dorsiventral "shoot," which implies that the categorical status remained static.

Process morphology uncovers the obscured dynamics. Since in process morphology a structure *is* the process combination, any change in the processes changes the structure. And if a change occurs in a process that defines the structure as a category, then the change affects the categorization. It is exactly this influence on categorization that may be overlooked when structure is not identified with process. One can say then that a structure changed without noticing that the categorical status of that structure also has changed and therefore is no longer the same kind of structure. Thus, in the above example process morphology makes it explicit that a change from radial to dorsiventral growth brings the new combination closer to that which represents a typical leaf (for other examples see Sattler, 1992, p. 709; see also Veilex 1973-1976, 1979-1981). As a result of this change, the new structure no longer fits any category and therefore categorical thinking has to be transcended. This means that homology concepts rooted in 1:1 correspondence, which is categorical, have to be broadened to include partial correspondences provided one is interested in the comparison of entire structures (process combinations).

If, however, one is content to compare only invariant properties (including processes) then a qualitative homology concept as 1:1 correspondence or equivalence, as discussed in Chapter 6, is still possible at least in those cases where invariant properties can be found. This kind of homology or equivalence, however, is much more abstract than morphological or morphogenetic distance because it disregards all partial correspondences and thus may fail to refer to entire structures or process combinations (Roth, 1988, p. 17). A discussion of which approach is correct appears misguided. The two approaches are complementary to each other: 1:1 correspondence or equivalence underlines invariant properties (including processes) in a logical transformation (see Chapter 6) or historical transformation, whereas morphological or morphogenetic distance emphasizes the relation of entire structures (process combinations) in terms of abstracted variables.

One might argue that the approach of process morphology reduces evolutionary change to change in similarity. This is correct if similarity is understood broadly as a matter of degree with regard to process combinations and their sequences. In contrast, homology, it is often argued, must go beyond mere similarity to imply phylogenetic or cladistic relationship (e.g., Patterson, 1987, p. 14; Crisp and Weston, 1987; Gould, 1988, p. 26). There is, however, no agreement on this issue (Woodger, 1945, p. 109; Jardine, 1969, p. 329; Sneath and Sokal 1973, pp. 75 and 76; Kaplan 1984, pp. 52 and 53; Goodwin, 1984, 1988; and see Chapters 1 and 6 in this volume). In any case, regardless of which position one endorses, and again one may hold both because they can be seen as complementary, it appears that all homology concepts (historical as well as ahistorical) are rooted in similarity. Qualitative homology concepts reflect only a very narrow range of similarity, namely 1:1 correspondence. Stevens (1984, p. 396) called it "simple similarity." In view of the enormous complexity of organic diversity, an insistence on only this simple similarity leads to a rather impoverished view.

There are many oversimplifications in evolutionary biology and phylogenetic reconstruction. In the present context, the idea of evolutionary character (state) transformation is especially pertinent (Hennig, 1967, p. 93). According to this idea, characters (or their states) are transformed during evolution. As a result, a 1:1 correspondence is seen between the original character (state) and the character (state) into which it was transformed. This 1:1 correspondence is fundamental in cladistic analysis even in cases where the morphological distance between the original and the transformed characters (or their states) is rather great, so that in terms of the above quantitative analyses the two characters (or their states) would only partially correspond to each other.

As pointed out by Hay and Mabberley (1993), the idea of character (state) transformation is at best a metaphor that does not appear quite appropriate. First of all, like all concepts, characters and their states are abstractions (Hay and Mabberley, 1993). Second, they are not really transformed into each other. In each generation, organisms are (re)constructed (Goodwin, 1984, p. 113; Oyama, 1988, 1989), and in this (re)construction more or less similar or different characters or character states may appear. This is indicated in highly diagrammatic form in Fig. 7. To the left (Fig. 7b), an axillary structure **AB** is reconstructed as in the ancestor at the bottom of Figure 7a. To the right (Fig. 7c), the axillary structure is identified by the processes **BC**. This means that during the construction of this plant, processes were combined that in the ancestor belonged to different combinations. Consequently, **BC** is a hybrid structure and therefore is neither a transformation of **AB**, nor of **CD** in the ancestor. It is a novel combination. The search for 1:1 correspondence (through transformation) of entire structures is misguided in such a situation where the new structure is related to both **AB** and **CD** (see also Minelli and Peruffo, 1991, p. 431). Not even primary homology (*sensu* de Pinna, 1991) can be postulated. Hence, there is no basis of potential homology (1:1 correspondence) of entire structures whose congruence could be tested. Process morphology, including the notions of partial correspondence, morphogenetic distance, and multiple relations, helps us understand such cases. Homology concepts based on 1:1 correspondence of entire structures will lead to distortions and futile debates of whether the novel structure is homologous to **BC** or **CD** [for concrete examples see Cooney-Sovetts and Sattler, 1987; Sattler, 1984, 1988a; Bell, 1991, pp. 206-215; and the literature on many homeotic mutants (see Section III)].

I conclude that in view of the dynamics and complexity of evolutionary change, our methodology has to be adapted to this situation. Greater emphasis will have to be placed on the multitude and nonhierarchical network of changes in the structural dynamics during development and evolution (Minelli and Peruffo, 1991; Sattler, 1992).

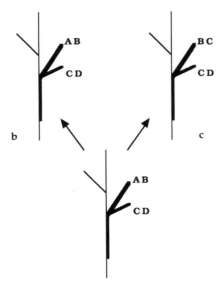

Fig. 7.   Diagrammatic representation of an ancestral plant (a) and two derivative plants (b and c). In (b), the ancestral condition is reconstructed, i.e., the root-stem axis with two leaves, one of which (**CD**) has an axillary branch (**AB**). In (c), the axillary branch is replaced by a hybrid structure (**BC**) which combines property **B** of the axillary branch (**AB**) with property **C** of the subtending leaf (**CD**). Instead of indicating a single property, **A**, **B**, **C**, **D** may each represent a set of properties. For concrete examples see Cooney-Sovetts and Sattler (1987).

VII. SUMMARY AND CONCLUSIONS

All homology concepts appear to be (at least methodologically) rooted in correspondence or relation. For some concepts such as the cladistic one, correspondence provides only the basis on which by means of additional criteria synapomorphy is constructed or tested. In any case correspondence or relation appear to be basic and therefore the focus in this chapter has been on these concepts. Although my illustrations have been from plants, especially flowering plants, the basic ideas I discussed apply also to animals (e.g., van der Klaauw, 1966; Sneath and Sokal, 1973; Ouweneel, 1976; Roth, 1984, 1988; Hinchliffe, 1990; Hall, 1992; Fig. 3 in Chapter 10 in this volume).

Correspondence *sensu stricto* is 1:1 correspondence, whereas correspondence *sensu lato* also includes partial correspondence. The latter can be seen as a multiple relation as is evident from examples of clinal variation such as **AB**, **BC**, **CD** (see also Sattler, 1984, p. 384): **BC** is related to both **AB** and **CD**. There is a wealth of actual relations that constitute partial correspondences. Such partial correspondences occur not only with regard to whole structures, but also with respect to properties of structures. Therefore, homologizing properties instead of entire structures (Roth, 1988, p. 17) does not completely eliminate problems of homologization in terms of 1:1 correspondence, although it may reduce or abolish conflicts between homology criteria in many cases.

Homeosis, implying known genealogy or hypothetical phylogeny, also has been defined in a narrow and broad sense. If it is broadly defined, it includes partial replacements and, furthermore, the structures or features that replace each other need not be totally homologous. Therefore, when one part of an organism is (nearly) totally or partially replaced by a nonhomologous part, a homology transgression occurs. Such transgressions may be saltational and therefore highly significant for macroevolution.

Homeosis, a phenomenon reported in many artifically induced mutants, teratologies, cultivars, natural populations, and natural taxa, undermines many traditional homologies in at least two ways: (1) It shows that partial correspondence plays an important role and that an insistence on only 1:1 correspondence may lead to distortions of relationships. (2) It weakens the usefulness of the criterion of relative position that often has been considered the most important homology criterion. This weakening is due to strong evidence that positional relations are not always conserved during evolution. Even structurally nonhomologous parts may occupy the same relative position. If in such cases we insist on the general reliability of the position criterion, we may reach absurd conclusions, such as, for example, the conclusion that a structure that on the basis of its special quality is generally recognized as a root corresponds or is homologous to a shoot.

Homeosis and other phenomena demonstrate an extraordinary flexibility and variation of structural relations. These relations can be quantified by means of multivariate analysis such as principal component analysis and discriminant analysis (Sattler and Jeune, 1992). The analysis of what are generally considered typical structures places these structures into discrete points or clusters [root, shoot, caulome (stem), phyllome (leaf), and trichome (hair)]. However, the additional analysis of atypical or controversial structures (i.e., structures whose homology is debated) reveals these structures as more or less intermediate between the so-called typical structures. As a result a heterogeneous continuum is obtained in which one cannot draw a sharp demarcation line between typical and atypical structures (Fig. 4B). There is no indication of a hierarchy of mutually exclusive structural categories and no basis for a fundamental opposition of homology and analogy (in the sense of similarity due to similar function).

Any structure within the heterogeneous continuum can be related to any other. This relation can be expressed quantitatively as the linear distance between the structures. It may be called *morphological distance*. As this distance approaches zero, the partial correspondences approximate 1:1 correspondence. A distance of zero constitutes complete 1:1 correspondence (with regard to the selected parameters). It is evident that from this perspective 1:1 correspondence represents only a subset of the whole set of correspondences *sensu lato* that also comprises partial correspondences. Basing homology of entire structures only on 1:1 correspondence, as is commonly done, entails two oversimplifications: (1) All partial correspondences are ignored which greatly impoverishes our view of nature, and/or (2) partial correspondences are forced into the mold of 1:1 correspondence, which violates and distorts nature (see below).

Nonetheless, homology or equivalence reflecting 1:1 correspondence may be retained if they are based only on invariant properties in logical or historical transformations. This restriction entails, however, that in many cases homology or equivalence can refer only to *properties* of structures

(Roth, 1988, p. 17). Homologization of entire structures may no longer be possible in terms of 1:1 correspondence.

In general, the notions of homology, based on 1:1 correspondence, and morphological distance are complementary to each other: 1:1 correspondence underlines invariable properties in a logical transformation (Chapter 6) or historical transformation, whereas morphological distance emphasizes the relation of entire structures in terms of abstracted variables.

Process morphology deals with morphological relations dynamically (Sattler 1990, 1992): structure is seen as process. In this process, four aspects can be distinguished: growth and decay, differentiation and dedifferentiation. Each of these can be analyzed in terms of parameters with two or three values that represent more specific processes such as radial growth or dorsiventral growth. Each particular structure can then be seen as a particular process combination. Diversity of form is a diversity of process combinations. Evolution of form is the evolution of process combinations (Sattler, 1992).

When process combinations are compared by means of multivariate analysis, the result is again a heterogeneous continuum, that differs, however, in some respects from the continuum based on static characters that are often used in comparative morphology, systematics, and evolutionary biology (Hay and Mabberley, 1993). If morphology is broadly defined so that it comprises morphogenesis, the distance between process combinations could be referred to as morphological distance. If, however, morphology is more narrowly defined, it would be more appropriate to refer to this distance as morphogenetic distance, since it is based on morphogenetic processes such as growth processes.

One of my central conclusions is that if we want to gain a more comprehensive understanding of the homology of entire structures we will have to take into consideration partial correspondences and multiple relations, preferably in a dynamic sense, since nature is basically dynamic (Whitehead, 1978; Birch and Cobb, 1981, pp. 84-91; Ho, 1988; Sattler, 1993). One might argue that correspondences *sensu lato* and multiple relations are a matter of similarity. This is correct, but the study of similarity is fundamental because synapomorphy is based on similarity or some sort of correspondence. Unfortunately, cladists

restrict this similarity to simple similarity or 1:1 correspondence. As a result, we obtain an impoverished view of
organic diversity and relationships of organisms and taxa.

The reason for a widespread insistence on 1:1 correspondence (i.e., either-or thinking) may be a fallacious
transfer of genealogical reasoning to parts of organisms.
Disregarding some complexities, we can say that two organisms *either* have a common ancestor at a certain level *or*
they do not. However, this kind of reasoning does not apply
to parts of organisms or characters (or character states)
because there is no genealogical relation between parts or
characters (and their states) (see, e.g., Sattler, 1984, p.
386; Wagner, 1989a, p. 1159; Sattler and Jeune, 1992, p.
257). As is well known, parts or characters (and their
states) "are built anew in each generation" (Wagner, 1989a,
p. 1159). During this (re)construction processes from nonhomologous structures in the ancestor may be combined.
This process of *developmental hybridization* gives rise to a
hybrid or mosaic structure. Such a structure cannot be
traced back to one ancestral structure as an organism can be
traced back to an ancestral organism. A hybrid structure
must be traced to more than one structure in the ancestral
organism (Fig. 7). Therefore it is inappropriate to ask
whether it corresponds to either this or that structure. It
partially corresponds to both, just as a mixture of a salt and
sugar solution is partially salt and partially sugar and not
either one or the other.

Instead of remaining caught in either-or thinking, I
propose that we look at organic diversity, phylogeny, and
evolution in terms of morphological or morphogenetic distances of structures seen as process combinations. A
comparison of whole organisms seen as sequences of
process combinations would be even more appropriate.
Regardless of whether whole organisms or parts of organisms are compared, I do not wish to claim that morphological or morphogenetic distances correspond to phylogenetic
distances. In other words, degree of similarity does not
necessarily reflect phylogenetic relationship. Degree of
similarity in terms of the structural dynamics is only the
basis for phylogenetic reconstruction. To obtain phylogenetic inferences, additional assumptions and criteria are

necessary just as additional procedures are required to pass from a 1:1 correspondence to a postulate of a synapomorphy.

In this chapter the focus has been on methodology, not on definitions. Definitions are, however, important for communication. If homology of structures is defined so broadly as to include partial correspondences in a process morphological sense, then homology might remain the central concept of comparative morphology of structures. If, however, homology is understood narrowly as based on 1:1 correspondence, then it will have to be complemented by morphological or morphogenetic distances between structures or, preferably, whole organisms.

A comparison of the dynamics of whole organisms would allow us to avoid fragmenting organisms into parts, characters, and character states. Such fragmentation is the first step in the establishment of structural homologies and thus all homologization in the traditional sense is somewhat removed from nature because of the abstraction that leads to fragmentation (e.g., Chapters 4 and 5 in Sattler, 1986; Hay and Mabberley, 1993). Processes are also abstractions. However, they appear less removed from nature than structural fragments. In other words: processes appear more adequate (for the notion of adequacy see Sattler, 1986, p. 76). Since they interlace, a dynamic view of organisms is less fragmenting than structural decomposition. As a result, the dynamic view such as that of process morphology allows us glimpses of the fluidity of nature that are beyond the scope of the traditional homologization of body parts.

ACKNOWLEDGMENTS

Preparation of this chapter was supported by Grant A2594 from the Natural Sciences and Engineering Research Council of Canada. For valuable comments on a first draft I am grateful to Fred L. Bookstein, Brian K. Hall, Alistair Hay, Bernard Jeune, Naida Lehmann, V. Louise Roth, Rolf Rutishauser, Neil H. Shubin, and Peter H. Weston. I thank contributors to this volume for kindly sending me copies of their manuscripts.

REFERENCES

Anthony, M., and Sattler, R. (1990). Pathological ramification of leaves and the pyramid model of plant construction. *Acta Biotheor.* **38**, 165-170.

Bateman, R. M., DiMichele, W. A., and Willard, D. A. (1992). Experimental cladistic analysis of anatomically preserved arborescent Lycopsids from the carboniferous of Euramerica: An essay on paleobotanical phylogenetics. *Ann Mo. Bot. Gard.* **79**, 50-59

Bateson, G. (1972). "Steps to an Ecology of Mind." Ballantine Books, Random House, New York.

Bateson, W. (1894). "Materials for the Study of Variation." Macmillan, London.

Bell, A. D. (1991). "Plant Form: An Illustrated Guide to Flowering Plant Morphology." Oxford Univ. Press, Oxford.

Birch, C., and Cobb, J. B., Jr. (1981). "The Liberation of Life — from Cell to the Community." Cambridge Univ. Press, Cambridge.

Bohm, D. (1980). "Wholeness and the Implicate Order." Routledge & Kegan Paul, London.

Bowman, J. L., Smyth, D. R., and Meyerowitz, E. M. (1989). Genes directing flower development in *Arabidopsis. Plant Cell* **1**, 37-52.

Bowman, J. L., Smyth, D. R., and Meyerowitz, E. M. (1991). Genetic interactions among floral homeotic genes of *Arabidopsis. Development* **112**, 1-20.

Brady, R. H. (1987). Form and cause in Goethe's morphology. *In* "Goethe and the Sciences: A Reappraisal" (F. Amrine, F. J. Zucker, and H. Wheeler, eds.), pp. 257-300. Reidel, Dordrecht, The Netherlands.

Broadhvest, J., Daigle, N., Martin, M., Haughn, G. W., and Bernier, F. (1992). Appendix: A novel type of homeotic mutation affecting floral morphology. *Plant J.* **2**, 991-997.

Brückner, C. (1991). Zur Interpretation des Karpells — eine Übersicht. *Gleditschia* **19**, 3-14.

Brugger, J., and Rutishauser, R. (1989). Bau und Entwicklung landbewohnender *Utricularia* — Arten. *Bot. Helv.* **99**(2), 91-146.

Champagnat, M., and Blatteron, S. (1966). Ontogénie des organes axillaires du Cresson (*Nasturtium officinale* R. Br.). *Rev. Gén. Bot.* **73**, 85-102.

Charlton, W. A. (1991). Homeosis and shoot construction in *Azara microphylla* Hook. (Flacourtiaceae). *Acta Bot. Neerl.* **40**, 329-337.

Coen, E. S. (1991). The role of homeotic genes in flower development and evolution. *Annu. Rev. Plant Physiol. Plant Mol. Biol.* **42**, 241-279.

Cooney-Sovetts, C., and Sattler, R. (1987). Phylloclade development in the Asparagaceae: An example of homoeosis. *Bot. J. Linn. Soc.* **94**, 327-371.

Cresens, E. M., and Smets, E. F. (1989). The carpel — a problem child of floral morphology and evolution. *Bull. Jard. Bot. Nat. Belg./Bull. Nat. Plant. Belg.* **59**, 377-409.

Crisp, M. D., and Weston, P. H. (1987). Cladistics and legume systematics, with an analysis of the Bossiaeeae, Brongniartieae and Mirbelieae. *In* "Advances in Legume Systematics" (C. H. Stirton, ed.), Part 3, pp. 65-130. Royal Botanic Gardens, Kew.

Croizat, L. (1960). "Principia Botanica," 2 vols. Published by the author, Caracas.

Croizat, L. (1962). "Space, Time, Form: the Biological Synthesis." Published by the author, Caracas.

Cusset, G. (1982). The conceptual bases of plant morphology. *Acta Biotheor.* **31A**, 8-86.

Cusset, G. (1991). Modélisation de la croissance de la tige feuillée. *Can. J. Bot.* **69**, 1810-1818.

Cusset, G., and Ferrand, A. (1988). La morphogenèse du calice de *Fittonia verschaffeltii* Lemaire. III. Orientations des mitoses. *Beitr. Biol. Pflanz.* **62**, 441-476.

Cutter, E. G. (1966). Patterns of organogenesis in the shoot. *In* "Trends in Plant Morphogenesis" (E. G. Cutter, ed.), pp. 220-234. Wiley, New York.

de Pinna, M. C. C. (1991). Concepts and tests of homology in the cladistic paradigm. *Cladistics* **7**, 367-394.

Donoghue, M. J. (1992). Homology *In* "Keywords in Evolutionary Biology" (E. F. Keller and E. A. Lloyd, eds), pp. 170-177. Harvard Univ. Press, Cambridge, MA.

Eckardt, T. (1964). Das Homologieproblem und Fälle strittiger Homologien. *Phytomorphology* **14**, 79-92.

Fries, E. M., and Endress, P. K. (1990). Origin and evolution of angiosperm flowers. *Adv. Bot. Res.* **17**, 99-162.

Froebe, H. A. (1982). Homologiekriterien oder Argumentationsverfahren. *Ber. Dtsch. Bot. Ges.* **95**, 19-34.

Geoffroy Saint-Hilaire, E. (1830). "Principes de Philosophie Zoologique, discutés en Mars 1830, au Sein de l'Académie Royale des Sciences." Pichon et Didier, Paris.

Goodwin, B. C. (1984). Changing from an evolutionary to a generative paradigm in biology. In "Evolutionary Theory: Paths into the Future" (J. W. Pollard, ed.), pp. 99-120. Wiley, New York.

Goodwin, B. C. (1988). Morphogenesis and heredity. In "Evolutionary Processes and Metaphors" (M.-W. Ho and S. W. Fox, eds.), pp. 145-162. Wiley, New York.

Gould, S. J. (1988). The heart of terminology. Nat. Hist. **97**(2), 24-31.

Greyson, R. I. (1993). "The Development of Flowers." Oxford Univ. Press, New York (in press).

Guédès, M. (1979). "Morphology of Seed-Plants." Cramer, Vaduz.

Hagemann, W. (1976). Sind Farne Kormophyten? Eine Alternative zur Telomtheorie. Plant Syst. Evol. **124**, 251-277.

Hagemann, W. (1984). "Baupläne der Pflanzen." Lect. Notes. Univ. of Heidelberg, Heidelberg.

Hagemann, W. (1989). Acrogenous branching in pteridophytes. In "Proceedings of the International Symposium on Systematic Pteridology" (K. H. Shing and K. U. Kramer, eds.), pp. 245-258. China Science and Technology Press, Beijing.

Hall, B. K. (1992). "Evolutionary Developmental Biology." Chapman & Hall, London and New York.

Hassenstein, B. (1954). Abbildende Begriffe. Verh. Dtschen Zool. Ges., 197-202.

Hassenstein, B. (1971). Injunktion. In "Historisches Wörterbuch der Philosophie" (J. Ritter, ed.), Vol. 4, pp. 367-368. Schwabe, Basel.

Hay, A., and Mabberley, D. J. (1991). 'Transference of Function' and the origin of aroids: Their significance in early angiosperm evolution. Bot. Jahrb. Syst. **113**, 339-428.

Hay, A., and Mabberley, D. J. (1993). On perception of plant morphology: some implications for phylogeny. In "Shape and Form in Plants and Fungi" (D. Ingram, ed.). Academic Press, London. (in press).

Hennig, W. (1967). "Phylogenetic Systematics." Univ. of Illinois Press, Urbana.

Hill, J. P., and Lord, E. M. (1989). Floral development in *Arabidopsis thaliana*: A comparison of the wild type and the homeotic *pistillata* mutant. *Can. J. Bot.* **67**, 2922-2936.

Hinchliffe, R. (1990). Towards a homology of process: Evolutionary implications of experimental studies on the generation of skeletal pattern in avian limb development. *In*: "Organizational Constraints on the Dynamics of Evolution" (J. Maynard Smith and G. Vida, eds.), pp. 119-131. Manchester Univ. Press, Manchester and New York.

Ho, M.-W. (1988). On not holding nature still: evolution by process, not by consequence. *In* "Evolutionary Processes and Metaphors" (M.-W. Ho and S. W. Fox, eds.), pp. 117-144. Wiley, New York.

Holmes, S. (1979). "Henderson's Dictionary of Biological Terms." Van Nostrand-Reinhold, New York.

Innes, R. L., Remphrey, W. R., and Lenz, L. M. (1989). An analysis of the development of single and double flowers in *Potentilla fruticosa. Can. J. Bot.* **67**, 1071-1079.

Jäger-Zürn, I. (1992). Morphologie der Podostemaceae II. *Indotristicha ramosissima* (Wight) van Royen (Tristichoideae). *In* "Tropische und subtropische Pflanzenwelt" (W. Rauh, ed.). Vol. 80, pp. 1-48. Akademie der Wissenschaften und Literatur, Mainz. Steiner Verlag, Stuttgart.

Jardine, N. (1967). The concept of homology in biology. *Br. J. Philos. Sci.* **18**, 125-139.

Jardine, N. (1969). The observational and theoretical components of homology: A study based on the morphology of the dermal skull-roofs of rhipidistian fishes. *Biol. J. Linn. Soc.* **1**, 327-361.

Jeune, B. (1984). Etude biometrique comparée de la croissance de feuilles et de tiges de Dicotyledones. *Bull. Soc. Bot. Fr. Lett. Bot.* **131**, 111-120.

Jeune, B., and Sattler, R. (1992). Multivariate analysis in process morphology of plants. *J. Theor. Biol.* **156**, 147-167.

Kaplan, D. R. (1984). The concept of homology and its central role in the elucidation of plant systematic relationships. *In* "Cladistics: Perspectives on the Reconstruction of Evolutionary History" (T. Stuessy and T. Duncan, eds.), pp. 51-70. Columbia Univ. Press, New York.

Kirchoff, B. K. (1991). Homeosis in the flowers of the Zingiberales. *Am. J. Bot.* **78**, 833-837.

Korzybski, A. (1948). "Science and Sanity." International Non-Aristotelian Library, Lakeville.

Lacroix, C., and Sattler, R. (1988). Phyllotaxis theories and pepal-stamen superposition in *Basella rubra. Am. J. Bot.* **75**, 906-917.

Leavitt, R. G. (1909). A vegetative mutant, and the principle of homeosis in plants. *Bot. Gaz. (Chicago)* **47**, 30-68.

Lehmann, N. L., and Sattler, R. (1992). Irregular floral development in *Calla palustris* (Araceae) and the concept of homeosis. *Am. J. Bot.* **79**, 1145-1157.

Lehmann, N. L., and Sattler, R. (1993). Homeosis in floral development of *Sanguinaria canadensis* and *S. canadensis* 'multiplex'. *Am. J. Bot.* (in press).

Leins, P. (1964). Die frühe Blütenentwicklung von *Hypericum hookerianum* Wight et Arn, und *H. aegypticum* L. *Ber. Dtsch. Bot. Ges.* **77**, 112-123.

Leins, P. (1983). Muster in Blüten. *Bonn. Universitätsbl.*, pp. 21-33.

Leins, P., and Erbar, C. (1991). Fascicled androecia in Dilleniidae and some remarks on the *Garcinia* androecium. *Bot. Acta* **104**, 257-344.

Leins, P., Tucker, S. C., and Endress, P. K., eds. (1988). "Aspects of Floral Development." Cramer, Berlin.

Lord, E. (1991). The concepts of heterochrony and homeosis in the study of floral morphogenesis. *Flower. Newsl.* **11**, 4-13.

Lyndon, R. F. (1978). Phyllotaxis and the initiation of primordia during flower development in *Silene. Ann. Bot. (London)* [N.S.] **42**, 1349-1360.

Mauseth, J. D. (1992). Possible examples of homeotic gene activity during morphogenesis in plants. *Am. J. Bot., Suppl.* **79**, 40 (abstr).

Mauseth, J. D., and Sajeva, M. (1992). Cortical bundles in the persistent, photosynthetic stems of cacti. *Ann. Bot. (London)* [N.S.] **70**, 317-324.

McCook, L. M., and Bateman, R. M. (1990). Homeosis in orchid flowers: a potential mechanism for saltational evolution. *Am. J. Bot.* **77**, Suppl., 145 (abstr.).

Meeuse, A. D. J. (1986). "Anatomy of Morphology." Brill/Backhuys, Leiden.

Meeuse, A. D. J. (1990). "Flowers and Fossils." Eburon, Delft.

Meyen, S. V. (1973). Plant morphology in its nomothetical aspects. *Bot. Rev.* **39**, 205-260.

Meyen, S. V. (1987). "Fundamental of Palaeobotany." Chapman & Hall, London.

Meyer, V. G. (1966). Flower abnormalities. *Bot. Rev.* **32**, 165-195.

Meyerowitz, E. M., Smyth, D. R., and Bowman, J. L. (1989). Abnormal flowers and pattern formation in floral development. *Development (Cambridge, UK)* **106**, 209-217.

Minelli, A., and Peruffo, B. (1991). Developmental pathways, homology, and homonomy in metameric animals. *J. Evol. Biol.* **3**, 429-445.

Negrutiu, I., Installé, P., and Jacobs, M. (1991). On flower design: A compilation. *Plant Sci.* **80**, 7-18.

Ouweneel, W. J. (1976). Developmental genetics of homeosis. *Adv. Genet.* **18**, 179-248.

Owen, R. (1843). "Lectures on the Comparative Anatomy and Physiology of the Invertebrate Animals, Delivered at the Royal College of Surgeons in 1843." Longman, London.

Oyama, S. (1988). Stasis, development and heredity. *In* "Evolutionary Processes and Metaphors" (M.-W. Ho and S. W. Fox, eds.), pp. 255-274. Wiley, New York.

Oyama, S. (1989). Ontogeny and the central dogma: Do we need the concept of genetic programming in order to have an evolutionary perspective? *In* "Systems and Development" (M. R. Gunnar and E. Thelen, eds.). Minn. Symp. Child Psychol., Vol. 22, pp. 1-34. Erlbaum, Hilldale, NJ.

Patterson, C. (1982). Morphological characters and homology. *In* "Problems of Phylogenetic Reconstruction" (K. A. Joysey and A. E. Friday, eds.), Syst. Assoc. Spec. Vol., No. 21, pp. 21-74. Academic Press, New York.

Patterson, C. (1987). Introduction. *In* "Molecules and Morphology in Evolution: Conflict or Compromise?" (C. Patterson, ed.), pp. 1-22. Cambridge Univ. Press, Cambridge.

Penzig, O. (1921/1922). "Pflanzenteratologie," 2nd. ed., 3 vols. Borntraeger, Berlin.

Polowick, P. L., and Sawhney, V. K. (1987). A scanning electron microscopic study on the influence of temperature on the expression of cytoplasmic male sterility in *Brassica napus. Can. J. Bot.* **65**, 807-814.

Remane, A. (1952). "Die Grundlagen des natürlichen Systems, der vergleichenden Anatomie und der Phylogenetik." Geest & Portig, Leipzig.

Remane, A. (1956). "Die Grundlagen des natürlichen Systems, der vergleichenden Anatomie und der Phylogenetik," 2nd ed. Geest & Portig, Leipzig.

Reynolds, J., and Tampion, J. (1983). "Double Flowers. A Scientific Study." Pembridge, London.

Riedl, R. (1978). "Order in Living Organisms." Wiley, New York.

Rieppel, O. C. (1988). "Fundamentals of Comparative Biology." Birkhäuser, Basel and Boston.

Rieppel, O. C. (1992). Homology and logical fallacy. *J. Evol. Biol.* **5**, 701-715.

Ronse Decraene, L.P. (1992). The Androecium of the Magnoliophytina: Characterisation and Systematic Importance. Ph.D. Thesis, Katholieke Universiteit Leuven, Belgium.

Ronse Decraene, L.-P., and Smets, E. F. (1992). Complex polyandry in the Magnoliatae: Definition, distribution and systematic value. *Nord. J. Bot.* **12**, 621-649.

Roth, V. L. (1984). On homology. *Biol. J. Linn. Soc.* **22**, 13-29.

Roth, V. L. (1988). The biological basis of homology. *In* "Ontogeny and Systematics" (C. J. Humphries, ed.), pp. 1-26. Columbia Univ. Press, New York.

Roth, V. L. (1991). Homology and hierarchies: Problems solved and unresolved. *J. Evol. Biol.* **4**, 167-194.

Rutishauser, R. (1989). A dynamic multidisciplinary approach to floral morphology. *Prog. Bot.* **51**, 54-69.

Rutishauser, R. (1993a). Reproductive development in seed plants: Research activities at the intersection of molecular genetics and systematic botany. *Prog. Bot.* (in press).

Rutishauser, R. (1993b). Unusual developmental morphology of leaves in aquatic angiosperms: a lesson in fuzzy morphology. *Am. J. Bot.* (abstr.) (in press).

Rutishauser, R., and Huber, K. A. (1991). The developmental morphology of *Indotristicha ramosissima* (Podostemaceae, Tristichoideae). *Plant Syst. Evol.* **178**, 195-223.

Rutishauser, R., and Sattler, R. (1985). Complementarity and heuristic value of contrasting models in structural botany. I. General considerations. *Bot. Jahrb. Syst.* **107**, 415-455.

Rutishauser, R. and Sattler, R. (1986). Architecture and development of the phyllode-stipule whorls of *Acacia longipedunculata*: Controversial interpretations and continuum approach. *Can. J. Bot.* **64**, 1987-2019.

Sachs, T. (1982). A morphogenetic basis for plant morphology. *Acta Biotheor.* **31A**, 121-137.

Sattler, R. (1966). Towards a more adequate approach to comparative morphology. *Phytomorphology* **16**, 417-429.

Sattler, R. (1984). Homology — a continuing challenge. *Syst. Bot.* **9**, 382-394.

Sattler, R. (1986). "Biophilosophy. Analytic and Holistic Perspectives." Springer-Verlag, Berlin and New York.

Sattler, R. (1988a). Homeosis in plants. *Am. J. Bot.* **75**, 1606-1617.

Sattler, R. (1988b). A dynamic multidimensional approach to floral morphology. *In* "Aspects of Floral Development" (P. Leins, S. C. Tucker, and P. K. Endress, eds.), pp. 1-6. Cramer, Berlin.

Sattler, R. (1990). Towards a more dynamic plant morphology. *Acta Biotheor.* **38**, 303-315.

Sattler, R. (1992). Process morphology: Structural dynamics in development and evolution. *Can. J. Bot.* **70,** 708-714.

Sattler, R. (1993). Why do we need a more dynamic study of morphogenesis? Descriptive and comparative aspects. In: "Morphogenèse et Dynamique" (D. Barabé and R. Brunet, eds.). Orbis, Frelighsburg (in press).

Sattler, R., and Jeune, B. (1992). Multivariate analysis confirms the continuum view of plant form. *Ann. Bot. (London)* [N.S.] **69**, 249-262.

Sattler, R., and Rutishauser, R. (1990). Structural and dynamic descriptions of the development of *Utricularia foliosa* and *U. australis. Can. J. Bot.* **68**, 1989-2003.

Sattler, R., and Rutishauser, R. (1992). Partial homology of pinnate leaves and shoots. Orientation of leaflet inception. *Bot. Jahrb. Syst.* **114**, 61-79.

Sattler, R., Luckert, D., and Rutishauser, R. (1988). Symmetry in plants: Phyllode and stipule development in *Acacia longipedunculata. Can. J. Bot.* **68**, 1989-2003.

Sawhney, V. K. (1992). Floral mutants in tomato: Development, physiology, and evolutionary implications. *Can. J. Bot.* **70**, 701-707.

Schultz, E. A. Pickett, F. B., and Haughn, G. W. (1991). The FLO10 gene product regulates the expression domain of homeotic genes AP3 and PI in *Arabidopsis flowers. Plant Cell* **3**, 1221-1237.

Smith-Huerta, N. L. (1992). A comparison of floral development in wild type and a homeotic sepaloid petal mutant of *Clarkia tembloriensis* (Onagraceae). *Am. J. Bot.* **79**, 1423-1430.

Sneath, P. H. A., and Sokal, R. R. (1973). "Numerical Taxonomy." Freeman, San Francisco.

Stebbins, G. L. (1974). "Flowering Plants: Evolution above the Species Level." Harvard Univ. Press (Belknap), Cambridge, MA.

Stevens, P. F. (1984). Homology and phylogeny: Morphology and systematics. *Syst. Bot.* **9**, 395-409.

Stevens, P. F. (1991). Character states, morphological variation, and phylogenetic analysis: A review. *Syst. Bot.* **16**, 553-583.

Stidd, B. M. (1987). Telomes, theory change, and the evolution of vascular plants. *Rev. Palaeobot. Palynol.* **50**, 115-126.

Tomlinson, P. B. (1984). Homology: An empirical view. *Syst. Bot.* **9**, 374-381.

Troll, W. (1954). "Praktische Einführung in die Pflanzenmorphologie," Part 1. Fischer, Jena.

Tucker, S. C. (1989). Evolutionary implications of floral ontogeny in legumes. *Adv. Legume Biol.* **29**, 59-75.

Tucker, S.C. (1992). The role of floral development in studies of legume evolution. *Can. J. Bot.* **70**, 692-700.

van der Klaauw, C. J. (1966). Introduction to the philosophic backgrounds and prospects of the supraspecific comparative anatomy of conservative characters in the adult stages of conservative elements of Vertebrata with an enumeration of many examples. Verh. K. Ned. Akad. Wet., Afd. Natuurkd., Reeks (2), **57**, 1-196.

Van Valen, L.M. (1982). Homology and causes. *J. Morphol.* **173**, 305-312.

Veilex, J. (1973-1976). Essais à travers l'étude de l'état articulé, d'une approche plus rationelle des problèmes posés par la morphologie des angiospermes. Première partie. *Bull. Soc. Etud. Sci. Nat., Vaucluse* **37-39**, 31-60.

Veilex, J. (1979-1981). Essais à travers l'étude de l'état articulé d'une approche plus rationelle des problèmes posés par la morphologie des angiospermes. Deuxième partie. *Bull. Soc. Etud. Sci. Nat., Vaucluse* **43-46,** 19-64.

Wagner, G. P. (1989a). The origin of morphological characters and the biological basis of homology. *Evolution (Lawrence, Kans.)* **43,** 1157-1171.

Wagner, G. P. (1989b). The biological homology concept. *Annu. Rev. Ecol. Syst.* **20,** 51-69.

Weston, P. H. (1988). Indirect and direct methods in systematics. *In* "Ontogeny and Systematics" (C. J. Humphries, ed.), pp. 27-56. Columbia Univ. Press, New York.

Whitehead, A. N. (1967). "Science and the Modern World." Macmillan, New York (first published 1925).

Whitehead, A. N. (1978). "Process and Reality." Corrected Edition by D. R. Griffin and D. W. Sherburne, Free Press, New York.

Wiley, E. O. (1975). Karl R. Popper, systematics, and classification: A reply to Walter Bock and other evolutionary taxonomists. *Syst. Zool.* **24,** 233-243.

Wiley, E. O. (1981). "Phylogenetics. The Theory and Practice of Phylogenetic Systematics." Wiley, New York.

Woodger, J. H. (1929). "Biological Principles: A Critical Study." K. Paul, Trench, Trubner & Co., London. (Reissued in 1967 with a new Introduction, Humanities Press, New York.

Woodger, J. H. (1945). On biological transformations. *In* "Essays on Growth and Form Presented to D'Arcy W. Thompson" (W. E. Le Gros Clark and P. B. Medawar, eds.), pp. 95-120. Oxford Univ. Press, Oxford.

Zimmermann, W. (1961). Phylogenetic shifting of organs, tissues and phases in pteridophytes. *Can. J. Bot.* **39,** 1547-1533.

INDEX

A